Symbol	Meaning
$X/\sim$	equivalence classes, quotient space
X/A	quotient space of X modulo A
(a, b)	bounded, open interval
$(-\infty, a), (a, \infty)$	unbounded, open intervals
$[a, b]$	closed, bounded interval
$(-\infty, a], [a, \infty)$	unbounded, closed intervals
$(a, b], [a, b)$	half-open, half-closed intervals
$|x|$	absolute value
$\|x\|$	norm
$d(x, y)$	distance from x to y
$d(x, A)$	distance from a point to a set
$D(A)$	diameter of a set
$B(a, r)$	open ball
$B[a, r]$	closed ball
S^n	n-dimensional sphere
I	closed unit interval $[0, 1]$
I^n	unit n-cube
$\mathbb{R}$	set of real numbers
$\mathbb{R}^n$	n-dimensional Euclidean space
θ	origin in $\mathbb{R}^n$
bdy A	boundary of a set
int A	interior of a set
$\bar{A}$	closure of a set
∂X	boundary of a manifold
Int X	interior of a manifold
$\mathcal{C}[a, b]$	space of continuous, real-valued functions on $[a, b]$
$C(X, \mathbb{R})$	space of continuous, bounded, real-valued functions on X
$\Pi_1(X, x_0)$	fundamental group
$\Pi_n(X, x_0)$	nth homotopy group
X_∞	one-point compactification
$\beta(X)$	Stone-Čech compactification
$[x]$	equivalence class
$[\alpha]$	homotopy class of a loop
$\mathbb{Z}$	set of integers
$\mathbb{Z}^+$	set of positive integers

PRINCIPLES OF
TOPOLOGY

FRED H. CROOM
The University of the South

Dover Publications
Garden City, New York

Bibliographical Note

This Dover edition, first published in 2016, is an unabridged and slightly corrected republication of the work originally published in 1989 by Saunders College Publishing, Philadelphia, and by Cengage Learning Asia in 2002.

Library of Congress Cataloging-in-Publication Data

Croom, Fred H., 1941–
  Principles of topology / Fred H. Croom. — Dover edition.
    pages cm
  Originally published: Philadelphia : Saunders College Publishing, 1989; slightly corrected.
  Includes bibliographical references and index.
  ISBN 978-0-486-80154-4
  ISBN 0-486-80154-3
    1. Topology. I. Title.
QA611.C76 2016
514—dc23

2015030379

Manufactured in the United States of America
80154309   2023
www.doverpublications.com

PREFACE

This text is designed for a one-semester introduction to topology at the undergraduate and beginning graduate levels. It is accessible to junior mathematics majors who have studied multivariable calculus.

The text presents the fundamental principles of topology rigorously but not abstractly. It emphasizes the geometric nature of the subject and the applications of topological ideas to geometry and mathematical analysis. The following basic premise motivated the writing of this book: Topology is a natural, geometric, and intuitively appealing branch of mathematics which can be understood and appreciated by undergraduate students as they begin their study of advanced mathematical topics. A course in topology can even be an effective vehicle for introducing students to higher mathematics. Topology developed in a natural way from geometry and analysis, and it is not an obscure, abstract, or intangible subject to be reserved only for graduate students.

The usual topics of point-set topology, including metric spaces, general topological spaces, continuity, topological equivalence, basis, subbasis, connectedness, compactness, separation properties, metrization, subspaces, product spaces, and quotient spaces, are treated in this text. In addition, the text contains introductions to geometric, differential, and algebraic topology. Each chapter has historical notes to put important developments into an historical framework and a supplementary reading list for those who want to go beyond the text in particular areas.

Chapter 1 introduces topology from an intuitive and historical point of view. This chapter also contains a brief summary of a modest amount of prerequisite material on sets and functions required for the remainder of the course. Chapter 2 initiates the rigorous presentation of topological concepts in the familiar setting of the real line and the Euclidean plane. Chapter 3 takes an additional step toward general topology with the introduction of metric spaces and treats such topics as open sets, closed sets, interior, boundary, closure, continuity, convergence, completeness, and subspaces in the metric context. Euclidean spaces and Hilbert space are emphasized.

The core of the course is Chapters 4 through 8. Chapter 4 extends the ideas of Chapter 3 to general topological spaces and introduces the additional concepts of basis, subbasis, topological equivalence, and topological invariants. The topological invariants discussed in Chapter 4 include separability, first and second countability, the Hausdorff property, and metrizability. Chapter 5 treats connectedness, with particular attention to the connected subsets of the real line and to applications in analysis. This chapter also introduces the related concepts of local connectedness, path connectedness, and local path connectedness. Chapter 6 deals with compactness and such related properties as countable compactness, local compactness, and the Bolzano-Weierstrass property. Particular attention is paid to compactness in Euclidean and metric spaces, the one-point compactification, and the Cantor set. Chapter 7 treats product and quotient spaces and introduces geometric and differential topology through an analysis of surfaces and manifolds. Chapter 8 deals with separation properties, metrization, and the Stone-Čech compactification.

Chapter 9, the final chapter of the text, provides an introduction to algebraic topology through the fundamental group and a brief encounter with categories. Like the other chapters, the final chapter emphasizes geometric applications, such as the Brouwer Fixed Point Theorem and the fundamental groups of familiar surfaces. A brief appendix on groups is included for the benefit of those who have not studied the necessary algebraic prerequisites for Chapter 9.

Most of the factual information about topology presented in this text is stated in the theorems and illustrated in the accompanying examples, figures, and exercises. Theorems are proved to provide a logical and rigorous framework, to show the development of the subject, and to illustrate important techniques. From the student's point of view, topology is an excellent subject for learning to prove theorems correctly, for learning the concepts of mathematical rigor, and for developing the mathematical maturity and sophistication that are required for higher level courses. The reader should keep in mind, however, that the examples are extremely important for bringing abstract concepts into more concrete form.

The exercises are the most important part of the text, since we all learn better by doing than simply by watching. This book contains many exercises of varying degrees of difficulty. Most of the exercises provide practice in applying the material from each section or ask the reader to supply arguments either omitted from the text or given only in outline form. Some of the exercises go considerably beyond the text and are worthy projects for undergraduate research and independent study.

No introductory course can cover all areas of topology, and many topics have necessarily been omitted or given only cursory treatment in this text. There is a supplementary reading list at the end of each chapter that points the way for the interested reader to the more advanced aspects of topology.

For reference within the text, theorems are numbered consecutively within each chapter. For example, "Theorem 5.6" refers to the sixth theorem of Chapter 5. In addition, the name by which a theorem is known in the mathematical literature is given whenever applicable. This is usually a descriptive name, like "the Mean Value Theorem," or a name that recognizes the discoverer, like "the Urysohn Metrization Theorem" or "Cantor's Nested Intervals Theorem." Examples are numbered consecutively within each section of each chapter. For example, "Example 3.7.4" is the fourth example of the seventh section of Chapter 3.

The notation used in this text is reasonably standard; a list of symbols with definitions appears on the front endsheets.

The following reviewers read several preliminary versions of this text and made many thoughtful and helpful suggestions for improvements: Paul J. Bankston,

Marquette University; Bruce P. Conrad, Temple University; Doug W. Curtis, Louisiana State University; Robert J. Daverman, University of Tennessee; and Dennis M. Roseman, University of Iowa. It is a pleasure to acknowledge their contributions to this project.

Fred H. Croom
Sewanee, Tennessee

CONTENTS

1 Introduction

1.1 THE NATURE OF TOPOLOGY

The word "topology" is derived from the Greek word "$\tau \acute{o} \pi o s$," which means "position" or "location." This name is appropriate, for topology deals with geometric properties which are dependent only upon the relative positions of the components of figures and not upon such concepts as length, size, or magnitude. Topology deals with properties which are not destroyed by continuous transformations like bending, shrinking, stretching, and twisting. Discontinuous transformations, such as cutting, tearing, and puncturing, are not allowed.

This section presents several examples to illustrate the fundamental concepts of topology. The examples are necessarily based on intuition and are intended only to give a heuristic introduction to the subject. The ideas sketched here will be made precise as the subject is developed in the succeeding chapters.

Example 1.1.1

Consider a circle in the ordinary Euclidean plane, as shown in Figure 1.1. Point A is inside the circle, B is on the circle, and C is outside the circle. Imagine that the entire plane undergoes a continuous deformation or transformation, that is, a stretching or twisting. For definiteness, imagine that horizontal distances are doubled and vertical distances are halved, so that the circle is transformed into

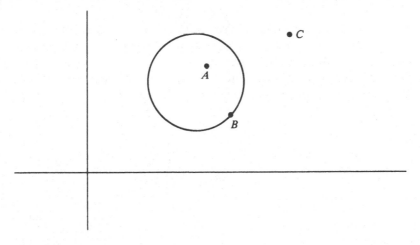

FIGURE 1.1

1

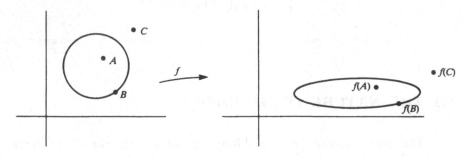

FIGURE 1.2

an ellipse and points *A*, *B*, and *C* are mapped by the transformation to points $f(A), f(B)$, and $f(C)$, as shown in Figure 1.2.

Note that point $f(A)$ is inside the image curve, $f(B)$ is on the image curve, and $f(C)$ is outside the image curve, the same relative positions held by their predecessors *A*, *B*, and *C* with respect to the original circle. Thus we note that the property of being inside, on, or outside a closed plane curve is not altered by this continuous transformation.

Let us now restrict out attention to the circle and the ellipse obtained from it in Figure 1.2 and omit consideration of the inside and outside points. Think of the ellipse as being obtained from the circle by a continuous transformation. By reversing the transformation to reduce horizontal distances by a factor of 1/2 and increase vertical distances by a factor of 2, we can imagine the ellipse being transformed back into the original circle. This is the basic idea of topological equivalence; we would say that the circle and the ellipse are "topologically equivalent" or "homeomorphic." For two figures to be topologically equivalent, there must be a reversible transformation between them which is continuous in both directions.

By defining suitable transformations, one can see that the following pairs of figures in the plane are topologically equivalent. Objects from different pairs are not topologically equivalent, however.

The preceding discussion illustrates why topology has often been called "rubber geometry." One imagines geometric objects made of rubber which undergo the

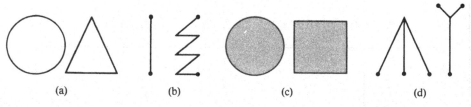

    (a)   (b)   (c)   (d)

FIGURE 1.3

continuous deformations of shrinking, stretching, bending, and twisting. The reader should be aware, however, that the rubber figure idea is too narrow for an accurate understanding of topology and somewhat misleading as well. For example, as the reader will easily be able to demonstrate later, the two objects illustrated in Figure 1.4 are topologically equivalent, but neither can be shrunk, stretched, bent, or twisted to match the other. The point here is that the definition of topological equivalence does not require that the stick "pass through" the spherical surface.

Example 1.1.2

The number 0 is the limit of the sequence $1/2, 1/3, 1/4, \ldots, 1/n, \ldots$. Students of calculus know the reason; the given sequence has limit 0 because no matter how small a positive number ϵ is given, there is a positive integer N such that all terms of the sequence from the Nth term on are within distance ϵ of 0.

   Here it is understood that distances on the number line are to be measured in the standard way. The distance $d(a, b)$ between real numbers a and b is the absolute value of their difference:

$$d(a, b) = |a - b|.$$

Suppose that we define a new distance function d' by

$$d'(a, b) = 20d(a, b).$$

In other words, the new distance from a to b is 20 times the usual distance. It should not take the reader long to realize that the sequence $1/2, 1/3, 1/4, \ldots,$ $1/n, \ldots$ still has 0 as its limit with this new method of measuring distances. Later we will see that the distance functions d and d' produce the same topological structure; for now it is sufficient to realize that multiplication of distances by a positive constant does not alter the convergence of sequences.

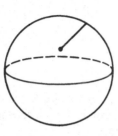

(a) Sphere with
stick inside

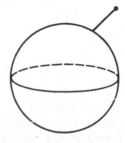

(b) Sphere with
stick outside

FIGURE 1.4

Topology replaces distance with a weaker, more general concept of nearness. Different distance functions may produce the same concept of nearness and, in this sense, be equivalent for the purposes of topology. This point will be made precise in Chapter 3.

Example 1.1.3

The *open unit interval* (0, 1) on the real number line ℝ consists of all real numbers x with $0 < x < 1$. The *closed unit interval* [0, 1] consists of all real numbers x with $0 \le x \le 1$. Thus [0, 1] contains the endpoints 0 and 1 while (0, 1) does not. The intervals (0, 1) and [0, 1] are topologically different, for the following reasons:

   (a) The open interval (0, 1) contains sequences of points which converge to limits not in (0, 1). For example, the sequence $1/2, 1/3, 1/4, \ldots, 1/n, \ldots$ converges to 0. The closed interval [0, 1], on the other hand, has the property that every convergent sequence of its points converges to a point in [0, 1].

   (b) If any point of (0, 1) is removed, the remaining points make up two disjoint or disconnected intervals. In other words, every point of (0, 1) is a *cut point*, since removing any point "cuts" the interval into two disjoint pieces. The closed interval [0, 1], however, has two points, the endpoints 0 and 1, which are not cut points. We shall see later that the number of cut points and the number of non-cut points of a figure are unaltered by topological transformations.

Explain in terms of cut points why the geometric objects in the following figure are not topologically equivalent:

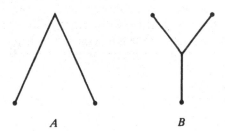

A   B

FIGURE 1.5

Answer   One reason is that A has two non-cut points while B has three. Another is that each cut point of A separates A into two components while B has one special cut point which separates it into three components.

   The reader has probably seen the following theorem and been told that its proof would be given in a later course:

Theorem 1.1:   The Intermediate Value Theorem. *Let f:* $\mathbb{R} \to \mathbb{R}$ *be a continuous function on the set* $\mathbb{R}$ *of real numbers and suppose that there are real numbers a and b for which f(a) < 0 and f(b) > 0. Then there is a real number c between a and b for which f(c) = 0.*

This theorem is made very plausible by considering the possibilities for the graph of the function $y = f(x)$. Since the point $(a, f(a))$ lies below the x-axis and $(b, f(b))$ lies above it, and since the graph of the function must connect $(a, f(a))$ and $(b, f(b))$ with a continuous unbroken curve, then the graph must cross the x-axis at some point $(c, f(c))$ with c between a and b. Then c is the desired real number with $f(c) = 0$.

The main problems with the argument of the preceding paragraph are that it does not make precise what is meant by a "continuous unbroken curve," it does not establish that the graph of a continuous function is a continuous unbroken curve, and it does not give any reason beyond intuition why the curve from $(a, f(a))$ to $(b, f(b))$ must intersect the x-axis. By topological considerations, the above argument will be made precise in Chapter 5 and will be used to prove a more general version of the Intermediate Value Theorem (Theorem 5.8).

The Intermediate Value Theorem is the type of result with which topology has been most successful. The theorem is called an "existence theorem" because it asserts the existence of a real number c with $f(c) = 0$ without, however, giving any method for determining the value of c in particular cases. Since existence theorems usually do not give methods for finding solutions, they may appear to be of little value. Precisely the opposite is true. Existence theorems are the basis on which calculus and real analysis rest, and in differential equations and functional analysis, for example, there are many applications in which the existence of a solution and not its particular form is the most important factor.

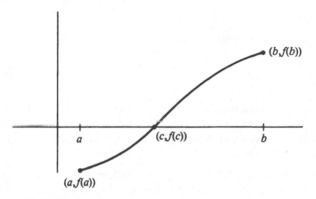

FIGURE 1.6

EXERCISE 1.1

1. Tell whether the following pairs of figures are topologically equivalent. Give reasons for your answers.

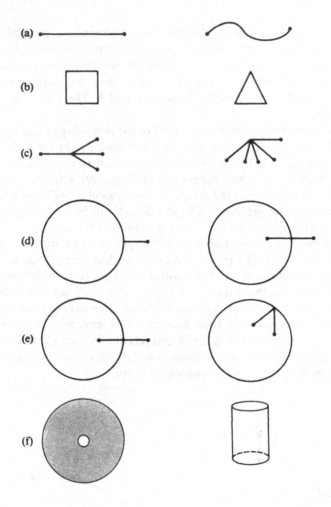

(a)

(b)

(c)

(d)

(e)

(f)

2. It is sometimes said that a topologist is a person who can't tell the difference between a doughnut and a coffee cup. By imagining a (solid) doughnut made of rubber, explain intuitively how to make a topological transformation of the doughnut to a coffee cup by bending, twisting, and stretching.

3. Separate the letters of the alphabet, as printed below, into groups in such a way that members of the same group are topologically equivalent and members of different groups are not.

A B C D E F G H I J K L M N O P Q R S T U V W X Y Z

1.2 THE ORIGIN OF TOPOLOGY

Topology emerged as a well-defined mathematical discipline during the early years of the twentieth century, but isolated instances of topological problems and precursors of the theory can be traced back several centuries. Gottfried Wilhelm Leibniz (1646–1716) was the first to foresee a geometry in which position, rather than magnitude, was the most important factor. In 1676 Leibniz used the term "geometria situs" (geometry of position) in predicting the development of a type of vector calculus somewhat similar to topology as it is known today. Leibniz is now best known as one of the independent inventors of calculus, along with Isaac Newton (1642–1727).

The first practical application of topology was made in the year 1736 by the Swiss mathematician Leonhard Euler (1707–1783) and is explained in the following example.

Example 1.2.1   The Königsberg Bridge Problem

In the eighteenth century the German city of Königsberg was located on an island in the Pregel River and on the surrounding banks, at the point where the river divided into the Old Pregel and New Pregel. Island and mainland were joined by a network of seven bridges as shown in Figure 1.7.

The problem, which was of interest to Sunday strollers, was to cross each of the seven bridges exactly once in one continuous trip. This is clearly a topological problem, since it depends only upon the relative positions of bridges and land masses and not on the size of the island or the lengths of the bridges. Following Euler, let us replace each land mass by a point and each bridge by a line segment (Fig. 1.8). In the resulting configuration, called by Euler a "graph," each point is a "vertex" and each line segment is an "edge."

A vertex with an odd number of edges leading from it is an *odd vertex*, and *even vertex* is defined in the analogous way for an even number of edges. As can be seen in Figure 1.8, each vertex of the Königsberg Bridge problem is

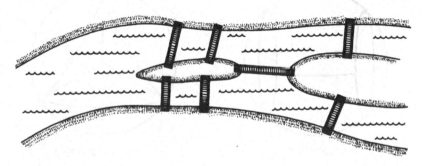

FIGURE 1.7

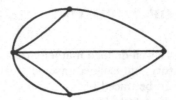

FIGURE 1.8   **Graph for the Königsberg Bridge problem.**

odd. Note that the bridges at any odd vertex can be crossed exactly once only if that vertex is either the beginning point or the ending point of the journey. But since there are more than two odd vertices in this case, Euler showed that the desired route is impossible. Euler went on to give a general solution for the number of continuous journeys required to traverse exactly once each edge of a connected graph. The number of odd vertices is always an even number, and if this number is 0 or 2, then the graph can be traversed, as required, in one continuous journey. If the number of odd vertices exceeds 2, then the number of continuous journeys required will be half the number of odd vertices.

Euler also proved that the first topological formula, which was conjectured earlier but not proved by René Descartes (1596–1650): For a connected graph drawn on the surface of a sphere,

$$V - E + F = 2$$

where V denotes the number of vertices, E the number of edges, and F the number of faces or areas into which the spherical surface is divided by the graph. This principle is illustrated in Figure 1.9.

Carl F. Gauss (1777–1855), who influenced so much of modern mathematics, predicted in 1833 that "geometry of location" would become a mathematical dis-

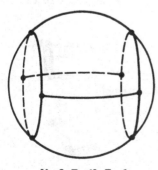

$V = 8, E = 12, F = 6$

$V = 4, E = 6, F = 4$

FIGURE 1.9

cipline of great importance. His study of closed surfaces such as the sphere and the torus (surface of a doughnut) and surfaces like those encountered in multidimensional calculus may be considered a harbinger of general topology. Gauss was also interested in knots, which are of current topological interest.

The word "topology" was coined by the German mathematician Joseph B. Listing (1808–1882) for the title of his book *Vorstudien zur Topologie (Introductory Studies in Topology)*, a textbook published in 1847. Listing's book dealt with knots and surfaces but failed to popularize either the subject or the name. Throughout the nineteenth and early twentieth centuries, the loosely defined area of geometry that was later to become topology was called *analysis situs* (analysis of position).

Example 1.2.2   The Möbius Strip

In 1858 the German mathematician A. F. Möbius (1790–1868) discovered a curious surface with only one side and one edge which, remarkably, can be easily constructed from a strip of paper. Cut a thin strip of paper and after giving the strip a twist through 180 degrees, join the two ends with glue or tape. The resulting surface is the Möbius strip.

As one can see by tracing with finger or pencil, the Möbius strip has only one continuous surface and one edge. Try drawing a closed curve along the length of the band and then cutting along the curve, as though cutting the band into two bands of half the original width. You may be surprised at the result.

The first mathematician really to foresee topology in anything like the generality it has achieved today was Bernard Riemann (1826–1866). Riemann initiated the study of the connectivity of a surface (the arrangement of the holes in a surface). He also used concepts in which the number of dimensions exceeded three, which was generally conceded to be the maximum number of dimensions involved with any geometric object.

The mathematical research of the nineteenth century which eventually produced the field of topology can now be traced to two primary sources: the development of non-Euclidean geometry and the process of putting calculus on a firm mathematical foundation. For over 2000 years it was believed that the ordinary

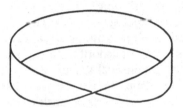

FIGURE 1.10   **Möbius Strip.**

two- and three-dimensional geometry of Euclid was the only geometry that pertained to the "real" world. Geometric research was restricted to the system satisfying Euclid's axioms and to projective geometry, which was of interest in Renaissance art because of its applications to perspective. The dominance of Euclidean ideas was especially reinforced by the philosopher Immanuel Kant (1724–1804), who taught that human intuition, perception, and experience were necessarily restricted to the realm of Euclidean geometry. Such beliefs discouraged free thinking and severely retarded the advancement of mathematics, particularly geometry.

Finally, around 1830, geometries which do not satisfy all the Euclidean axioms were invented independently by Janos Bolyai (1802–1860) and N. I. Lobachevsky (1792–1856). Their work involved an attempt to show that the parallel postulate of Euclid could be derived from the other axioms. Actually, they showed that it could not and that, indeed, there was a perfectly reasonable geometry, now called a non-Euclidean geometry, which does not satisfy the parallel postulate but which does satisfy the other Euclidean axioms. Interest generated by the discoveries of Bolyai and Lobachevsky stimulated free geometric thinking and led to other non-Euclidean geometries, for example the elliptic geometry of Riemann, and, eventually, to the abstraction of geometric ideas to form the subject of topology.

The second nineteenth century current that influenced the development of topology was the work done by many mathematicians, notably A. L. Cauchy (1784–1857) and Karl Weierstrass (1815–1897), in defining rigorously the real number system and the concept of limit in order to put the foundations of calculus on firm ground. Pathological examples like Weierstrass' function which is everywhere continuous but nowhere differentiable and the "space filling curves" of Guiseppe Peano (1858–1932), which mapped an interval onto a square or a cube, demanded that lines, planes, curves, and surfaces be defined rigorously and that loose arguments which appealed only to intuitive plausibility be thrown out. During the latter part of the nineteenth century, problems arose in functional analysis and differential equations which made it necessary to consider large collections of functions, curves, and surfaces as collections, not merely as individuals.

Point-set topology or set-theoretic topology, which is the branch of topology considered in this text, was decisively influenced by the work of Georg Cantor (1845–1918) during the years 1872 to 1890. Cantor and his coworkers discovered many properties of the real number line that are now considered the basic concepts of point-set topology. In addition, Cantor laid the foundations and posed many basic questions and paradoxes concerning the theory of infinite sets. With the introduction by Maurice Fréchet (1878–1973) in 1906 of general distance functions for abstract spaces whose "points" were not required to be the points of ordinary geometry, the groundwork for topology was laid. The subject emerged as a cohesive discipline with the publication of the textbook *Grundzüge der Mengenlehre* by Felix Hausdorff (1868–1942) in 1914. Hausdorff's classic treatise presented a list of defining axioms for the term "topological space," a very general concept which included the ordinary line, the plane, three-dimensional space, spaces of more than three dimensions, curves, surfaces, spaces of curves, spaces of functions, and even spaces of sets.

As a mathematical discipline, topology is divided into several overlapping areas. Chapters 1 through 8 deal primarily with point-set topology. Algebraic topology, which is the subject of Chapter 9, attempts to describe geometric objects in terms of algebraic structures. Algebraic topology is less general in terms of the objects studied than is point-set topology, and it is more specialized in its methods of attack. Differential topology, which is introduced in Chapter 7, is concerned with the properties of "smooth" spaces and surfaces which permit a concept of differentiability for functions. Geometric topology is also introduced in Chapter 7. As the subjects are taught today, an introduction to point-set topology is required before one can learn anything of significance about algebraic, differential, or geometric topology.

Progress in topology was greatly accelerated when Hausdorff carefully selected from the work of his predecessors those principles of greatest importance and presented them, in a general and abstract form, as an object for consideration. In the years immediately following the appearance of Hausdorff's book, point-set topology developed wide applicability. The power of the subject is derived from its generality; from a few simple axioms and definitions one can deduce principles which apply to problems in real and complex analysis, differential equations, functional analysis, and other areas where the relations of points to sets and continuity of functions are important.

Much progress in mathematics as a whole, not just in topology, has been the result of abstraction of ideas to their basic elements. Reduction to basic principles strips away superfluous and confusing information; it leads to simplification and to the unification of ideas once thought to be completely separate. Often, however, one can see the beauty and power of an abstract subject only after studying it for some time. For the beginner, abstraction is more often a bane than a blessing; it can make the subject appear artificial and obscure its utility; it can stifle the intuition and produce confusion instead of clarity. For these reasons, we shall enter the abstract world of topology gradually, avoiding the temptation to do everything in the smallest possible number of steps.

In the remaining sections of this chapter, we shall review some basic ideas on sets and functions which are useful in topology. Then, following Cantor and the early workers in topology, we shall undertake in Chapter 2 a study of the real number line and plane. After studying general distance functions in Chapter 3, we shall move on to general topological spaces in Chapter 4. Those who like to think in terms of an historical perspective may imagine that Chapter 2 corresponds roughly to the year 1890, Chapter 3 to 1906 when general distance functions were defined by Fréchet, and Chapter 4 to 1914 when Hausdorff defined general topological spaces. Succeeding chapters lead to the more modern aspects of the subject.

1.3 PRELIMINARY IDEAS FROM SET THEORY

It is assumed that the terms "set" and "member of a set" are familiar terms from the reader's previous training in mathematics. The terms "collection," "aggregate," and "family" are synonyms for set; the members of a set are often referred to as its "elements" or "objects." These terms are used in this book in a way which agrees with their customary usage.

Sets are usually denoted by capital letters and elements by lower case letters. The symbol "∈" (abbreviation for "belongs to") indicates set membership. Thus $a \in A$ means that object a is a member of set A, and $b \notin A$ means that b is not a member of A. Sets are often described using brackets: $\{x: \ldots\}$ denotes the set of all elements x satisfying the statement . . . .

Example 1.3.1

Let $A = \{x: x$ is an integer between 0 and 5 inclusive$\}$. Then set A has as its elements the integers 0, 1, 2, 3, 4, and 5 and could be expressed as $A = \{0, 1, 2, 3, 4, 5\}$. This method of defining a set by listing its members has obvious drawbacks for large sets. One could not, for example, list all members of the set $B = \{x: x$ is a real number larger than 10$\}$.

The set having no members is called the *empty set* and is denoted by the symbol $\varnothing$.

If A and B are sets for which each member of A is also a member of B, then A is a *subset* of B or is said to be *contained in B*. Such set inclusion is denoted by the symbol "⊂": $A \subset B$ provided that A is a subset of B.

Observe that $A \subset A$ and $\varnothing \subset A$ for every set A. The latter inclusion is true because $\varnothing$ has no members and therefore has no members outside A. Two sets A and B are equal precisely when each is a subset of the other. This fact is often used in showing set equality.

Note   The concept of set inclusion defined here allows for the possibility of equality: $A \subset B$ includes the possibility $A = B$. In some texts, this relation is expressed $A \subseteq B$ and is read A is *contained in or equal to B*.

The collection of all subsets of a given set A is called the *power set* of A and is denoted $\mathcal{P}(A)$. Thus $X \in \mathcal{P}(A)$ means that $X \subset A$. The subsets of A other than A itself and the empty set are the *proper* subsets of A.

Example 1.3.2

Let $A = \{0, 1, 2\}$. The subsets of A are $\varnothing$, $\{0\}$, $\{1\}$, $\{2\}$, $\{0, 1\}$, $\{0, 2\}$, $\{1, 2\}$, and A. The power set of A is the set whose elements are the eight subsets of A:

$$\mathcal{P}(A) = \{\varnothing, \{0\}, \{1\}, \{2\}, \{0, 1\}, \{0, 2\}, \{1, 2\}, \{0, 1, 2\}\}.$$

It is often desirable to discuss collections of sets and to name many sets in a systematic way. The standard method of doing this, called *indexing*, is defined next.

Definition: *Let A be a set for which, corresponding to each element $a \in A$, there is a set M_a. Then the collection of sets $\{M_a : a \in A\}$, also denoted $\{M_a\}_{a \in A}$, is said to be **indexed** by A or to have A as **index set**.*

Example 1.3.3

(a) For each real number a, let L_a denote the collection of all real numbers less than a. Then the family of sets

$$\mathcal{L} = \{L_a : a \in \mathbb{R}\}$$

is indexed by the set $\mathbb{R}$ of real numbers.

(b) As a simpler illustration, consider four sets A_1, A_2, A_3, and A_4. The family of sets

$$\mathcal{A} = \{A_i : i = 1, 2, 3, 4\}$$

is indexed by the set $\{1, 2, 3, 4\}$.

EXERCISE 1.3

1. Find the power set of $B = \{a, b, c, d\}$.

2. Suppose that A and B are sets for which $\mathcal{P}(A) = \mathcal{P}(B)$. Show that $A = B$.

3. Explain why each of the following statements is false. Alter each one to make a true statement:

    (a) $a \subset \{a, b, c\}$.

    (b) $\{a\} \in \{a, b, c\}$.

    (c) $A \subset \mathcal{P}(A)$.

4. Tell whether each of the following statements is true or false for a given set A.

    (a) $\varnothing \subset \mathcal{P}(A)$.   (d) $\varnothing = \{\varnothing\}$.

    (b) $\varnothing \in \mathcal{P}(A)$.   (e) $\varnothing = \mathcal{P}(\varnothing)$.

    (c) $A \in \mathcal{P}(A)$.   (f) $\{\varnothing\} = \mathcal{P}(\varnothing)$.

5. Prove the following *transitive property* of set inclusion: If $A \subset B$ and $B \subset C$, then $A \subset C$.

6. Prove that if set A has n members, n a positive integer, then $\mathcal{P}(A)$ has 2^n members. (*Hint:* A subset B of A associates with each member of A one of the two words "in" and "out.")

1.4 OPERATIONS ON SETS: UNION, INTERSECTION, AND DIFFERENCE

A note on word usage is in order before defining the standard set operations. The conjunction "or" is used in mathematics and logic in the inclusive sense: If p and q are statements, then the statement "p or q" is true whenever at least one of p, q is true. The only case for which "p or q" is false is the case in which p is false and q is also false.

A similar interpretation applies to the indefinite articles "a" and "an." These articles indicate at least one object of a specified type. Thus, "There is a real number between 0 and 100" is a perfectly correct statement even though there are many real numbers between 0 and 100.

Definition: *If A and B are sets, the **union** $A \cup B$ of A and B is the set consisting of all elements x which belong to at least one of the sets A, B:*

$$A \cup B = \{x: x \in A \text{ or } x \in B\}.$$

*The **intersection** $A \cap B$ of A and B is the set of all elements x which belong to both A and B, that is, the set of elements common to A and B:*

$$A \cap B = \{x: x \in A \text{ and } x \in B\}.$$

*Sets A and B are said to be **disjoint** if $A \cap B = \varnothing$.*

Example 1.4.1

Let $A = \{0, 1, 2, 3\}$ and $B = \{2, 3, 5, 7, 8\}$. Then

$$A \cup B = \{0, 1, 2, 3, 5, 7, 8\}, A \cap B = \{2, 3\}.$$

The set operations of union and intersection are represented pictorially in the Venn diagram on the following page (Figure 1.11).

Note the following elementary properties of $\cup$ and $\cap$ for any sets A and B:

(a) $A \cup A = A \cap A = A.$

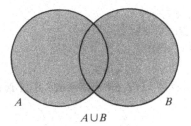
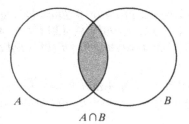

$A \cup B$ $A \cap B$

FIGURE 1.11

(b) Both A and B are subsets of $A \cup B$.

(c) $A \cap B$ is a subset of both A and B.

Theorem 1.2:   *The following statements are equivalent for any sets A and B.*

(a) $A \subset B$;

(b) $A \cup B = B$;

(c) $A \cap B = A$.

Proof: *The following argument is for the equivalence of (a) and (b); the analogous argument for the equivalence of (a) and (c) is left to the reader.*
Suppose $A \subset B$ and consider $A \cup B$. Since $A \subset B$, then

$$A \cup B \subset B \cup B = B$$

so $A \cup B$ is a subset of B. But $B \subset A \cup B$ for any sets A and B, so it follows that $A \cup B = B$. Thus (a) implies (b).
To reverse the implication, suppose $A \cup B = B$. Then

$$A \subset A \cup B = B$$

so $A \subset B$.   □

Theorem 1.3:   The Distributive Properties for Union and Intersection.
For any sets A, B, and C,

(a) $A \cup (B \cap C) = (A \cup B) \cap (A \cup C)$;

(b) $A \cap (B \cup C) = (A \cap B) \cup (A \cap C)$.

Proof of (a): $A \cup (B \cap C) = \{x: x \in A \text{ or } x \in B \cap C\}$
$= \{x: x \text{ belongs to } A \text{ or } x \text{ belongs to both } B \text{ and } C\}$

= {x: x belongs to A or B and x belongs to A or C}
= {x: x ∈ A ∪ B and x ∈ A ∪ C} = (A ∪ B) ∩ (A ∪ C).
The analogous argument for (b) is left to the reader.   □

Definition: *Let A and B be subsets of a set X. The **set difference** B\A is the set of all points of B which do not belong to A:*

$$B \backslash A = \{x: x \in B \text{ and } x \notin A\}.$$

*The difference X\A is called the **complement** of A relative to X.*

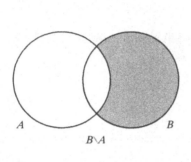
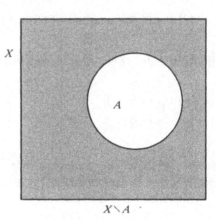

FIGURE 1.12

Note that the complement of *A* relative to *X* is used only when *A* is a subset of *X*, but the difference *B\A* is defined whether *A* is a subset of *B* or not.

Theorem 1.4:   De Morgan's Laws.   *If A and B are subsets of a set X, then*

(a)   *X\(A ∪ B) = (X\A) ∩ (X\B);*
(b)   *X\(A ∩ B) = (X\A) ∪ (X\B).*

Proof of (a):   *X\(A ∪ B) = {x: x ∈ X and x ∉ A ∪ B}*
= {x: x ∈ X and x belongs to neither A nor B}
= {x: x ∈ (X\A) and x ∈ (X\B)} = (X\A) ∩ (X\B).
The analogous proof of (b) is left as an exercise.   □

Definition: *Let {A_i: i ∈ I} be a family of sets indexed by index set I. The **union** and **intersection** of this family of sets are defined respectively by*

$$\bigcup_{i \in I} A_i = \{x: x \in A_i \text{ for some } i \in I\}$$

$$\bigcap_{i \in I} A_i = \{x: x \in A_i \text{ for all } i \in I\}.$$

The reader will no doubt observe that the definitions of union and intersection of two sets are special cases of the above definitions in which the index set is a set of two elements.

It is customary to denote a family of sets A_i indexed by the integers 1 through n by $\{A_i\}_{i=1}^n$ and to express the union and intersection of such a family by

$$\bigcup_{i=1}^n A_i \text{ and } \bigcap_{i=1}^n A_i.$$

Similarly, $\{A_i\}_{i=1}^\infty$ denotes a family of sets indexed by the set of positive integers, and the union and intersection of such a family are denoted by

$$\bigcup_{i=1}^\infty A_i \text{ and } \bigcap_{i=1}^\infty A_i.$$

A family $\{A_i: i \in I\}$ of sets is *pairwise disjoint* provided that no two of the sets have any member in common:

$$A_i \cap A_j = \emptyset \quad \text{for} \quad i \neq j.$$

The next two theorems are natural extensions of Theorems 1.3 and 1.4 to the general case.

Theorem 1.5:   Distributive Laws for Union and Intersection.   *Let A be a set and $\{B_i: i \in I\}$ a family of sets indexed by set I. Then*

(a) $A \cap \bigcup_{i \in I} B_i = \bigcup_{i \in I} (A \cap B_i)$;

(b) $A \cup \bigcap_{i \in I} B_i = \bigcap_{i \in I} (A \cup B_i)$.

Proof: *This theorem can be proved in a manner similar to that used for Theorem 1.3. For a slightly different approach that may be easier to follow, a more detailed argument is given here for (a), showing that the sets $A \cap \bigcup_{i \in I} B_i$ and $\bigcup_{i \in I} (A \cap B_i)$ are subsets of each other.*

Suppose $x \in A \cap \bigcup_{i \in I} B_i$. Then $x \in A$ and x belongs to B_i for some $i \in I$. Thus $x \in A \cap B_i$ for some i, so $x \in \bigcup_{i \in I} (A \cap B_i)$. From this we conclude

$$A \cap \bigcup_{i \in I} B_i \subset \bigcup_{i \in I} (A \cap B_i).$$

For the reverse inclusion, suppose $x \in \bigcup_{i \in I} (A \cap B_i)$. Then $x \in A \cap B_i$ for some $i \in I$. This means that $x \in A$ and $x \in B_i$ for some $i \in I$. Thus $x \in A$ and $x \in \bigcup_{i \in I} B_i$, so $x \in A \cap \bigcup_{i \in I} B_i$. Thus

$$\bigcup_{i \in I} (A \cap B_i) \subset A \cap \bigcup_{i \in I} B_i$$

and the proof of (a) is complete.

The proof of (b) is left as an exercise.   □

Theorem 1.6:   De Morgan's Laws.   Let X be a set and $\{A_i : i \in I\}$ a family of subsets of X. Then

(a) $X \backslash (\bigcup_{i \in I} A_i) = \bigcap_{i \in I} (X \backslash A_i)$;

(b) $X \backslash (\bigcap_{i \in I} A_i) = \bigcup_{i \in I} (X \backslash A_i)$.

Proof of (a):   Let $x \in X \backslash (\bigcup_{i \in I} A_i)$. Then $x \in X$ and x does not belong to A_i for any $i \in I$. Thus $x \in (X \backslash A_i)$ for all $i \in I$ and $x \in \bigcap_{i \in I} (X \backslash A_i)$.

For the opposite inclusion, let $x \in \bigcap_{i \in I} (X \backslash A_i)$. Then x belongs to the complement of each set A_i, so x does not belong to the union of the sets A_i. In other words, $x \in X \backslash (\bigcup_{i \in I} A_i)$. Since we have demonstrated that the sets $X \backslash (\bigcup_{i \in I} A_i)$ and $\bigcap_{i \in I} (X \backslash A_i)$ are subsets of each other, we conclude that they are equal.

The proof of (b) is left as an exercise.   □

EXERCISE 1.4

1.   Prove part (b) of Theorems 1.3 and 1.5.

2.   Prove part (b) of Theorems 1.4 and 1.6.

3.   Prove that for any sets A and B, $A \subset B$ if and only if $A \cap B = A$.

4.   Let X be a set with subsets A and B. Prove:

(a) $X \backslash (X \backslash A) = A$.

(b) If $A \subset B$, then $X \backslash B \subset X \backslash A$.

(c) $A \subset B$ if and only if $X \backslash B \subset X \backslash A$.

(d) $X \backslash A \subset B$ if and only if $A \cup B = X$.

(e) $A \subset X \backslash B$ if and only if $A \cap B = \emptyset$.

(f) $A \backslash B = A \cap (X \backslash B)$.

(g) $X \backslash (A \backslash B) = B \cup (X \backslash A)$.

5.   Assume that A, B, and C are subsets of a set X. Express each of the following using the symbols $\cup$, $\cap$, and $\backslash$.

(a)   The elements of X which belong to both A and B but do not belong to C.

(b)   The elements of X which belong to C and to either A or B.

(c)   The elements of X which belong to A but not to both B and C.

(d)   The elements of X which do not belong to all three sets A, B, and C.

(e)   The elements of X which fail to belong to at least two of the sets A, B, and C.

1.5 CARTESIAN PRODUCTS

An *ordered pair* of objects is a set of two objects in which one element is designated as the first term and the other element as the second term. For objects x and y, (x, y) denotes the ordered pair whose first term is x and whose second term is y. If n is a positive integer, an *ordered n-tuple* $(a_1, a_2, \ldots, a_n)$ is an ordered arrangement of n objects.

Definition: *Let A and B be sets. The **Cartesian product** or simply **product** of A and B is the set $A \times B$ (read A cross B) of all ordered pairs whose first terms belong to A and whose second terms belong to B:*

$$A \times B = \{(a, b): a \in A \text{ and } b \in B\}.$$

Example 1.5.1

The number plane of ordinary analytic geometry is the product $\mathbb{R} \times \mathbb{R}$, where $\mathbb{R}$ denotes the set of real numbers.

The definition of Cartesian product is easily extended to more than two factors. If $\{A_i\}_{i=1}^n$ is a finite sequence of sets, then their Cartesian product, denoted by $A_1 \times A_2 \times \cdots \times A_n$ or by $\prod_{i=1}^n A_i$, is defined by

$$\prod_{i=1}^n A_i = \{(a_1, a_2, \ldots, a_n): a_i \in A_i \text{ for each } i = 1, 2, \ldots, n\}.$$

For an infinite sequence of sets $\{A_i\}_{i=1}^\infty$, the Cartesian product is defined by

$$\prod_{i=1}^\infty A_i = \{(a_1, a_2, a_3, \ldots): a_i \in A_i \text{ for each positive integer } i\}.$$

Notation The symbol $\mathbb{R}^n$ denotes the Cartesian product formed by taking the set $\mathbb{R}$ of real numbers as a factor n times. Thus $\mathbb{R}^n = \{(x_1, x_2, \ldots, x_n): x_i \text{ is a real number for each } i = 1, 2, \ldots, n\}$. In particular, $\mathbb{R}^2$ and $\mathbb{R}^3$ are the ordinary plane and three-dimensional spaces encountered in calculus. The set $\mathbb{R}^n$ is called *Euclidean n-dimensional space*. Naturally, we consider the real line $\mathbb{R} = \mathbb{R}^1$ to be one-dimensional Euclidean space.

EXERCISE 1.5

1. If A has m members and B has n members, prove that $A \times B$ has mn members.

2. Prove that $\varnothing \times B = \varnothing$ for each set B.

3. Sketch each of the following sets on the number plane.

   (a) $\{0\} \times \mathbb{R}$ (d) $\{(x, y): 0 \leq y \leq 5\}$

   (b) $\mathbb{R} \times \{0, 1\}$ (e) $\{x: 0 \leq x \leq 1\} \times \{y: 1 \leq y \leq 2\}$

   (c) $\{(x, y): x \text{ and } y \text{ are both integers}\}$

4. Find a subset of $\mathbb{R}^2$ which does not equal $A \times B$ for any subsets A and B of $\mathbb{R}$.

5. Suppose that X and Y are sets having at least two members. Prove that $X \times Y$ has a subset which is not the product of a subset of X with a subset of Y.

6. Show that $(A \times B) \times C$, $A \times (B \times C)$, and $A \times B \times C$ are identical for any sets A, B, and C except for placement of parentheses. (In practice, no distinction is made for products of sets associated in different ways.)

1.6 FUNCTIONS

The following sequence of definitions involving functions is probably familiar from the study of calculus.

Definition: *A function f from set X to set Y, denoted f: X → Y or X $\xrightarrow{f}$ Y, is a rule which assigns to each member x of X a unique member y = f(x) of Y. If y = f(x), then y is called the **image** of x and x is called a **preimage** of y. The set X is the **domain** of f and Y is the **codomain** or **range** of f.*

Note that for a function $f: X \to Y$, each element x in X has a unique image $f(x)$ in Y. However, the number of preimages of a point y in Y is not necessarily one; the number of preimages may be zero, one, or more than one.

Definition: *Let f: X → Y be a function: For a subset A of X, the set*

$$f(A) = \{y \in Y: y = f(x) \text{ for some } x \in A\}$$

is called the **image** of *A* under *f*. The set *f(X)*, the image of the domain under *f*, is sometimes called the **image** of the function.

For a subset *C* of *Y*, the set

$$f^{-1}(C) = \{x \in X : f(x) \in C\}$$

is the **inverse image** of *C* under *f*. The set of points $\{(x, y) \in X \times Y : y = f(x)\}$ is called the **graph** of the function *f*.

Example 1.6.1

Consider the function $f : \mathbb{R} \to \mathbb{R}$ defined by the rule $f(x) = x^2$. The graph of this function appears in Figure 1.13.

Since $f(2) = 4$, then 4 is the image of 2 under *f* and 2 is a preimage of 4. However, -2 is also a preimage of 4. Note that a member *y* of the range has two, one, or zero preimages as the value of *y* is positive, zero, or negative.

For an example of an image set and inverse image, note that

$$f(\{0, 2, -3\}) = \{0, 4, 9\}$$
$$f^{-1}(\{0, 4, 9\}) = \{0, 2, -2, 3, -3\}.$$

For any subset *D* of $\mathbb{R}$ consisting only of negative values, $f^{-1}(D) = \varnothing$. The image of *f* is the set of non-negative real numbers:

$$f(\mathbb{R}) = \{y \in \mathbb{R} : y \geq 0\}.$$

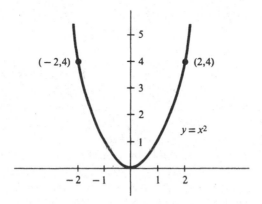

FIGURE 1.13

Definition: *A function f: X → Y is* **one-to-one** *or* **injective** *means that for distinct elements x_1, x_2, of X, $f(x_1) \neq f(x_2)$. In other words, f is one-to-one provided that no* *two distinct points in the domain have the same image. In contrapositive form this can be stated as:* $f(x_1) = f(x_2)$ *implies that* $x_1 = x_2$.

A function f for which $f(X) = Y$, that is, for which the image $f(X)$ equals the codomain, is said to map X **onto** Y or to be **surjective.**

A one-to-one function from X onto Y is called a **one-to-one correspondence** or a **bijection.** Thus $f: X \to Y$ is a one-to-one correspondence provided that each member of Y is the image under f of exactly one member of X. In this case there is an **inverse function** $f^{-1}: Y \to X$ which assigns to each y in Y its unique preimage $x = f^{-1}(y)$ in X.

Example 1.6.2

Let $X = \{a, b, c, d, e\}$, $Y = \{1, 2, 3, 4, 5\}$. The function $f: X \to Y$ defined by

$$f(a) = 1, \quad f(b) = 2, \quad f(c) = 3, \quad f(d) = 4, \quad f(e) = 5$$

is a one-to-one correspondence with inverse function $f^{-1}: Y \to X$ defined by

$$f^{-1}(1) = a, \quad f^{-1}(2) = b, \quad f^{-1}(3) = c, \quad f^{-1}(4) = d, \quad f^{-1}(5) = e.$$

Theorem 1.7:   *Let* $f: X \to Y$ *be a function,* A_1 *and* A_2 *subsets of* X, *and* B_1 *and* B_2 *subsets of* Y. *Then*

(a) $f(A_1 \cup A_2) = f(A_1) \cup f(A_2)$,
(b) $f(A_1 \cap A_2) \subset f(A_1) \cap f(A_2)$,
(c) $f^{-1}(B_1 \cup B_2) = f^{-1}(B_1) \cup f^{-1}(B_2)$,
(d) $f^{-1}(B_1 \cap B_2) = f^{-1}(B_1) \cap f^{-1}(B_2)$.

Proof: (a) *Since* A_1 *and* A_2 *are subsets of* $A_1 \cup A_2$, *it follows easily that* $f(A_1)$ *and* $f(A_2)$ *are subsets of* $f(A_1 \cup A_2)$. *Thus* $f(A_1) \cup f(A_2) \subset f(A_1 \cup A_2)$. *For the reverse inclusion, consider a member* $y \in f(A_1 \cup A_2)$. *Then* $y = f(x)$ *for some* $x \in A_1 \cup A_2$. *Thus* $y = f(x)$ *for some* x *in either* A_1 *or* A_2. *Hence* $y \in f(A_1) \cup f(A_2)$, *so* $f(A_1 \cup A_2) \subset f(A_1) \cup f(A_2)$ *and the proof of (a) is complete.*

(b) *Since* $A_1 \cap A_2$ *is a subset of both* A_1 *and* A_2, *it follows from the definition of image that* $f(A_1 \cap A_2)$ *is a subset of both* $f(A_1)$ *and* $f(A_2)$. *Thus* $f(A_1 \cap A_2) \subset f(A_1) \cap f(A_2)$. *The reverse inclusion is not true in general. It is left as an exercise for the reader to find an example for which* $f(A_1) \cap f(A_2)$ *is not a subset of* $f(A_1 \cap A_2)$.

(c) $f^{-1}(B_1 \cup B_2) = \{x \in X: f(x) \in B_1 \cup B_2\}$
$= \{x \in X: f(x) \in B_1 \text{ or } f(x) \in B_2\} = \{x \in X: f(x) \in B_1\} \cup \{x \in X: f(x) \in B_2\}$
$= f^{-1}(B_1) \cup f^{-1}(B_2)$.
The proof of part (d) is similar.   □

Definition: *The **identity function** $i_X: X \to X$ from a set X into itself is the function defined by*

$$i_X(x) = x, \quad x \in X.$$

Definition: *If $f: X \to Y$ and $g: Y \to Z$ are functions on the indicated sets, then the **composite function** $g \cdot f: X \to Z$ is defined by*

$$g \cdot f(x) = g(f(x)), \quad x \in X.$$

The composite function $g \cdot f$ is sometimes denoted simply gf.

The idea of composition of functions is illustrated in Figure 1.14.

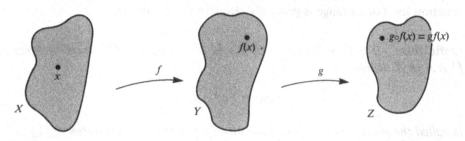

FIGURE 1.14

Example 1.6.3

Consider the functions $f: \mathbb{R} \to \mathbb{R}$ and $g: \mathbb{R} \to \mathbb{R}$ defined by $f(x) = x^2$, $g(x) = x + 1$. Then the composite functions $g \cdot f$ and $f \cdot g$ are both defined and have the following formulas:

$$g \cdot f(x) = g(f(x)) = g(x^2) = x^2 + 1,$$
$$f \cdot g(x) = f(g(x)) = f(x + 1) = (x + 1)^2.$$

It is left as an exercise for the reader to show that if $f: X \to Y$ is a one-to-one correspondence with inverse function $f^{-1}: Y \to X$, then the composite function $f^{-1} \cdot f$ is the identity function on X and $f \cdot f^{-1}$ is the identity function on Y.

Definition: *A **sequence** is a function whose domain is the set $\mathbb{Z}^+$ of all positive integers or the set of positive integers less than or equal to some given positive integer*

*N. The sequence is called **infinite** if its domain is $\mathbb{Z}^+$ and **finite** if its domain is $\{1, 2, \ldots, N\}$. A sequence is usually represented in the form $\{x_n\}_{n=1}^{\infty}$ or $\{x_n\}_{n=1}^{N}$ where x_n denotes the value of the sequence at the integer n.*

Example 1.6.4

The sequence $\{1/n\}_{n=1}^{\infty}$ is the function f from $\mathbb{Z}^+$ into the set of real numbers whose values are $f(1) = 1, f(2) = 1/2, f(3) = 1/3, \ldots, f(n) = 1/n, \ldots$.

In practice, the word *sequence* is often used to refer to what is properly the range or image of the sequence. For example, one often refers to the sequence of points $\{1/n\}_{n=1}^{\infty}$, meaning the set $\{1/n: n \in \mathbb{Z}^+\}$ rather than the function which maps 1 to 1, 2 to 1/2, 3 to 1/3, and so on. Whether the object of interest is the function itself or its range is generally clear from the context.

Definition: *If $f: X \to Y$ is a function and A is a subset of X, then the function $f|_A: A \to Y$ defined by*

$$f|_A(a) = f(a), \ a \in A,$$

*is called the **restriction** of f to A. Equivalently, f is called an **extension** of $f|_A$ to X.*

EXERCISE 1.6

1. Give an example to show that $f(A_1 \cap A_2)$ may not equal $f(A_1) \cap f(A_2)$. Show that equality does hold if f is one-to-one.

2. Let $f: X \to Y$ be a one-to-one correspondence with inverse function $f^{-1}: Y \to X$. Show that $f^{-1} \cdot f$ and $f \cdot f^{-1}$ are the identity functions on X and Y, respectively.

3. Let $f: X \to Y$ be a function, $\{A_i: i \in I\}$ a family of subsets of X, and $\{B_j: j \in J\}$ a family of subsets of Y. Prove that
    (a) $f(\bigcup_{i \in I} A_i) = \bigcup_{i \in I} f(A_i)$;
    (b) $f(\bigcap_{i \in I} A_i) \subset \bigcap_{i \in I} f(A_i)$;
    (c) $f^{-1}(\bigcup_{j \in J} B_j) = \bigcup_{j \in J} f^{-1}(B_j)$;
    (d) $f^{-1}(\bigcap_{j \in J} B_j) = \bigcap_{j \in J} f^{-1}(B_j)$.

4. Let $f: X \to Y$ be a function, A a subset of X and B a subset of Y. Prove:
    (a) $A \subset f^{-1}(f(A))$.
    (b) $f(f^{-1}(B)) \subset B$.
    (c) If f is one-to-one, then $f^{-1}(f(A)) = A$.

(d) If f maps X onto Y, then $f(f^{-1}(B)) = B$.

(e) $f^{-1}(Y\backslash B) = X\backslash f^{-1}(B)$.

(f) $f(A \cap f^{-1}(B)) = f(A) \cap f(f^{-1}(B)) = f(A) \cap B$.

5. Let $f: X \to Y$ and $g: Y \to Z$ be functions on the indicated sets. For a subset A of Z, show that $(g \cdot f)^{-1}(A) = f^{-1}(g^{-1}(A))$.

6. Let $f: X \to Y$ and $g: Y \to Z$ be functions on the indicated sets. Prove:

    (a) If f and g are one-to-one, then $g \cdot f$ is one-to-one.

    (b) If f and g are surjective, then $g \cdot f$ is surjective.

    (c) If f and g are bijections, then $g \cdot f$ is a bijection and $(g \cdot f)^{-1} = f^{-1} \cdot g^{-1}$.

7. **Definition.** *Suppose $f: X \to Y$ is a function. A **left inverse** for f is a function $g: Y \to X$ for which $g \cdot f$ is the identity function on X. A **right inverse** for f is a function $h: Y \to X$ for which $f \cdot h$ is the identity function on Y.* Prove the following statements for a function $f: X \to Y$.

    (a) f is one-to-one if and only if f has a left inverse.

    (b) f maps X onto Y if and only if f has a right inverse.

    (c) f is a one-to-one correspondence if and only if there is a function $k: Y \to X$ such that $k \cdot f$ and $f \cdot k$ are the identity functions on X and Y respectively. Show in addition that if $k \cdot f = i_X$ and $f \cdot k = i_Y$, then $k = f^{-1}$. (Do not assume that the left inverse from (a) and right inverse from (b) are equal without proper proof.)

8. Let X and Y be sets and A a subset of X.

    (a) Explain how it is possible for two different functions $f: X \to Y$ and $g: X \to Y$ to have identical restrictions to A.

    (b) Explain how it is possible for a function $h: A \to Y$ to have more than one extension to X.

9. Let X and Y be sets, A a subset of X, and $f: A \to Y$ and $F: X \to Y$ functions on the sets indicated. Show that F is an extension of f if and only if the graph of f is a subset of the graph of F.

1.7 EQUIVALENCE RELATIONS

Definition: *Let X be a set. A **relation** R on X is a subset of $X \times X$. If $(x, y) \in R$, it is customary to say that x is **related** to y by R and to write xRy.*

Example 1.7.1

The usual order for real numbers determines a relation R, called the *less than relation*, on the set of real numbers as follows: $(x, y) \in R$ or xRy provided that

x is less than y. It is customary to denote this relation by $x < y$ instead of xRy. Note in this example that no element is related to itself by R and that xRy ($x < y$) is definitely not the same as yRx ($y < x$).

Definition: *A relation R on a set X is called **reflexive, symmetric,** or **transitive** if it satisfies the corresponding property stated below:*

  (a)   *The Reflexive Property: xRx for all $x \in X$.*
  (b)   *The Symmetric Property: If xRy, then yRx.*
  (c)   *The Transitive Property: If xRy and yRz, then xRz.*

The relation $<$ on the set of real numbers is transitive, but it is not reflexive or symmetric. The relation $\leq$ is reflexive and transitive but not symmetric. Equality is reflexive, symmetric, and transitive.

Definition: *An **equivalence relation** on a set X is a relation on X which is reflexive, symmetric, and transitive.*

Notation   Equivalence relations are usually denoted by symbols like $\sim$, $\approx$, and $\equiv$ rather than by letters of the alphabet. A slanted bar through the symbol indicates that the relation does not hold: $x \not\approx y$ indicates that x is not related to y by $\approx$.

Example 1.7.2   Examples of Equivalence Relations

  (a)   The relation of equality on any set.
  (b)   Congruence of geometric figures.
  (c)   Similarity of plane triangles.
  (d)   The relation $\sim$ defined on the set $\mathbb{Z}$ of integers as follows: $x \sim y$ means that $x - y$ is an even integer. Under this relation any two even integers are related to each other, and any two odd integers are related to each other. Anticipating the next definition, we say that there are two "equivalence classes" for this relation; one class is the set of even integers, and the other is the set of odd integers.

Definition: *Let $\approx$ denote an equivalence relation on X. For $x \in X$, the set $[x]$ of all elements of X to which x is related by $\approx$ is called the **equivalence class** of x:*

$$[x] = \{y \in X : y \approx x\}.$$

Proofs to establish the following properties of equivalence classes are left as exercises:

  (a) $x \in [x]$ for each $x \in X$.

(b) $x \approx y$ if and only if $[x] = [y]$.

(c) $x \napprox y$ if and only if $[x] \cap [y] = \varnothing$.

(d)   For x, y in X, $[x]$ and $[y]$ are either identical or disjoint.

Example 1.7.3

(a) *Congruence modulo 5*, denoted $\equiv$ mod 5, is the relation on $\mathbb{Z}$ defined as follows: $x \equiv y$ mod 5 means that $x - y$ is divisible by 5. Congruence modulo 5 is an equivalence relation with the following five equivalence classes:

$$[0] = \{0, 5, -5, 10, -10, \ldots\} \ (\text{multiples of 5})$$

$$[1] = \{1, 6, -4, 11, -9, \ldots\} \quad (\text{integers of the form } 5k + 1)$$

$$[2] = \{2, 7, -3, 12, -8, \ldots\} \quad (\text{integers of the form } 5k + 2)$$

$$[3] = \{3, 8, -2, 13, -7, \ldots\} \quad (\text{integers of the form } 5k + 3)$$

$$[4] = \{4, 9, -1, 14, -6, \ldots\} \quad (\text{integers of the form } 5k + 4)$$

(b) *Congruence modulo n*, $\equiv$ mod n, is the relation on $\mathbb{Z}$ defined as follows: $x \equiv y$ mod n means that $x - y$ is divisible by n. This is an equivalence relation with n equivalence classes $[0], [1], \ldots, [n - 1]$ corresponding to the possible remainders when dividing by n.

EXERCISE 1.7

1. Let $\sim$ be the so-called sibling relation defined on the set of all people: $x \sim y$ means that y is a sibling of x. For the purposes of this problem, assume that each person is a sibling of himself or herself. Prove that $\sim$ is an equivalence relation.

2. Determine whether the following relations satisfy the reflexive, symmetric, or transitive properties:

    (a) $>$ on the set $\mathbb{R}$ of real numbers,

    (b) $\geq$ on $\mathbb{R}$,

    (c) $\approx$ defined on $\mathbb{Z}$ by $x \approx y$ if and only if $x - y$ is odd,

    (d) $\sim$ defined on $\mathbb{Z}$ by $x \sim y$ if and only if x is a divisor of y.

3. Let $f \colon X \to Y$ be a function, and define a relation $\sim$ on X as follows: $x \sim y$ if and only if $f(x) = f(y)$.

    (a)   Prove that $\sim$ is an equivalence relation.

    (b)   Prove that the equivalence class $[x]$ is the set of preimages of $f(x)$:

    $$[x] = f^{-1}(\{f(x)\}).$$

4. Show that congruence modulo n is an equivalence relation on the set of integers.

5. Consider the unit circle C with equation $x^2 + y^2 = 1$ in the plane. Thus

$$C = \{(x, y) \in \mathbb{R} \times \mathbb{R} : x^2 + y^2 = 1\}.$$

Define a relation $\sim$ on C as follows: For $(x, y) \in C$, (x, y) is related to itself and to its *antipodal point* $(-x, -y)$. In symbols,

$$(x, y) \sim (x, y); \quad (x, y) \sim (-x, -y).$$

(a) Show that $\sim$ is an equivalence relation.

(b) Think of C as a rubber band and imagine gluing together the equivalent antipodal points of C. Describe the resulting figure. (Much more will be said on this topic in Chapter 7.)

SUGGESTIONS FOR FURTHER READING

For intuitive examples illustrating topological concepts, see *Intuitive Concepts in Elementary Topology* by B. H. Arnold, *First Concepts of Topology* by Steenrod and Chinn, and *What is Mathematics?* by Courant and Robbins. For an interesting historical account of the development of point-set topology, see *The Genesis of Point-Set Topology* by J. Manheim.

The papers "Elementary Point-Set Topology" by R. H. Bing and "Topology" by Tucker and Bailey may also be of interest. (Complete bibliographical information on suggested books and papers appears in the Bibliography.)

2 The Line and the Plane

In this chapter we shall examine some topological properties of the real number line $\mathbb{R}$ and the Euclidean plane $\mathbb{R}^2$. Some of the ideas presented here will be familiar from algebra and calculus, but others may be new. In later chapters, the ideas introduced here will be carried over to a much more general family of sets. Section 2.1 reviews properties of the real number system, and Section 2.2 uses the real number system to introduce the concept of the number of elements in a set. Topological considerations begin formally with consideration of open and closed sets in Section 2.3.

2.1 UPPER AND LOWER BOUNDS

In order to establish some necessary terminology, let us review the various types of intervals studied in calculus.

Let a and b be real numbers with $a < b$. The set

$$(a, b) = \{x \in \mathbb{R} : a < x < b\}$$

is the *open interval* with *endpoints* a and b and

$$[a, b] = \{x \in \mathbb{R} : a \le x \le b\}$$

is the corresponding *closed interval*. Both of the sets

$$(a, b] = \{x \in \mathbb{R} : a < x \le b\}$$
$$[a, b) = \{x \in \mathbb{R} : a \le x < b\}$$

are called *half-open and half-closed intervals*. The latter two types are distinguished by the fact that $(a, b]$ is *open on the left and closed on the right*, while $[a, b)$ is *closed on the left and open on the right*. The sets (a, b), $[a, b]$, $(a, b]$, and $[a, b)$ are called *bounded intervals*, since they do not extend indefinitely in either the positive or negative direction. We shall also have occasion to refer to the *unbounded open intervals*

$$(a, \infty) = \{x \in \mathbb{R} : a < x\}, \quad (-\infty, a) = \{x \in \mathbb{R} : x < a\}, \quad (-\infty, \infty) = \mathbb{R}$$

and to the *unbounded closed intervals*

$$[a, \infty) = \{x \in \mathbb{R} : a \le x\}, \quad (-\infty, a] = \{x \in \mathbb{R} : x \le a\}, \quad (-\infty, \infty) = \mathbb{R}.$$

Note that intervals from a to b have been defined only for $a < b$. For the case of a closed interval, the definition is extended to allow equality of the endpoints by defining $[a, a]$ to be the singleton set $\{a\}$. The empty set $\varnothing$ is sometimes called the *empty interval*. Intervals that are either empty or have only one point are called *degenerate intervals*.

The term "interval" is used to refer to sets of the type (a, b), $[a, b]$, $(a, b]$, $[a, b)$, (a, ∞), $(-\infty, a)$, $[a, \infty)$, $(-\infty, a)$, $\mathbb{R}$, and $\varnothing$ when no further specialization in terms of openness, closedness, or boundedness is needed. Open intervals may be of the bounded type (a, b) and the unbounded types (a, ∞), $(-\infty, a)$, and $\mathbb{R}$. Closed intervals may be of the bounded type $[a, b]$ and the unbounded types $[a, \infty)$, $(-\infty, a]$, and $\mathbb{R}$. A further characterization of intervals will be undertaken in Chapter 5.

Definition: *A number u is an **upper bound** for a set A of real numbers provided that $a \leq u$ for all $a \in A$. If there is a smallest upper bound u_0 for A, that is, an upper bound u_0 less than all other upper bounds for A, then u_0 is called the **least upper bound** or **supremum** of A. The least upper bound for a set A is denoted by lub A or sup A.*

Example 2.1.1

(a)   Any real number greater than or equal to 3 is an upper bound for the set $\{0, 1, 2, 3\}$. The least upper bound is 3.

(b)   The least upper bound of $[a, b]$ is b.

(c)   The least upper bound of (a, b) is b.

(d)   The least upper bound of the set $(-\infty, 0)$ of negative numbers is 0.

(e)   The set $(0, \infty)$ of positive numbers has no upper bounds and therefore has no least upper bound.

Note that some sets of real numbers do not have upper bounds and that a set which has upper bounds may or may not contain its least upper bound. If a set contains its least upper bound, the least upper bound is simply the largest member of the set.

Definition: *A number l is a **lower bound** for a set A of real numbers provided that $l \leq a$ for all $a \in A$. If there is a largest lower bound l_0 for A, that is, a lower bound greater than all other lower bounds for A, then l_0 is called the **greatest lower bound** or **infimum** of A. The greatest lower bound of a set A is denoted by glb A or inf A.*

Example 2.1.2

   (a)   The greatest lower bound of $\{0, 1, 2, 3\}$ is 0.

   (b)   The greatest lower bound of $[a, b]$ is a.

   (c)   The greatest lower bound of (a, b) is a.

   (d)   The interval $(-\infty, 0)$ has no lower bounds and therefore has no greatest lower bound.

   (e)   The greatest lower bound of the interval $(0, \infty)$ is 0.

A set may or may not have lower bounds. A set which has lower bounds may or may not contain the greatest lower bound. If a set contains its greatest lower bound, then the greatest lower bound is the smallest member of the set.

The Least Upper Bound Property: Every non-empty set of real numbers which has an upper bound has a least upper bound.

We shall not prove the Least Upper Bound Property but rather accept it as one of the defining axioms of the real number system. A complete list of the axioms for the real number system, with corresponding constructive definition of $\mathbb{R}$, can be found in many textbooks on real analysis. Some references are given in the suggested reading list at the end of the chapter. It is left as an exercise for the reader to use this property to prove the corresponding Greatest Lower Bound Property.

The Greatest Lower Bound Property: Every non-empty set of real numbers which has a lower bound has a greatest lower bound.

Example 2.1.3

   (a)   Consider the sequence $\{1/n\}_{n=1}^{\infty}$ of real numbers 1, 1/2, 1/3, 1/4. . . . The least upper bound of this set is 1, the largest member. To see that 0 is the greatest lower bound, we reason as follows: Certainly 0 is a lower bound for the set since $0 < 1/n$ for each positive integer n. Now, a number $\epsilon > 0$ cannot be a lower bound for $\{1/n\}_{n=1}^{\infty}$ because $1/n < \epsilon$ when n is an integer greater than $1/\epsilon$. Thus 0 is a lower bound, and no number greater than 0 is a lower bound. This means that 0 is greater than every other lower bound and is hence the greatest lower bound of the sequence.

   (b)   The sequence $\{1 - 1/n\}_{n=1}^{\infty}$ of real numbers 0, 1/2, 2/3, 3/4 . . . has greatest lower bound 0 (the smallest term) and least upper bound 1.

To see that 1 is the least upper bound, first note that $1 - 1/n < 1$ for every positive integer n, so 1 is an upper bound. It now must be shown that no number less than 1 is an upper bound. To this end, consider a number $1 - \epsilon$, where $\epsilon > 0$, and let n be a positive integer greater than $1/\epsilon$. Then

$$1 - \epsilon < 1 - 1/n.$$

Thus $1 - \epsilon$ is not an upper bound for $\{1 - 1/n\}_{n=1}^{\infty}$, so 1 must be its least upper bound.

The next theorem is the first example of an important phenomenon called *denseness*, which will appear many times in later chapters. The theorem shows that there are rational numbers "very close" to every real number. (Recall that a *rational number* is a real number which can be expressed as p/q for some integers p, q. A real number which is not rational is called *irrational*.)

Theorem 2.1:   *Between any two real numbers there is a rational number.*

Proof: *Let a and b be real numbers with $a < b$. It must be shown that there is a rational number r with $a < r < b$. Intuitively, the argument proceeds as follows: Let q be a positive integer and consider the rational numbers p/q for $p = 0, \pm 1, \pm 2, \ldots$. The numbers p/q, when arranged in order on the number line, have successive terms separated by distance $1/q$. If $1/q$ is less than the distance from a to b, it seems reasonable that at least one number of the form p/q must fall between a and b.*

This intuitive idea is made precise by the upper bound concept. Let q be a positive integer for which $1/q$ is less than $b - a$. The set $P = \{p/q: p \in \mathbb{Z}\}$ has no bounds, either upper or lower. In particular, a is neither an upper bound nor a lower bound for P. Thus there is an integer p_0 such that $p/q \leq a$ when $p \leq p_0$ and $p/q > a$ when $p > p_0$. Then the rational number $r = (p_0 + 1)/q$ is greater than a. The fact that r is less than b follows from properties of p_0 and q: $p_0/q \leq a$ and $1/q < b - a$ so

$$r = \frac{p_0 + 1}{q} = p_0/q + 1/q < a + (b - a) = b.$$

Thus r is a rational number between a and b.   □

Theorem 2.1 can be extended to show that there is an unending sequence of rational numbers between any two real numbers a and b. There is a rational number r_1 between a and b. There must be, in addition, a rational number r_2 between a and r_1 and a rational number r_3 between r_1 and b. This process can always be repeated to show the existence of another rational number between any two already determined.

Theorem 2.1 also shows that there are rational numbers within any prescribed positive distance of a given real number. For a real number a and distance ϵ, there must be a rational number between $a - \epsilon$ and $a + \epsilon$. Such a rational number will be at a distance less than ϵ from a. The fact that there are rational numbers arbitrarily close to every real number is expressed by saying that the set of rational numbers is *dense* in $\mathbb{R}$.

Example 2.1.4

Let A denote the set of rational numbers between 0 and $\sqrt{2}$. It should be clear from the preceding discussion that A has least upper bound $\sqrt{2}$ and greatest lower bound 0.

EXERCISE 2.1

1. Find the least upper bound and greatest lower bound, if they exist, for the following sets:

   (a) The set $\mathbb{Z}$ of integers.

   (b) The set $\mathbb{Z}^+$ of positive integers.

   (c) The set of rational numbers greater than π.

   (d) $(-1, 2) \cup (3, 7)$.

2. Explain why a set of real numbers cannot have more than one least upper bound or more than one greatest lower bound.

3. Prove that a set cannot contain more than one of its upper bounds or more than one of its lower bounds.

4. Prove the Greatest Lower Bound Property assuming the Least Upper Bound Property as an axiom. (*Hint:* There is a natural correspondence between upper bounds of a set A and lower bounds of the set $-A$ of negatives of members of A.)

2.2 FINITE AND INFINITE SETS

Section 2.1 dealt with several subsets of the set of real numbers: integers, intervals, sequences, rational numbers, and irrational numbers. It was shown that between any two real numbers there is a rational number. In view of this property, how many rational numbers are there? Is the number of rational numbers equal to the number of real numbers? How does one compare sets to determine which one has the greater number of members anyway?

The definitions and theorems of this section apply to sets in general, not just to subsets of the real line. The real line is simply used in this section as the primary source of examples.

Definition: *A set A is **finite** provided that A is empty or there is a positive integer N for which there is a one-to-one correspondence between A and the integers 1 through N. In the latter case it is said that A has N members. A set which is not finite is called **infinite**.*

Example 2.2.1

Each of the following sets is infinite:

(a)   the set $\mathbb{Z}$ of integers,

(b)   the set $\mathbb{Z}^+$ of positive integers,

(c)   the set $\mathbb{R}$ of real numbers,

(d)   any interval with endpoints a and b for which $a < b$,

(e)   the set of rational numbers.

Definition: *Two sets A and B are **equipotent** or **have the same cardinal number** provided that there is a one-to-one correspondence from A onto B. This relation is expressed by $A \approx B$ or card A = card B. The terms "cardinally equivalent" and simply "equivalent" are sometimes used synonymously with "equipotent."*

The definition of equipotence of sets is more fundamental than the principle of counting. A small child, for example, learns that right and left hands have equal numbers of fingers by pressing corresponding fingers together in a one-to-one correspondence long before he or she can count to five. Furthermore, we shall soon see that the principle of counting does not apply to a large class of infinite sets.

Theorem 2.2:   *Equipotence of sets is an equivalence relation.*

Proof: *It must be shown that the relation $\approx$ of set equipotence is reflexive, symmetric, and transitive.*

The Reflexive Property: For any set A, $A \approx A$ because the identity function on A is one-to-one correspondence.

The Symmetric Property: *Suppose $A \approx B$ by a one-to-one correspondence f from A onto B. Then the inverse function f^{-1} is a one-to-one correspondence from B onto A, so $B \approx A$.*

The Transitive Property: *Suppose $A \approx B$ and $B \approx C$ by one-to-one correspondences $f: A \to B$ and $g: B \to C$. Then the composite map $g \cdot f$ is a one-to-one correspondence from A onto C, so $A \approx C$.*   □

Definition: *A set A is **denumerable** or **countably infinite** provided that A is equivalent to the set $\mathbb{Z}^+$ of positive integers. A set which is either finite or countably infinite is called **countable**; a set which is not countable is called **uncountable**.*

We thus have the following hierarchy of sets, according to size: finite sets, countably infinite sets, and uncountable sets. Sets of the first two types are referred to collectively as *countable sets*. Demonstration of the following important properties is left as an exercise for the reader:

(a) Each subset of a finite set is finite.

(b) Each subset of a countable set is countable.

(c) Each set which contains an infinite set is infinite.

(d) Each set which contains an uncountable set is uncountable.

Example 2.2.2

(a) The set $\mathbb{Z}^+ \cup \{0\}$ of all non-negative integers is countably infinite. In fact, the function $f: \mathbb{Z}^+ \cup \{0\} \to \mathbb{Z}^+$ defined by

$$f(n) = n + 1, \quad n \in \mathbb{Z}^+ \cup \{0\},$$

is a one-to-one correspondence. Thus $\mathbb{Z}^+$ has "the same number of members" as the set $\mathbb{Z}^+ \cup \{0\}$ formed by adjoining to $\mathbb{Z}^+$ another element.

(b) The set $\mathbb{Z}$ of all integers is countably infinite. It is left as an exercise for the reader to show that the following function g from $\mathbb{Z}$ to the set of non-negative integers is a one-to-one correspondence:

$$g(n) = \begin{cases} 2n - 1 & \text{if } n \text{ is positive,} \\ -2n & \text{if } n \text{ is negative or zero.} \end{cases}$$

This function was obtained by starting with the following correspondences: $0 \to 0, 1 \to 1, -1 \to 2, 2 \to 3, -2 \to 4$, etc.

(c)   The Cartesian product $\mathbb{Z}^+ \times \mathbb{Z}^+$ is countably infinite. The product $\mathbb{Z}^+ \times \mathbb{Z}^+$ of all ordered pairs (m, n) of positive integers is represented by the points having integral coordinates in the first quadrant of the Euclidean plane.

Figure 2.1 suggests a method of defining a one-to-one correspondence $f: \mathbb{Z}^+ \rightarrow \mathbb{Z}^+ \times \mathbb{Z}^+$ by working successively across the diagonals.

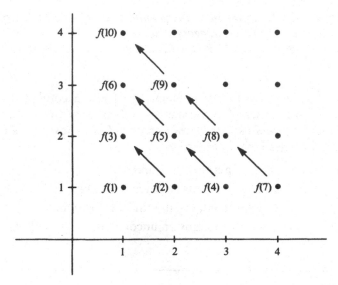

FIGURE 2.1

Another proof that $\mathbb{Z}^+ \times \mathbb{Z}^+$ is countably infinite can be made as follows: Define $g: \mathbb{Z}^+ \times \mathbb{Z}^+ \rightarrow \mathbb{Z}^+$ by

$$g(m, n) = 2^m 3^n, \quad (m, n) \in \mathbb{Z}^+ \times \mathbb{Z}^+.$$

Then g is not surjective, but the Fundamental Theorem of Arithmetic (unique factorization into primes) does guarantee that it is one-to-one. Thus $\mathbb{Z}^+ \times \mathbb{Z}^+$ is equivalent to a subset of $\mathbb{Z}^+$. Since every subset of a countable set is countable, then $\mathbb{Z}^+ \times \mathbb{Z}^+$ is countable. Since $\mathbb{Z}^+ \times \mathbb{Z}^+$ is clearly not finite, then it must be countably infinite.

Proofs of the following facts about unions and products of finite and countable sets are left as exercises:

Theorem 2.3:

(a)   If $\{A_i\}_{i=1}^{N}$ is a finite sequence of sets and each set A_i is finite, then $\bigcup_{i=1}^{N} A_i$ and $\prod_{i=1}^{N} A_i$ are finite.

Proof: *Every rational number can be expressed uniquely in lowest terms m/n, where m and n are integers with no common positive divisor other than 1, and n is positive. Thus we consider the function*

$$m/n \rightarrow (m, n)$$

from the set of rational numbers into $\mathbb{Z} \times \mathbb{Z}$. This function is one-to-one since the ordered pair (m,n) determines only one rational number m/n. The range $\mathbb{Z} \times \mathbb{Z}$ is countable since it is the product of two countable sets. Thus the set of rational numbers is equivalent to a subset of a countable set and hence is countable. Since the set of rational numbers is obviously not finite, then it must be countably infinite. □

Theorem 2.5: *The set $\mathbb{R}$ of real numbers is uncountable.*

Proof: *Recall that all terminating and non-terminating decimals represent real numbers. Since there is an obvious one-to-one correspondence between decimals and sequences of digits,*

$$.a_1a_2a_3 \ldots \leftrightarrow (a_1, a_2, a_3 \ldots),$$

Example 2.2.3 shows that the set of all decimals of the form $.a_1a_2a_3 \ldots$ where each a_i is 0 or 1 is uncountable. Note also that the two distinct decimals whose terms can be only 0 or 1 cannot represent the same real number. Thus $\mathbb{R}$ contains an uncountable set and must itself be uncountable. □

Corollary: *The set of irrational numbers is uncountable.*

Proof: *If the set I of irrational numbers were countable, then, since the set R of rational numbers is countable, it would follow that their union*

$$\mathbb{R} = I \cup R$$

is countable. Since $\mathbb{R}$ is uncountable, then I must be uncountable. □

EXERCISE 2.2

Explain why every set which contains an uncountable set must be uncountable.

Prove that the function g of Example 2.2.2 (b) is a one-to-one correspondence.

Prove Theorem 2.3.

4. Prove that every subset of a countable set is countable.

5. Prove that any two non-degenerate closed and bounded intervals have the same cardinal number. (*Hint:* Find a one-to-one correspondence from [0, 1] onto an arbitrary non-degenerate closed interval [a, b].)

6. Prove that (0, 1) is equivalent to ℝ.

7. Prove that every (non-degenerate) open interval is equivalent to ℝ.

8. (a) Prove that any two non-degenerate intervals have the same cardinal number.

   (b) Prove that every non-degenerate interval is uncountable.

   (c) Prove that every non-degenerate interval contains both rational and irrational numbers.

9. **Definition:** *For sets A and B, we say that **card A** ≤ **card B** provided that A is equivalent to a subset of B. The inequality **card A** < **card B** means that card A ≤ card B and card A ≠ card B.*

   (a) Restate the meanings of card A ≤ card B and card A < card B in terms of functions.

   (b) Show that if card A ≤ card B, then card $\mathcal{P}(A)$ ≤ card $\mathcal{P}(B)$.

   (c) Show that if A is a countable set, then card $\mathcal{P}(A)$ ≤ card ℝ.

   Note: If A is a countably infinite set, then card $\mathcal{P}(A)$ = card ℝ, but the proof goes beyond the amount of set theory developed in this course.

   The *continuum hypothesis,* originally proposed by Georg Cantor, says that if X is an infinite set with card X ≤ card ℝ, then X is either countably infinite or is equivalent to ℝ. Thus the continuum hypothesis asserts that there is no cardinal number "between" the cardinal numbers of the set of positive integers and the set of real numbers. Surprisingly, the continuum hypothesis has been shown to be independent of the usual axioms for set theory; both the continuum hypothesis and its negation are consistent with the usual axioms. Further information on the continuum hypothesis and axiomatic set theory can be found in the suggestions for further reading at the end of the chapter.

10. Show that for any set A, card A < card $\mathcal{P}(A)$. (*Hint:* The function $a \rightarrow \{a\}$, $a \in A$, establishes a one-to-one correspondence between A and the family of all singleton sets in $\mathcal{P}(A)$. Conclude from this that card A ≤ card $\mathcal{P}(A)$. Then, proceeding by contradiction, suppose that $f: A \rightarrow \mathcal{P}(A)$ is a one-to-one correspondence from A onto $\mathcal{P}(A)$. Let $Q = \{a \in A: a \notin f(a)\}$ and consider the member q of A for which $f(q) = Q$. Show that neither of the relations $q \in f(q)$ or $q \notin f(q)$ is possible.)

2.3 OPEN SETS AND CLOSED SETS ON THE REAL LINE

Distances between points on the real line ℝ are measured by the absolute value function, with $|a - b|$ being defined as the distance from a to b. The reader is probably familiar with the following distance properties:

(a) $|a - b| \geq 0$, and $|a - b| = 0$ only when $a = b$;

(b) $|a - b| = |b - a|$;

(c) $|a - c| \leq |a - b| + |b - c|$

for any real numbers a, b, and c.

Topology deals with distances between points and sets and, more generally, with a concept of "closeness" of points and sets which can be defined independently of any distance function. In this section, we shall explore the concept of closeness for the case of points and subsets of the real line. A similar property will be defined without recourse to a distance function in Chapter 4.

Definition: *Let a be a real number and B a non-empty subset of $\mathbb{R}$. The **distance** from a to B, $d(a, B)$, is the greatest lower bound of all distances $|a - b|$ for $b \in B$:*

$$d(a, B) = glb\{|a - b|: b \in B\}.$$

Note in the preceding definition that $\{|a - b|: b \in B\}$ is a set of non-negative real numbers and thus has 0 (and any negative number) as a lower bound. The Greatest Lower Bound Property insures that $d(a, B)$ is well-defined and non-negative.

Example 2.3.1

(a) $d(0, [1, 2]) = d(0, (1, 2)) = 1$.

(b) $d(1, [1, 2]) = d(1, (1, 2)) = 0$.

(c)   For the set R of rational numbers, the denseness property of R in $\mathbb{R}$ (Theorem 2.1) insures that $d(x, R) = 0$ for every real number x.

Definition: *Let A be a subset of $\mathbb{R}$. If $\{|x - y|: x, y \in A\}$ has an upper bound, then A is a **bounded set** and the least upper bound of all distances $|x - y|$ for x, y in A is called the **diameter** $D(A)$ of A.*

Example 2.3.2

(a) $D([a, b]) = D((a, b)) = |b - a|$.

(b)   If B is the set of rational numbers in (a, b), then $D(B) = |b - a|$.

(c) $\mathbb{R}$, intervals of the form (a, ∞), $[a, \infty)$, $(-\infty, a)$, and $(-\infty, a]$, the set of rational numbers, and the set of irrational numbers are all unbounded sets.

Definition: *A subset O of the real line $\mathbb{R}$ is an **open set** provided that O is the union of some family of open intervals.*

Theorem 2.6:   *The following statements are equivalent for a subset O of $\mathbb{R}$:*

(a)   *O is an open set.*

(b)   *For each $x \in O$, there is an open interval I_x centered at x and contained in O.*

For $O \neq \mathbb{R}$, (a) and (b) are equivalent to:

(c)   *For each $x \in O$, $d(x, \mathbb{R} \setminus O) > 0$.*

Proof: *The proof will be accomplished by showing that (a) is equivalent to (b) and (b) is equivalent to (c). For condition (c) we assume that $O \neq \mathbb{R}$, since otherwise $\mathbb{R} \setminus O$ would be the empty set and the distance from x to $\mathbb{R} \setminus O$ would be undefined.*

To see that (a) implies (b), suppose O is an open set and $x \in O$. Since O is an open set, then x belongs to some open interval (a, b) contained in O. Thus $|x - a|$ and $|x - b|$ are both positive numbers. If ϵ is a positive number less than or equal to both $|x - a|$ and $|x - b|$, then $I_x = (x - \epsilon, x + \epsilon)$ is an open interval centered at x and contained in O.

The argument that (b) implies (a) is easier. Assuming (b), there exists for each x in O an open interval I_x centered at x and contained in O. Thus

$$\bigcup_{x \in O} I_x \subset O.$$

Since each x in O is a member of the interval I_x, then

$$O \subset \bigcup_{x \in O} I_x.$$

Thus

$$O = \bigcup_{x \in O} I_x,$$

so O is a union of open intervals.

To see that (b) implies (c), consider an open interval $I_x = (x - \epsilon, x + \epsilon)$ centered at x and contained in O. Then any point within distance ϵ of x must be in O, so the distance from x to any point outside O must be at least ϵ. Thus $d(x, \mathbb{R} \setminus O)$ is positive for each x in O.

Assuming that (c) holds, $d(x, \mathbb{R} \setminus O)$ is a positive number α_x. This means that the distance from x to any point outside O is at least α_x, so any real number within distance α_x of x must be in O. Thus the open interval $(x - \alpha_x, x + \alpha_x)$, which contains only points within distance α_x of x, is contained in O. Thus we conclude that (c) implies (b), and the proof is complete. $\square$

Theorem 2.7:   *The open subsets of $\mathbb{R}$ have the following properties:*

(a) $\mathbb{R}$ *and* $\emptyset$ *are open sets.*

(b)   *If* $\{O_\alpha: \alpha \in A\}$ *is a collection of open sets, then* $\bigcup_{\alpha \in A} O_\alpha$ *is open.*

(c)   *If* $\{O_i\}_{i=1}^n$ *is a finite collection of open sets, then* $\bigcap_{i=1}^n O_i$ *is open.*

Proof:

(a)   *The real line* $\mathbb{R}$ *is open since it can be expressed in many different ways as a union of open intervals. For example,*

$$\mathbb{R} = (-\infty, \infty) = \bigcup_{n=1}^{\infty} (-n, n) = \bigcup_{n=-\infty}^{\infty} (n, n+2).$$

The empty set $\emptyset$ *is open since it is the union of the empty collection of open intervals.*

(b)   *If* $\{O_\alpha: \alpha \in A\}$ *is a family of open sets indexed by a set A, then for each $\alpha \in A$, O_α is a union of open intervals. Then $\bigcup_{\alpha \in A} O_\alpha$ is the union of all the open intervals of which the open sets O_α are composed and is, therefore, an open set. Property (b) is sometimes paraphrased by saying that the union of any family of open sets is open.*

(c)   *We shall prove (c) first for $n = 2$ and then complete the proof by induction. Suppose that O_1 and O_2 are open sets and that point x belongs to $O_1 \cap O_2$. By Theorem 2.6, there exist open intervals I_1 and I_2 centered at x and contained in O_1 and O_2, respectively. But then $I_1 \cap I_2$ is an open interval centered at x and contained in $O_1 \cap O_2$, and it follows from Theorem 2.6 that $O_1 \cap O_2$ is open.*

Proceeding inductively, suppose that the intersection of every family of $n - 1$ open sets is open and consider a family $\{O_i\}_{i=1}^n$ of n open sets. Then $\bigcap_{i=1}^{n-1} O_i$ and O_n are open sets, so what has just been proved shows that their intersection $\bigcap_{i=1}^n O_i$ is open also. Property (c) is paraphrased by saying that the intersection of any finite family of open sets is open.

□

Definition: *A subset C of $\mathbb{R}$ is a **closed set** if its complement $\mathbb{R} \setminus C$ is open.*

Note in the preceding definition that the term "closed set" does *not* mean a set which is not open. We shall see many examples to illustrate that "closed" and "not open" are very different properties for sets.

Theorem 2.8:   *The closed subsets of $\mathbb{R}$ have the following properties:*

(a) $\mathbb{R}$ *and* $\emptyset$ *are closed sets.*

(b)   If $\{C_\alpha : \alpha \in A\}$ *is a family of closed sets, then* $\bigcap_{\alpha \in A} C_\alpha$ *is closed.*

(c)   If $\{C_i\}_{i=1}^n$ *is a finite family of closed sets, then* $\bigcup_{i=1}^n C_i$ *is closed.*

Proof: (a) ℝ *and* ∅ *are closed because their respective complements* ∅ *and* ℝ *are open.*

Statements (b) and (c) are proved from Theorem 2.7 and De Morgan's Laws (Theorem 1.6). The method is illustrated here for (b), and the analogous proof of (c) is left to the reader. For any collection $\{C_\alpha : \alpha \in A\}$ *of closed subsets of* ℝ, *the family of complements* $\{\mathbb{R} \backslash C_\alpha : \alpha \in A\}$ *is a family of open sets. Theorem 2.7 insures that*

$$\bigcup_{\alpha \in A} (\mathbb{R} \backslash C_\alpha) = \mathbb{R} \backslash \bigcap_{\alpha \in A} C_\alpha$$

is an open set. Then, by definition, $\bigcap_{\alpha \in A} C$ *is a closed set.*   □

Note the duality between open sets and closed sets: A set is closed if and only if its complement is open. The union of any family of open sets is open; the intersection of any family of closed sets is closed. The intersection of any finite family of open sets is open; the union of any finite family of closed sets is closed.

Example 2.3.3

(a)   It is obvious from the definition that an open interval is an open set. It is true, but not quite so obvious, that a closed interval $[a, b]$ is a closed set:

$$\mathbb{R} \backslash [a, b] = (-\infty, a) \cup (b, \infty)$$

is open so $[a, b]$ is closed. Closed intervals of the form $(-\infty, a]$ and $[a, \infty)$ are also closed sets.

(b)   Since a singleton set $\{a\}$ can be regarded as a closed interval $[a, a]$, then each singleton subset of ℝ is closed. Combining this fact with statement (c) of Theorem 2.8, we conclude that every finite subset of ℝ is closed.

(c)   Many subsets of ℝ are neither open nor closed. Examples of such sets are the rational numbers, the irrational numbers, and half-open half-closed intervals. (The interval $[a, b)$ is not open because it contains no open interval centered at a; it is not closed because its complement

$$\mathbb{R} \backslash [a, b) = (-\infty, a) \cup [b, \infty)$$

contains no open interval centered at b.)

The announced idea of closeness of a point to a set appears in the next definition.

Definition: *A point x in* $\mathbb{R}$ *is a **limit point** or **accumulation point** of a subset A of* $\mathbb{R}$ *provided that every open set containing x contains a point of A distinct from x.*

Note in the definition of limit point that it is not required that every open set containing x contain the same point of A; different open sets may contain different points of A. Note also that the possibility is left open for a limit point of A to be either in A or outside it. Limit points are sometimes called *cluster points*.

Theorem 2.9:   *A real number x is a limit point of a subset A of* $\mathbb{R}$ *if and only if* $d(x, A\backslash\{x\}) = 0.$

Proof: *Suppose first that x is a limit point of A and let* ϵ *be a positive number. Then the open set* $(x - \epsilon, x + \epsilon)$ *contains a point y of A distinct from x. Since* $y \in (x - \epsilon, x + \epsilon)$, *then* $d(x, y) < \epsilon$ *so* $d(x, A\backslash\{x\}) < \epsilon$. *Since the latter inequality holds for all* $\epsilon > 0$, *then* $d(x, A\backslash\{x\}) = 0.$
    Suppose now that $d(x, A\backslash\{x\}) = 0$ *and consider an open set O containing x. Then O contains an open interval* $(x - \delta, x + \delta)$ *for some positive number* δ. *Since* $d(x, A\backslash\{x\}) < \delta$ *and the interval* $(x - \delta, x + \delta)$ *consists precisely of all points at a distance less than* δ *from x, then* $(x - \delta, x + \delta)$ *must contain a point z in* $A\backslash\{x\}$. *Thus*

$$z \in (x - \delta, x + \delta) \subset O$$

and $z \neq x$ *since* $z \in A\backslash\{x\}$. *Hence every open set containing x contains a point of A distinct from x, and x is a limit point of A.* $\square$

Theorem 2.9 is interpreted intuitively as saying that x is a limit point of A if and only if x is arbitrarily close to points of A other than x itself. (The phrase "other than x itself" is applicable only in case x belongs to A.) This interpretation of limit point is illustrated in the following examples.

Example 2.3.4

(a)   0 is the only limit point of $\{1/n\}_{n=1}^{\infty}$. Note in this case that the limit point is outside the set.

(b)   The set of limit points of a closed interval $[a, b]$, $a < b$, is precisely the same interval.

(c) The set of limit points of an open interval (a, b), $a < b$, is the corresponding closed interval $[a, b]$. In the latter case, the set is a proper subset of its set of limit points.

(d) A finite subset of $\mathbb{R}$ has no limit points.

(e) The set of limit points of the set of rational numbers is the entire real line. This is a consequence of Theorem 2.1.

The following theorem explains the relationship between closed sets and limit points.

Theorem 2.10: *A subset A of $\mathbb{R}$ is closed if and only if A contains all its limit points.*

Proof: *Suppose first that A is closed and consider a limit point x of A. If x were outside A, then $\mathbb{R} \backslash A$ would be an open set containing x but containing no point of A, and we would be forced to conclude that x is not a limit point of A. Thus, if x is a limit point of A, then it must be a member of A.*

Now suppose that A contains all its limit points. We shall show that A is closed by showing that $\mathbb{R} \backslash A$ is open. If $y \in \mathbb{R} \backslash A$, then y is not a limit point of A so there is an open set O_y containing y but containing no point of A. Then $\mathbb{R} \backslash A$ is the union of such open sets O_y and is an open set. Thus A is closed. □

The reader should be aware of the distinction between limit point and limit of a sequence. A real number x is the *limit* of a sequence $\{a_n\}_{n=1}^{\infty}$, or $\{a_n\}_{n=1}^{\infty}$ *converges* to x, provided that given $\epsilon > 0$ there is a positive integer N such that if $n \geq N$, then $|a_n - x| < \epsilon$. As the reader will see in Problem 9 of Exercise 2.3, saying that x is a limit point of A is equivalent to saying that there is a sequence $\{a_n\}_{n=1}^{\infty}$ of distinct members of A which has limit x. It should be noted, however, that if a sequence does not have distinct terms, then its limit is not necessarily a limit point of the range of the sequence. Keep in mind for future reference that sequences are not adequate to determine limit points in the more general context in which limit points will be considered later in the course. In a general topological space, if a sequence of distinct points of a set A converges to a point x, then x is a limit point of A. However, A may have limit points which are not limits of sequences of distinct points of A. We shall return to this subject in Section 4.1.

EXERCISE 2.3

1. Let x be a real number, A a subset of $\mathbb{R}$, and ϵ a positive number. Prove that $(x - \epsilon, x + \epsilon) \subset A$ if and only if $d(x, \mathbb{R} \backslash A) \geq \epsilon$.

2. Let x be a real number and A a subset of $\mathbb{R}$.

   (a) Prove that if $d(x, A) > 0$, then $d(x, y) > 0$ for all $y \in A$.

   (b) Give an example for which $d(x, y) > 0$ for all $y \in A$ but $d(x, A) = 0$.

3. Prove that a subset of $\mathbb{R}$ is bounded if and only if it has both upper and lower bounds.

4. Complete the proof of Theorem 2.8.

5. Prove that a non-empty subset C of $\mathbb{R}$ is closed if and only if $d(x, C) > 0$ for each point x in the complement of C.

6. Let a be a real number and let B, C be subsets of $\mathbb{R}$. Prove that $d(a, B \cup C)$ is the smaller of $d(a, B)$ and $d(a, C)$.

7. Give examples to show that $D(A \cup B)$ may be larger or smaller than $D(A) + D(B)$.

8. Show that if x is the limit of the sequence $\{a_n\}_{n=1}^{\infty}$ of real numbers and all the terms of the sequence are distinct, then x is a limit point of the range of the sequence. Give an example to show that the limit of a sequence may not be a limit point of the range of the sequence if the terms of the sequence are not distinct.

9. Let x be a real number and A a subset of $\mathbb{R}$.

   (a) Prove that x is a limit point of A if and only if there is a sequence of distinct points of A which converges to x.

   (b) Prove that x is a limit point of A if and only if every open set containing x contains infinitely many points of A.

2.4 THE NESTED INTERVALS THEOREM

The closed intervals $[0, 2]$, $[1/2, 3/2]$, $[2/3, 4/3]$, $\ldots$ $[(n - 1)/n, (n + 1)/n]$, $\ldots$ have precisely one point in common, the number 1. The main result of this section shows that any such collection of "shrinking" or "nested" closed intervals with diameters approaching zero must have exactly one point in common. This may seem intuitively obvious or even uninteresting, but it is a property of great importance in mathematics. It is one of the early topological discoveries of Georg Cantor in his work on $\mathbb{R}$. We shall prove Cantor's Nested Intervals Theorem and formulate two related but less obvious properties in this section. The importance of Cantor's theorem will become clear in Chapter 6 and in the reader's study of real analysis.

Definition: *A sequence $\{S_n\}_{n=1}^{\infty}$ of sets is **nested** if $S_{n+1} \subset S_n$ for each positive integer n.*

Theorem 2.11: Cantor's Nested Intervals Theorem *If $\{[a_n, b_n]\}_{n=1}^{\infty}$ is a nested sequence of closed and bounded intervals, then $\bigcap_{n=1}^{\infty} [a_n, b_n]$ is not empty. If, in* *addition, the diameters of the intervals converge to 0, then $\bigcap_{n=1}^{\infty} [a_n, b_n]$ has precisely one member.*

Proof: *Since $[a_{n+1}, b_{n+1}] \subset [a_n, b_n]$ for each positive integer n, the sequences $\{a_n\}_{n=1}^{\infty}$ and $\{b_n\}_{n=1}^{\infty}$ of left and right endpoints have the following properties:*

(i) *$\{a_n\}_{n=1}^{\infty}$ is an increasing sequence $(a_1 \leq a_2 \leq \cdots \leq a_n \leq a_{n+1} \leq \cdots)$;*

(ii) *$\{b_n\}_{n=1}^{\infty}$ is a decreasing sequence $(b_1 \geq b_2 \geq \cdots \geq b_n \geq b_{n+1} \geq \cdots)$;*

(iii) *Each left endpoint is less than or equal to each right endpoint.*

Let c denote the least upper bound of the left endpoints and d the greatest lower bound of the right endpoints. Note that the existence of c and d is guaranteed by the Least Upper Bound Property and the Greatest Lower Bound Property, respectively. Then, by property (iii), $c \leq b_n$ for each n so $c \leq d$. Since $a_n \leq c \leq d \leq b_n$, then $[c, d] \subset [a_n, b_n]$ for each n. Thus $\bigcap_{n=1}^{\infty} [a_n, b_n]$ contains the closed interval $[c, d]$ and must therefore be non-empty.

If we assume further that the diameters of the intervals $[a_n, b_n]$ approach 0, then it follows that $c = d$ and that c is the one point in $\bigcap_{n=1}^{\infty} [a_n, b_n]$. $\square$

Example 2.4.1

The closed intervals $[a_n, b_n]$ of Cantor's Nested Intervals Theorem cannot be replaced by open intervals. Note, for example, that $\{(0, 1/n)\}_{n=1}^{\infty}$ has empty intersection.

The next two theorems are consequences of Theorem 2.11, but they are not so intuitively plausible.

Theorem 2.12: The Heine-Borel Theorem *Let $[a, b]$ be a closed and bounded interval and $\mathcal{O}$ a collection of open intervals whose union contains $[a, b]$. Then there is a finite subset $\{O_1, O_2, \ldots, O_N\}$ of $\mathcal{O}$ whose union contains $[a, b]$.*

Proof: *The following terminology will simplify the proof. The intervals $[c, (c + d)/2]$ and $[(c + d)/2, d]$ will be called the **left half** and **right half** of the interval $[c, d]$, respectively. If $[c, d]$ is contained in the union of a finite number of members of $\mathcal{O}$, then we shall say that $[c, d]$ is **finitely coverable** by $\mathcal{O}$.*

Proceeding with the proof of the theorem by the method of contradiction, suppose that $[a, b]$ is not finitely coverable by $\mathcal{O}$. Then either the left half or the right half of $[a, b]$ is not finitely coverable by $\mathcal{O}$. Let $[a_1, b_1]$ denote either half of $[a, b]$ which is not finitely coverable by $\mathcal{O}$.

The same reasoning applies to $[a_1, b_1]$. Since $[a_1, b_1]$ is not finitely coverable by $\mathcal{O}$, it has (at least) one half $[a_2, b_2]$ which is not finitely coverable by $\mathcal{O}$. Applying *this reasoning inductively, there exists a nested sequence $\{[a_n, b_n]\}_{n=1}^{\infty}$ of closed intervals, none of which is finitely coverable by $\mathcal{O}$, such that $[a_{n+1}, b_{n+1}]$ is either the left half or the right half of $[a_n, b_n]$. The diameters of the intervals $[a_n, b_n]$ have limit 0 since the length decreases by a factor of 1/2 at each stage; i.e., the length of $[a_n, b_n]$ is $(b - a)/2^n$.*

Cantor's Nested Intervals Theorem guarantees that there is precisely one point p common to all the intervals $[a_n, b_n]$. Since $p \in [a, b]$, there is some open interval O in $\mathcal{O}$ with $p \in O$. Let ϵ be a positive number such that $(p - \epsilon, p + \epsilon) \subset O$, and let n be a positive integer such that $(b - a)/2^n < \epsilon$. Then, since $p \in [a_n, b_n]$, it follows that

$$[a_n, b_n] \subset (p - \epsilon, p + \epsilon) \subset O.$$

*But this contradicts the fact that $[a_n, b_n]$ is not finitely coverable by $\mathcal{O}$: $[a_n, b_n]$ is contained in **one** member of $\mathcal{O}$. Assuming that $[a, b]$ is not finitely coverable by $\mathcal{O}$ has led to a contradiction, so we conclude that $[a, b]$ is finitely coverable by $\mathcal{O}$.* □

Example 2.4.2

The closed interval $[a, b]$ of the preceding theorem cannot be replaced by an open interval (a, b). Note for example, that $(0, 1)$ is contained in the union of the family of open intervals $\mathcal{O} = \{(1/n, 1)\}_{n=2}^{\infty}$ but is not finitely coverable by $\mathcal{O}$.

Theorem 2.13:   The Bolzano-Weierstrass Theorem   *Every bounded, infinite subset of $\mathbb{R}$ has a limit point.*

The proof of the Bolzano-Weierstrass Theorem is left as an exercise with the hint that a proof can be modeled after the proof of Theorem 2.12: A bounded, infinite set must be a subset of some closed interval $[a, b]$. Divide the interval $[a, b]$ into halves, and the halves into halves, and so on, with at least one half at each stage always containing an infinite number of members of the original infinite set.

EXERCISE 2.4

1. Give an example of an infinite subset of $\mathbb{R}$ which has no limit point.

2. Give an example of a nested sequence $\{[a_n, b_n)\}_{n=1}^{\infty}$ whose intersection is empty.

3. Consider $[0, 1]$ and the family of open intervals $\mathcal{O} = \{(-0.001, 0.001), (0.999, 1.001)\}$ $\cup \{(1/n, 1)\}_{n=1}^{\infty}$. Find a finite subcollection of $\mathcal{O}$ whose union contains $[0, 1]$.

4. Prove the Bolzano-Weierstrass Theorem (Theorem 2.13).

5. Prove the following generalization of Cantor's Nested Intervals Theorem (Theorem 2.11): If $\{A_n\}_{n=1}^{\infty}$ is a nested sequence of sets each of which is closed, bounded, and non-empty, then $\bigcap_{n=1}^{\infty} A_n$ is non-empty.

6. Prove the following generalization of the Heine-Borel Theorem (Theorem 2.12): Let $[a, b]$ be a closed, bounded interval and $\mathcal{U}$ a collection of open sets whose union contains $[a, b]$. Then there is a finite subcollection $\{U_1, U_2, \ldots, U_N\}$ of $\mathcal{U}$ whose union contains $[a, b]$.

7. Prove that the theorem stated in Problem 6 remains valid with $[a, b]$ replaced by an arbitrary closed and bounded subset A of $\mathbb{R}$. (*Hint:* $A \subset [a, b]$ for some closed interval $[a, b]$ and $\mathbb{R}\backslash A$ is an open set.)

8. Show that every uncountable subset of $\mathbb{R}$ has a limit point. (*Hint:* Show that such a set must have infinitely many of its members in an interval of the form $[n, n + 1]$ where n is an integer.)

2.5 THE PLANE

The purpose of this section is to show that the ideas presented for the real line in Sections 2.3 and 2.4 are also applicable to the Euclidean plane $\mathbb{R}^2$. Recall that the *distance* $d(a, b)$ between points $a = (a_1, a_2)$ and $b = (b_1, b_2)$ in $\mathbb{R}^2$ is defined by

$$d(a, b) = ((a_1 - b_1)^2 + (a_2 - b_2)^2)^{1/2}.$$

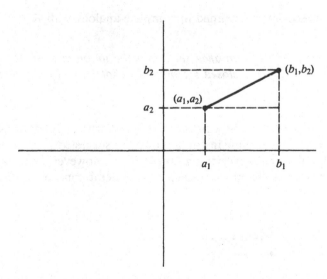

FIGURE 2.2   The distance from (a_1, a_2) to (b_1, b_2) is $((a_1 - b_1)^2 + (a_2 - b_2)^2)^{1/2}$.

This distance function satisfies the following properties:

(a) $d(a, b) \geq 0$, and $d(a, b) = 0$ only when $a = b$;

(b) $d(a, b) = d(b, a)$;

(c) $d(a, c) \leq d(a, b) + d(b, c)$

for any points $a = (a_1, a_2)$, $b = (b_1, b_2)$, $c = (c_1, c_2)$ in $\mathbb{R}^2$.

The basic object used to define open sets in $\mathbb{R}$ is the open interval. The generalization of open sets to $\mathbb{R}^2$ can easily be made once the subsets of $\mathbb{R}^2$ which are analogous to the open intervals of $\mathbb{R}$ are determined. An open interval in $\mathbb{R}$ centered at a is an interval $(a - r, a + r)$, where r is a positive number. In terms of distance, $(a - r, a + r)$ consists of all points x in $\mathbb{R}$ for which the distance from x to a is less than r:

$$(a - r, a + r) = \{x \in \mathbb{R}: |a - x| < r\}.$$

We naturally call a the *center* and r the *radius* of the interval.

It is now easy to formulate the analogue of an open interval for $\mathbb{R}^2$: It is a disk that excludes the circular edge in the same way that an open interval excludes the endpoints. The term "ball" is used rather than "disk" to anticipate the generalization to dimensions three and higher.

Definition: *Let* $a = (a_1, a_2) \in \mathbb{R}^2$ *and let* r *be a positive number. The **open ball** $B(a, r)$ with **center** a and **radius** r is the set*

$$B(a, r) = \{x = (x_1, x_2) \in \mathbb{R}^2: d(a, x) < r\}.$$

Open sets and closed sets are defined in complete analogy with $\mathbb{R}$.

Definition: *A subset O of $\mathbb{R}^2$ is an **open set** if it is the union of some family of open balls. A subset C of $\mathbb{R}^2$ is a **closed set** provided that its complement $\mathbb{R}^2 \backslash C$ is open.*

Theorems corresponding to those proved for $\mathbb{R}$ in Sections 2.3 and 2.4 can be formulated and proved for $\mathbb{R}^2$; some instructive ones are suggested in the exercises for this section. Rather than redo everything for the plane, however, let us set our sights a bit higher. The next chapter introduces the general concept of distance

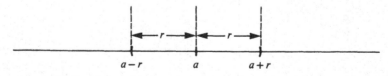

FIGURE 2.3   **The open interval with center** *a* **and radius** *r.*

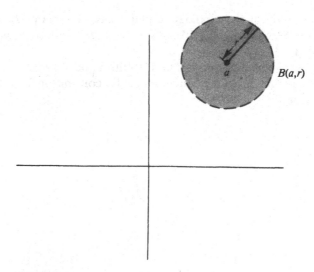

FIGURE 2.4

function, and the properties touched on here for the plane are presented in detail for any set which has a distance function.

EXERCISE 2.5

1. Give the definition of the following terms for $\mathbb{R}^2$:
   (a) Distance from a point to a set.
   (b) Bounded set.
   (c) Diameter of a set.
   (d) Limit point of a set.
2. Prove the analogues of Theorems 2.7 and 2.8 for $\mathbb{R}^2$.
3. State and prove the analogue of Cantor's Nested Intervals Theorem (Theorem 2.11) for a nested sequence $\{[a_n, b_n] \times [c_n, d_n]\}_{n=1}^{\infty}$ of closed rectangles in $\mathbb{R}^2$.
4. Prove the analogues of Theorems 2.9 and 2.10 for $\mathbb{R}^2$.

SUGGESTIONS FOR FURTHER READING

For a readable introduction to the real number system based on the axiomatic approach, see Apostol's *Calculus*. Rudin's *Principles of Mathematical Analysis* is recommended for a constructive definition of the real number system.

For further study of the topology of plane sets, Newman's *Elements of the Topology of Plane Sets of Points* and Wall's *A Geometric Introduction to Topology* are recommended.

Numbers, Sets and Axioms by A. G. Hamilton and *Infinity and the Mind* by R. Rucker both give readable introductions to the continuum hypothesis and to axiomatic set theory.

HISTORICAL NOTES FOR CHAPTER 2

The ideas introduced in this chapter were developed during the latter part of the nineteenth century as mathematicians examined the real number system and attempted to make rigorous the foundations of calculus. Those primarily responsible for coherent definitions of the real number system were Karl Weierstrass, Richard Dedekind (1831–1916), Charles Méray (1835–1911), and Georg Cantor (1845–1918). Least upper bound and greatest lower bound were defined by Weierstrass in his lectures during the 1880's, but they were used in less rigorous form by Bernard Bolzano (1781–1848) as early as 1817.

It is Cantor who deserves the credit for the theory of infinite sets. In a series of papers during the years 1872 to 1878, Cantor defined equivalence of sets, countable sets, and uncountable sets. He was led to these ideas by his work on the convergence of trigonometric series, examining the "exceptional points" where the series failed to converge. Cantor proved that the set of rational numbers is countable by the diagonal counting method of Example 2.2.2(c). He proved also that the set of algebraic numbers, which includes the rational numbers, is countable. (An *algebraic number* is a real number which is a root of a polynomial equation with integer coefficients.) Cantor showed that the union of a countable family of countable sets is countable. The argument of Example 2.2.3 was given by Cantor in 1890 to prove that the closed interval [0, 1] is uncountable.

The properties of open and closed sets in Section 2.3 are also due to Cantor. He defined limit points, dense sets, open sets, and closed sets and established their properties for the line, plane, and higher dimensional Euclidean spaces during the years 1872 to 1890. Cantor's Nested Intervals Theorem (Theorem 2.11) was proved in 1884. Considerations of limit points and related ideas were made independently and approximately concurrently with those of Cantor by Paul du Bois-Reymond (1831–1889).

The Heine-Borel Theorem (Theorem 2.12) was first proved by Emile Borel (1871–1956) in 1894 under the additional assumption that the collection of open intervals whose union contains [a, b] is countable. More will be said about the history and significance of Borel's theorem in the historical notes to Chapter 6.

The Bolzano-Weierstrass Theorem (Theorem 2.13) is usually credited to Weierstrass in the 1880's, but the method of proof used in the text for Theorem 2.12 and suggested for the Bolzano-Weierstrass Theorem was used by Bolzano in his lectures in 1817 and is clearly explained in his book *Paradoxien des Unendlichen* (*Paradoxes of the Infinite*), published in 1851.

3

Metric Spaces

In the preceding chapter we studied the notion of limit point, defined in terms of open sets, for the real line and plane. That investigation continues in this chapter with the line and plane replaced by an arbitrary set in which it is possible to measure distances. The properties required of a distance function, also called a *metric*, are given in the definition in Section 3.1. In analogy with $\mathbb{R}$ and $\mathbb{R}^2$, there is a natural definition of open sets for any set X on which a metric is defined. This chapter explores the fundamental and most useful aspects of metrics, which are of real importance in modern mathematics. Historically, the development of metric space led to the more general concept of topological space, which is introduced in Chapter 4.

3.1 THE DEFINITION AND SOME EXAMPLES

Definition: *Let X be a set and $d{:}X \times X \to \mathbb{R}^+$ a function from $X \times X$ to the set $\mathbb{R}^+$ of non-negative real numbers satisfying the following properties. For all x, y, z in X,*

(a) $d(x, y) = 0$ *if and only if* $x = y$;
(b) $d(x, y) = d(y, x)$;
(c) $d(x, z) \le d(x, y) + d(y, z)$.

*Then d is called a **metric** or **distance function** on X and $d(x, y)$ is called the **distance** from x to y. The set X with metric d is called a **metric space** and is denoted by (X, d).*

Notice that the properties required of a metric are parallel to those of the distance functions used for $\mathbb{R}$ and $\mathbb{R}^2$ in Chapter 2. In analogy with the plane, property (c) is often called the "Triangle Inequality." When the metric under consideration is clear or when the symbol for the metric is unimportant, we shall often omit mention of the metric and refer to metric space X instead of (X, d).

The real line $\mathbb{R}$ and the plane $\mathbb{R}^2$, with the metrics defined for them in Chapter 2, are special cases of the most important class of metric spaces, the *Euclidean spaces*

$$\mathbb{R}^n = \{x = (x_1, x_2, \ldots, x_n){:} \ x_i \text{ is a real number for } i = 1, 2, \ldots, n\}$$

with metric d defined for each n by

$$d(x, y) = \left(\sum_{i=1}^{n} (x_i - y_i)^2 \right)^{1/2}, \quad x = (x_1, \ldots, x_n), \quad y = (y_1, \ldots, y_n) \in \mathbb{R}^n.$$

Strictly speaking, we should use a symbol like d^n, indicating the dimension, for the metric on $\mathbb{R}^n$ since there is a different metric in each dimension. This notation is cumbersome, however, so we shall avoid it. The dimension in question will always be indicated by the superscript of $\mathbb{R}^n$. Anticipating the fact that d is a metric for each n, we call d the *usual metric* for $\mathbb{R}^n$. Unless stated otherwise, we shall assume that $\mathbb{R}^n$ is assigned the usual metric. To see that d is a metric for $\mathbb{R}^n$, it will be helpful to review some of the vector properties of $\mathbb{R}^n$.

For points $a = (a_1, \ldots, a_n)$ and $b = (b_1, \ldots, b_n)$ in $\mathbb{R}^n$, the *sum* $a + b$ and *difference* $a - b$ are defined by

$$a + b = (a_1 + b_1, \ldots, a_n + b_n)$$

$$a - b = (a_1 - b_1, \ldots, a_n - b_n)$$

It is often said that addition and subtraction in $\mathbb{R}^n$ are defined *coordinatewise*. The *dot product* or *scalar product* $a \cdot b$ is defined by

$$a \cdot b = \sum_{i=1}^{n} a_i b_i.$$

The *norm* or *length* $\|a\|$ of a vector a is the distance from a to the origin $\theta = (0, \ldots, 0)$ (the point all of whose coordinates are zero):

$$\|a\| = d(a, \theta) = \left(\sum_{i=1}^{n} a_i^2 \right)^{1/2}.$$

With this notation, the distance between two vectors is simply the norm of their difference: $d(a, b) = \|a - b\|$.

Theorem 3.1:   The Cauchy-Schwarz Inequality   *For any points $a = (a_1, \ldots, a_n)$ and $b = (b_1, \ldots, b_n)$ in $\mathbb{R}^n$,*

$$|a \cdot b| \le \|a\| \|b\|.$$

Proof: *If either a or b is the origin, the result is true because both sides of the inequality reduce to zero. Thus we may assume that both a and b have at least one non-zero coordinate and hence that $\|a\|$ and $\|b\|$ are both positive numbers.*

For each i = 1, . . . , n,

$$0 \le \left(\frac{|a_i|}{\|a\|} - \frac{|b_i|}{\|b\|} \right)^2$$

so

$$\frac{2|a_i b_i|}{\|a\|\,\|b\|} \le \frac{a_i^2}{\|a\|^2} + \frac{b_i^2}{\|b\|^2}.$$

Then

$$\sum_{i=1}^{n} \frac{2|a_i b_i|}{\|a\|\,\|b\|} \le \sum_{i=1}^{n} \left(\frac{a_i^2}{\|a\|^2} + \frac{b_i^2}{\|b\|^2} \right).$$

Splitting the right hand side into two sums and factoring constant terms from both sides gives

$$\frac{2}{\|a\|\,\|b\|} \sum_{i=1}^{n} |a_i b_i| \le \frac{1}{\|a\|^2} \sum_{i=1}^{n} a_i^2 + \frac{1}{\|b\|^2} \sum_{i=1}^{n} b_i^2 = \frac{\|a\|^2}{\|a\|^2} + \frac{\|b\|^2}{\|b\|^2} = 2.$$

Thus

$$\frac{1}{\|a\|\,\|b\|} \sum_{i=1}^{n} |a_i b_i| \le 1$$

so

$$|a \cdot b| = \left| \sum_{i=1}^{n} a_i b_i \right| \le \sum_{i=1}^{n} |a_i b_i| \le \|a\|\,\|b\|. \qquad \square$$

Theorem 3.2:   The Minkowski Inequality   *For any points* $a = (a_1, \ldots, a_n)$ *and* $b = (b_1, \ldots, b_n)$ *in* $\mathbb{R}^n$,

$$\|a + b\| \le \|a\| + \|b\|.$$

Proof: *The Cauchy-Schwarz Inequality applies to produce the following:*

$$\|a + b\|^2 = \sum_{i=1}^{n} (a_i + b_i)^2 = \sum_{i=1}^{n} (a_i^2 + 2a_i b_i + b_i^2) = \sum_{i=1}^{n} a_i^2 + 2 \sum_{i=1}^{n} a_i b_i + \sum_{i=1}^{n} b_i^2$$

$$= \|a\|^2 + 2a \cdot b + \|b\|^2 \le \|a\|^2 + 2\|a\|\,\|b\| + \|b\|^2 = (\|a\| + \|b\|)^2.$$

Taking square roots of the first and last terms gives the desired result $\|a + b\| \leq \|a\| + \|b\|$. $\qquad\qquad\qquad\qquad\qquad\qquad\qquad\qquad\qquad\qquad\qquad\qquad\quad$ □

We are now ready to prove that d is a metric for $\mathbb{R}^n$. It follows directly from the definition that $d(x, y) = 0$ precisely when $x = y$ and that $d(x, y) = d(y, x)$. The proof of the remaining metric property, the Triangle Inequality, will use the Minkowski Inequality. For points $x = (x_1, \ldots, x_n)$, $y = (y_1, \ldots, y_n)$ and $z = (z_1, \ldots, z_n)$ in $\mathbb{R}^n$,

$$d(x, z) = \|x - z\|$$
$$= \|(x - y) + (y - z)\| \leq \|x - y\| + \|y - z\| = d(x, y) + d(y, z).$$

Thus d is a metric and $(\mathbb{R}^n, d)$ is a metric space. $\qquad\qquad\qquad\qquad\qquad\qquad$ □

Example 3.1.1   The Taxicab Metric for $\mathbb{R}^n$

Define a function d' on $\mathbb{R}^n \times \mathbb{R}^n$ as follows: For $x = (x_1, \ldots, x_n)$ and $y = (y_1, \ldots, y_n)$ in $\mathbb{R}^n$,

$$d'(x, y) = \sum_{i=1}^{n} |x_i - y_i|.$$

The proof that d' is a metric is left to the reader. It is called the *taxicab metric* because, in the plane, the distance from x to y is the sum of the lengths of a horizontal segment and a vertical segment ("streets") joining x to y.

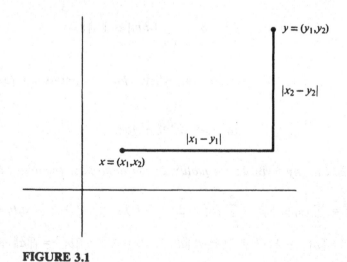

FIGURE 3.1

Example 3.1.2   The Max Metric for $\mathbb{R}^n$

Another metric d'' for $\mathbb{R}^n$ is defined by taking the largest or maximum of the absolute values of the differences of the coordinates of x and y:

$$d''(x, y) = \max \{|x_i - y_i|\}_{i=1}^n.$$

It should be clear that d'' satisfies properties (a) and (b) in the definition of metric. To see that it also satisfies the Triangle Inequality, consider points $x = (x_1, \ldots, x_n)$, $y = (y_1, \ldots, y_n)$, and $z = (z_1, \ldots, z_n)$ in $\mathbb{R}^n$:

$$d''(x, z) = \max \{|x_i - z_i|\}_{i=1}^n$$
$$= \max \{|(x_i - y_i) + (y_i - z_i)|\}_{i=1}^n \leq \max \{|x_i - y_i| + |y_i - z_i|\}_{i=1}^n$$
$$\leq \max \{|x_i - y_i|\}_{i=1}^n + \max \{|y_i - z_i|\}_{i=1}^n = d''(x, y) + d''(y, z).$$

Example 3.1.3

For an arbitrary set X, define $d(x, y)$ to be 0 when $x = y$ and 1 when $x \neq y$. The reader should check to see that d is a metric; it is called the *discrete metric* and is usually of little interest. It does demonstrate, however, that every set can be assigned a metric.

Example 3.1.4

Consider the set $\mathscr{C}[a, b]$ of all continuous real-valued functions defined on a given closed interval $[a, b]$. For f, g in $\mathscr{C}[a, b]$, define

$$\rho(f, g) = \int_a^b |f(x) - g(x)| \, dx.$$

The fact that ρ is a metric follows easily from properties of the Riemann integral. This metric measures the "distance" between two functions to be the area enclosed between their graphs from $x = a$ to $x = b$.

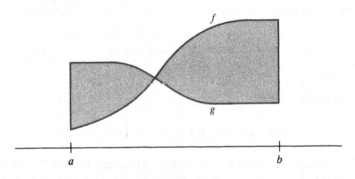

FIGURE 3.2

Example 3.1.5

For the set $\mathcal{C}[a, b]$ of Example 3.1.4, define ρ' by

$$\rho'(f, g) = \text{lub } \{ |f(x) - g(x)| : x \in [a, b] \}.$$

The proof that ρ' is a metric is left to the reader. (A proof of the Triangle Inequality can be made along the lines of Example 3.1.2.) The metric ρ' is called the *supremum metric* or the *uniform metric* for $\mathcal{C}[a, b]$. It measures the "distance" from f to g as the supremum (which, in this case, is the same as the maximum) of vertical distances from points $(x, f(x))$ to $(x, g(x))$ on the graphs of f and g.

Examples 3.1.4 and 3.1.5 suggest an important point about the value of generalization and abstraction. When people think of metric spaces, they usually conceive something like the plane with its usual distance function. Imagination and geometric intuition may suggest theorems whose proofs depend only on properties which the plane shares with other metric spaces. Such theorems would, of course, be true for all metric spaces, including the spaces of Examples 3.1.4 and 3.1.5. The process of generalization, which brings under one umbrella a large collection of apparently disparate objects, has been one of the great advances of modern mathematics. It gives much more than the simple economic benefit of not having to give a separate proof for each special situation; it also gives new and decisive insights into complicated phenomena and suggests new relationships where none had been seen before. In short, generalization and abstraction of mathematical concepts provide a deeper and more profound understanding than can be attained by considering each example in isolation.

Definition: *Let (X, d) be a metric space and A a non-empty subset of X. If $\{d(x, y): x, y \in A\}$ has an upper bound, then A is called a **bounded set** and lub $\{d(x, y): x, y \in A\}$ is called the **diameter** $D(A)$ of A. For completeness, we define the diameter of the empty set to be zero. If the set X is bounded, then (X, d) is called a **bounded metric space.***

Example 3.1.6

Consider the unit square

$$S = \{x = (x_1, x_2): 0 \leq x_i \leq 1; \quad i = 1, 2\}$$

in $\mathbb{R}^2$. With the usual metric d, this set has diameter $\sqrt{2}$; with the taxicab metric d', its diameter is 2; with the max metric d'', its diameter is 1; and with the discrete metric its diameter is 1.

Definition: *Let (X, d) be a metric space, A a non-empty subset of X, and x a point of X. The **distance** $d(x, A)$ from x to A is defined by*

$$d(x, A) = glb\{d(x, y): y \in A\}.$$

EXERCISE 3.1

1. For $a = (-2, 1)$ and $b = (3, 4)$ in $\mathbb{R}^2$, compute the distance from a to b in each of the following metrics: (a) usual, (b) taxicab, (c) max, (d) discrete.

2. Determine the distance from $(3, 4)$ to the unit square $[0, 1] \times [0, 1]$ in $\mathbb{R}^2$ with respect to each of the four metrics listed in Problem 1.

3. Prove that the taxicab metric d' is actually a metric for $\mathbb{R}^n$.

4. Prove that each of the following functions is a metric:

   (a) the discrete metric of Example 3.1.3

   (b) the function ρ of Example 3.1.4

   (c) the function ρ' of Example 3.1.5

5. Describe pictorially in $\mathbb{R}^2$ the set of points x whose distance from the origin is less than or equal to 1 with respect to each of the following metrics: (a) usual, (b) taxicab, (c) max, (d) discrete.

6. Repeat Problem 5 for the set of points whose distance from the origin is less than 1.

7. Describe pictorially (on a graph) the set of functions g in $\mathscr{C}[a, b]$ whose distance from a given function f is less than or equal to 1 for each of the following metrics: (a) the integral metric ρ of Example 3.1.4, (b) the supremum metric ρ' of Example 3.1.5, (c) the discrete metric.

8. Let $B = \{x = (x_1, x_2, x_3) \in \mathbb{R}^3: x_1^2 + x_2^2 + x_3^2 \le 1\}$ be the unit ball in $\mathbb{R}^3$. Compute the diameter of B for each of the following metrics: (a) usual, (b) taxicab, (c) max, (d) discrete.

9. Show that if (X, d) is a metric space with discrete metric d and A is a subset of X with at least two members, then the diameter of A is 1.

10. Let $A = \{x = (x_1, x_2) \in \mathbb{R}^2: x_1^2 + x_2^2 \le 1\}$ and let $b = (1, 1)$. Find the distance from b to A for the following metrics: (a) usual, (b) taxicab, (c) max, (d) discrete.

11. Let (X, d) be a metric space and A a subset of X. Prove that the diameter of A is zero if and only if A has fewer than two members.

3.2 OPEN SETS AND CLOSED SETS IN METRIC SPACES

There will be a rather obvious parallelism between the concepts defined in this section and those defined for the line and plane in Chapter 2.

Definition: *Let (X, d) be a metric space, a a member of X, and r a positive number. The **open ball** $B_d(a, r)$ with **center** a and **radius** r is the set*

$$B_d(a, r) = \{x \in X: d(a, x) < r\}.$$

*The corresponding **closed ball** $B_d[a, r]$ is defined by*

$$B_d[a, r] = \{x \in X: d(a, x) \le r\}.$$

When there is only one metric under consideration, the symbols for open balls and closed balls are sometimes simplified to $B(a, r)$ and $B[a, r]$.

The following example is for those who did not do Problem 5 of the preceding exercises.

Example 3.2.1

(a)   For the plane $\mathbb{R}^2$ with the usual metric d, $B_d(\theta, 1)$ is the region inside the circle with center at the origin θ and radius 1. The closed ball $B_d[\theta, 1]$ is the union of $B_d(\theta, 1)$ with the bounding circle.

(b)   For $\mathbb{R}^2$ with the taxicab metric d',

$$B_{d'}(\theta, 1) = \{(x, y) \in \mathbb{R}^2: |x| + |y| < 1\}$$

is the interior of the diamond shown in Figure 3.3. The closed ball $B_{d'}[\theta, 1]$ is the union of $B_{d'}(\theta, 1)$ with the four bounding line segments.

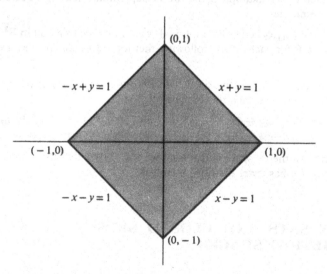

FIGURE 3.3

(c)   For $\mathbb{R}^2$ with the max metric d'',

$$B_{d''}(\theta, 1) = \{(x, y) \in \mathbb{R}^2 : \max \{|x|, |y|\} < 1\}$$

is the interior of the square of side 2 centered at θ; $B_{d''}[\theta, 1]$ is the union of $B_{d''}(\theta, 1)$ with the four bounding line segments.

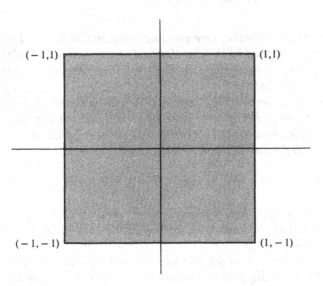

FIGURE 3.4

(d)   For any set X with the discrete metric,

$$B(a, r) = \{a\} \quad \text{if} \quad r \leq 1,$$
$$B[a, r] = \{a\} \quad \text{if} \quad r < 1,$$
$$B[a, r] = X \quad \text{if} \quad r = 1,$$
$$B(a, r) = B[a, r] = X \quad \text{if} \quad r > 1.$$

Definition: *A subset O of a metric space (X, d) is an **open set** with respect to the metric d provided that O is a union of open balls. The family of open sets defined in this way is called the **topology for X generated by** d. A subset C of X is a **closed set** with respect to d provided that its complement $X \backslash C$ is an open set with respect to d.*

As usual, when there is only one metric under consideration, repeated references to it will be omitted.

Theorem 3.3:   *The following statements are equivalent for a subset of O of a metric space (X, d).*

    (a)   O is an open set.

    (b)   For each $x \in O$, there is an open ball $B(x, \epsilon_x)$, for some positive radius ϵ_x, which is contained in O. For $O \neq X$, (a) and (b) are equivalent to:

    (c)   For each $x \in O$, $d(x, X \backslash O) > 0$.

Proof: *As was done for the corresponding result in Chapter 2 (Theorem 2.6), the proof will be accomplished by showing that (a) is equivalent to (b) and (b) is equivalent to (c). In condition (c) we again assume $O \neq X$ since the distance from a point to the empty set is not defined.*

    To see that (a) implies (b), suppose O is open and $x \in O$. Since O is a union of open balls, then x belongs to some open ball $B(a, r)$ contained in O. Then $d(x, a) < r$. Let ϵ_x be a positive number less than or equal to $r - d(x, a)$. Then $B(x, \epsilon_x) \subset B(a, r)$ for the following reason: If $y \in B(x, \epsilon_x)$,

$$d(y, a) \leq d(y, x) + d(x, a) < \epsilon_x + d(x, a) \leq r - d(x, a) + d(x, a) = r.$$

Thus $B(x, \epsilon_x)$ is an open ball of positive radius centered at x and contained in O.

    The proof that (b) implies (a) is immediate: Assuming (b), O must be the union of the balls $B(x, \epsilon_x)$.

    To see that (b) implies (c), consider an open ball $B(x, \epsilon_x)$ centered at x and contained in O. Then any point within distance ϵ_x of x is in O, so the distance from x to $X \backslash O$ must be at least ϵ_x. Thus $d(x, X \backslash O) > 0$ for each $x \in O$.

    Assuming that (c) holds, $d(x, X \backslash O)$ is a positive number α_x, depending on x. This means that the distance from x to a point outside O must be at least α_x, so any point within distance α_x of x must be in O. In other words, $B(x, \alpha_x) \subset O$.   □

Theorem 3.4:   *The open subsets of a metric space (X, d) have the following properties:*

    (a)   X and $\emptyset$ are open sets.

    (b)   The union of any family of open sets is open.

    (c)   The intersection of any finite family of open sets is open.

Proof:

    (a)   The entire space X is open since it is the union of all open balls of all possible centers and radii. The empty set $\emptyset$ is open since it is the union of the empty collection of open balls.

    (b)   If $\{O_\alpha : \alpha \in A\}$ is a collection of open sets in X, then for each α in the index set A, O_α is a union of open balls. Then $\bigcup_{\alpha \in A} O_\alpha$ is the union of all the open balls of which the open sets O_α are composed and is, therefore, an open set.

(c)   Let $\{O_i\}_{i=1}^n$ be a finite collection of open sets and let $x \in \cap_{i=1}^n O_i$. Then
by Theorem 3.3(b) there is for each $i = 1, \ldots, n$ a positive number ϵ_i
such that $B(x, \epsilon_i) \subset O_i$. Then

$$\bigcap_{i=1}^n B(x, \epsilon_i) \subset \bigcap_{i=1}^n O_i.$$

But the intersection of the balls $B(x, \epsilon_i)$ is simply the ball $B(x, \epsilon)$ where
$\epsilon = minimum \{\epsilon_i\}_{i=1}^n$, so $B(x, \epsilon)$ is an open ball centered at x and contained
in $\cap_{i=1}^n O_i$. Thus $\cap_{i=1}^n O_i$ is open.   □

In Chapter 4 we shall define a topology for an arbitrary set X by taking as the
defining properties the statements (a), (b), and (c) of Theorem 3.4.
    The proof of the following theorem, the analogue of Theorem 2.8, is left as
an exercise.

Theorem 3.5:   *The closed subsets of a metric space (X, d) have the following
properties:*

(a)   *X and $\emptyset$ are closed sets.*
(b)   *The intersection of any family of closed sets is closed.*
(c)   *The union of any finite family of closed sets is closed.*

Example 3.2.2

Whether a set is open or not open depends upon the space in which it is considered. For example, it is common practice to identify the real line $\mathbb{R}$ with the
horizontal axis $\{(x, 0) \in \mathbb{R}^2 : x \in \mathbb{R}\}$ in $\mathbb{R}^2$. Since $\mathbb{R}$ contains no open balls in $\mathbb{R}^2$,
then $\mathbb{R}$ is not open when considered as a subset of $\mathbb{R}^2$. Similarly, whether or not
a set is closed also depends upon the space in which it is being considered.

Example 3.2.3

In the plane with the usual metric, the set $A = \{(x_1, x_2) \in \mathbb{R}^2 : 0 \leq x_i < 1; i =
1, 2\}$ shown in Figure 3.5 is neither open nor closed.

Definition: *Let (X, d) be a metric space and A a subset of X. A point $x \in X$ is a*
limit point *or* ***accumulation point*** *of A provided that every open set containing x
contains a point of A distinct from x. The set of limit points of A is called its* ***de-***
rived *set.*

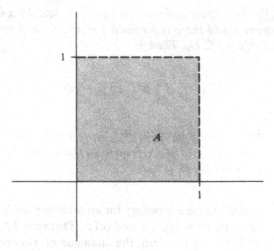

FIGURE 3.5

Theorem 3.6:   *Let (X, d) be a metric space and A a subset of X. A point $x \in X$ is a limit point of A if and only if $d(x, A\backslash\{x\}) = 0$.*

The proof of Theorem 3.6 is completely analogous to that of Theorem 2.9 and is left to the reader.

Example 3.2.4

For $\mathbb{R}^2$ with the usual metric d:

(a)   The origin is the only limit point of the sequence $\{(1/n, 1/n)\}_{n=1}^{\infty}$.

(b)   The derived set of the closed unit square $S = \{(x_1, x_2): 0 \leq x_i \leq 1; i = 1, 2\}$ is precisely the set S itself.

(c)   The derived set of the open unit square $U = \{(x_1, x_2): 0 < x_i < 1; i = 1, 2\}$ is the closed unit square. Note in this case that the set U is a proper subset of its derived set.

(d)   A finite set has no limit points.

(e)   The derived set of the set R of all points (x_1, x_2) having rational coordinates is the entire plane.

The proof of the following theorem is identical to that of Theorem 2.10, with $\mathbb{R}$ replaced by X.

Theorem 3.7:   *A subset A of a metric space (X, d) is closed if and only if A contains all its limit points.*

Definition: *Let (X, d) be a metric space and $\{x_n\}_{n=1}^{\infty}$ a sequence of points of X. Then $\{x_n\}_{n=1}^{\infty}$* **converges** *to the point $x \in X$, or x is the* **limit** *of the sequence, provided that given $\epsilon > 0$ there is a positive integer N such that if $n \geq N$, then $d(x_n, x) < \epsilon$. A sequence that converges is called a* **convergent sequence.**

Since $d(x_n, x) < \epsilon$ is equivalent to $x_n \in B(x, \epsilon)$, the definition of convergence can be restated as follows: *A sequence $\{x_n\}_{n=1}^{\infty}$ in a metric space X* **converges** *to $x \in X$ if and only if for each $\epsilon > 0$ the open ball $B(x, \epsilon)$ contains x_n for all but a finite number of positive integers n.*

Theorem 3.8: *A sequence in a metric space cannot converge to more than one limit.*

Proof: *Suppose to the contrary that $\{x_n\}_{n=1}^{\infty}$ converges to two distinct limits a and b in the metric space (X, d). Let $\epsilon = \frac{1}{2}d(a, b)$. By definition, there must exist integers N_a and N_b such that if $n \geq N_a$, then $d(x_n, a) < \epsilon$ and if $n \geq N_b$, then $d(x_n, b) < \epsilon$. This means that both $d(x_n, a)$ and $d(x_n, b)$ are less than ϵ when n is greater than or equal to the larger of N_a and N_b. Then*

$$d(a, b) \leq d(a, x_n) + d(x_n, b) < \epsilon + \epsilon = 2\epsilon = d(a, b)$$

so $d(a, b) < d(a, b)$, an obvious contradiction. Thus the assumption that $\{x_n\}_{n=1}^{\infty}$ converges to more than one limit must be false. □

Theorem 3.9: *Let (X, d) be a metric space and A a subset of X.*

(a) *A point x in X is a limit point of A if and only if there is a sequence of distinct points of A which converges to x.*

(b) *The set A is closed if and only if each convergent sequence of points of A converges to a point of A.*

Proof:

(a) *Suppose first that x is a limit point of A. Then there is a member x_1 of A distinct from x in the open ball $B(x, 1)$. Proceeding inductively, suppose that the first $n - 1$ terms $x_1, \ldots, x_{n-1}$ have been chosen, all distinct from each other and from x. It is left as an easy exercise to show that the finite set $\{x_i\}_{i=1}^{n-1}$ has no limit points and is therefore a closed set. Then the complement $X \setminus \{x_i\}_{i=1}^{n-1}$ is open, so $B(x, 1/n) \cap (X \setminus \{x_i\}_{i=1}^{n-1})$ is an open set containing x and must contain a point x_n of A distinct from x. The fact that $d(x, x_n) < 1/n$ insures that the resulting sequence $\{x_n\}_{n=1}^{\infty}$ converges to x. Since x_n is always chosen in the complement of $\{x_i\}_{i=1}^{n-1}$, it follows* that the terms of the sequence are all distinct. Thus there is a sequence of distinct points of A which converges to x.

For the reverse implication, suppose that there is a sequence $\{x_n\}_{n=1}^{\infty}$ of distinct points of A which converges to x. Let O be an open set containing x and ϵ a positive number for which $B(x, \epsilon) \subset O$. By the definition of convergent sequence, there is a positive integer N for which $x_n \in B(x, \epsilon)$ for all $n \geq N$. Since $B(x, \epsilon) \subset O$, we conclude that O contains points of A distinct from x and that x is a limit point of A.

(b)   To prove (b), suppose first that A is closed and consider a convergent sequence $\{y_n\}_{n=1}^{\infty}$ of points of A which converges to a point y in X. It must be shown that y is in A.

If the range of the sequence $\{y_n\}_{n=1}^{\infty}$ is infinite, it follows easily that y is a limit point of this set. Since A is closed, then y belongs to A. If, on the other hand, the range of $\{y_n\}_{n=1}^{\infty}$ is finite, then convergence of the sequence requires that it be constant from some point on, and this constant value $y_n = y$, $n \geq N$, is the limit of the sequence. Since each term of the sequence belongs to A, then y belongs to A in this case also.

To complete the proof, suppose that each convergent sequence of points of A converges to a point of A. We shall show that A is closed by showing that it contains all its limit points and invoking Theorem 3.7.

Let x be a limit point of A. By part (a), there is a sequence of distinct points of A which converges to x. By hypothesis, such a convergent sequence of points of A must converge to a point of A. Since the sequence cannot converge to two different limits, the one point x to which it does converge must be in A. We conclude that A contains all its limit points, so Theorem 3.7 guarantees that A is a closed set. This completes the proof.   □

Corollary: Let x be a limit point of a subset A of a metric space X. Then every open set containing x contains infinitely many members of A.

EXERCISE 3.2

1.   For metric space (X, d), $a \in X$, and $r > 0$, prove that the open ball $B(a, r)$ is an open set and the closed ball $B[a, r]$ is a closed set.

2.   Show that a finite subset of a metric space has no limit points and is therefore a closed set.

3.   Prove Theorem 3.5.

4.   Prove Theorem 3.6.

5.   Show that the limit of a convergent sequence of distinct points in a metric space is a limit point of the range of the sequence. Give an example to show that this is not true if the word "distinct" is omitted.

6. Determine whether the set A of Example 3.2.3 is open, closed, or neither for the taxicab and max metrics.

7. Prove that a non-empty subset C of a metric space (X, d) is a closed set if and only if $d(x, C) > 0$ for each $x \notin C$.

8. Prove that $d(a, B \cup C)$ is the smaller of $d(a, B)$ and $d(a, C)$ for a point a and subsets B, C of a metric space.

9. Let (X, d) be a metric space with the discrete metric.
    Prove:

    (a) Every subset of X is open.

    (b) Every subset of X is closed.

    (c) No subset of X has a limit point.

10. Let (X, d) be a metric space and x_1, x_2 distinct points of X. Prove that there are disjoint open sets O_1 and O_2 containing x_1 and x_2, respectively.

11. Show that the result of Problem 10 remains true when x_2 is replaced by a closed set C_2 which does not contain x_1.

12. Show that the result of Problem 10 remains true when x_1, x_2 are replaced by disjoint closed sets C_1, C_2.

13. Show that every open ball in $\mathbb{R}^2$ contains a point $x = (x_1, x_2)$ both of whose coordinates are rational.

14. Let R denote the subset of $\mathbb{R}^n$ consisting of points $x = (x_1, \ldots, x_n)$ all of whose coordinates are rational.

    (a) Prove that every non-empty open set in $\mathbb{R}^n$ contains a member of R.

    (b) Prove that every non-empty open set in $\mathbb{R}^n$ contains infinitely many members of R.

    (c) Prove that every point of $\mathbb{R}^n$ is a limit point of R.

3.3 INTERIOR, CLOSURE, AND BOUNDARY

This section introduces ideas closely related to open sets and closed sets.

Definition: *Let A be a subset of a metric space X. A point x in A is an **interior point** of A, or A is a **neighborhood** of x, provided that there is an open set O which contains x and is contained in A. The **interior** of A, denoted int A, is the set of all interior points of A.*

In the preceding definition, note that if O is an open set contained in A, then every point of O is an interior point of A. Hence the interior of A contains every open set contained in A and is the union of this family of open sets. This description

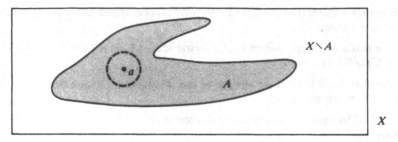

FIGURE 3.6   The point *a* is an interior point of *A*.

reveals two facts about interiors: (1) The interior of a set A is necessarily an open set. (2) Int A is the largest open set contained in A, in the sense that if U is an open set contained in A, then $U \subset$ int A.

Example 3.3.1

Consider $\mathbb{R}$ with the usual metric, as defined in Chapter 2.

  (a) For $a, b \in \mathbb{R}$ with $a < b$,

$$\text{int } (a, b) = \text{int } [a, b) = \text{int } (a, b] = \text{int } [a, b] = (a, b).$$

  (b) The interior of a finite set is empty since such a set contains no open interval.

  (c) The interior of the set of irrational numbers is empty since every open interval contains some rational numbers (Theorem 2.1). The interior of the set of rational numbers is also empty. (An open interval must be uncountable since it is equivalent to $\mathbb{R}$. Hence an open interval cannot contain only rational numbers because the set of rational numbers is countable.)

  (d) int $\varnothing = \varnothing$; int $\mathbb{R} = \mathbb{R}$.

Example 3.3.2

Consider $\mathbb{R}^2$ with the usual metric.

  (a) If $a \in \mathbb{R}^2$ and $r > 0$, then

$$\text{int } B(a, r) = \text{int } B[a, r] = B(a, r).$$

  (b) The interior of a finite set is empty.

  (c) The interior of the set of points with rational coordinates is empty. So is the interior of the set of points having at least one irrational coordinate.

  (d) int $\varnothing = \varnothing$; int $\mathbb{R}^2 = \mathbb{R}^2$.

It is left as an exercise for the reader to show that all parts of Example 3.3.2 generalize to $\mathbb{R}^n$.

Definition: *The **closure** $\bar{A}$ of a subset A of a metric space X is the union of A with its set of limit points:*

$$\bar{A} = A \cup A'$$

where A' is the derived set of A.

The preceding definition can be rephrased as follows: A point x is in $\bar{A}$ provided that either $x \in A$ or every open set containing x contains a point of A distinct from x. If $x \in A$, then every open set containing x contains a point of A, namely x itself. Thus if we omit the phrase "distinct from x" in the description of limit point, we may reformulate the definition of closure: $x \in \bar{A}$ if and only if every open set containing x contains a point of A.

Example 3.3.3

Consider $\mathbb{R}$ with the usual metric.

(a) For $a, b \in \mathbb{R}$ with $a < b$,

$$\overline{(a, b)} = \overline{[a, b)} = \overline{(a, b]} = \overline{[a, b]} = [a, b]$$

(b) If A is a finite set, then $\bar{A} = A$ because the derived set A' is empty.

(c) The closure of the set of rational numbers is $\mathbb{R}$. The closure of the set of irrational numbers is also $\mathbb{R}$. (Every open interval contains both rational and irrational numbers.)

(d) $\bar{\varnothing} = \varnothing$; $\bar{\mathbb{R}} = \mathbb{R}$.

Example 3.3.4

Consider $\mathbb{R}^n$ with the usual metric.

(a) If $a \in \mathbb{R}^n$ and $r > 0$, then

$$\overline{B(a, r)} = \overline{B[a, r]} = B[a, r].$$

(b) If A is a finite set, then $\bar{A} = A$.

(c) Let R be the subset of $\mathbb{R}^n$ consisting of all points having only rational coordinates. Then $\bar{R} = \mathbb{R}^n$. To see this, let $a = (a_1, \ldots, a_n) \in \mathbb{R}^n$ and let O be an open set containing a. By Theorem 3.3, there is an open ball $B(a, r)$ of positive radius r contained in O. By Theorem 2.1, there is for each $i = 1, \ldots, n$ a rational number x_i between $a_i - r/\sqrt{n}$ and $a_i + r/\sqrt{n}$. Then $x = (x_1, \ldots, x_n) \in R$ and

$$d(a, x) = \left(\sum_{i=1}^{n} (a_i - x_i)^2 \right)^{1/2} < \left(\sum_{i=1}^{n} (r/\sqrt{n})^2 \right)^{1/2} = (nr^2/n)^{1/2} = r$$

so

$$x \in B(a, r) \subset O.$$

Thus $a \in \bar{R}$ so $\bar{R} = \mathbb{R}^n$. The complement $I = \mathbb{R}^n \backslash R$ consisting of points having at least one irrational coordinate has the property $\bar{I} = \mathbb{R}^n$ by a similar argument.

(d) $\bar{\varnothing} = \varnothing$, $\overline{\mathbb{R}^n} = \mathbb{R}^n$.

The next two theorems explain the relations between closures and closed sets.

Theorem 3.10:   *If A is a subset of a metric space X, then $\bar{A}$ is a closed set and is a subset of every closed set containing A.*

Proof: *By Theorem 3.7, showing that $\bar{A}$ is closed can be accomplished by showing that it contains all its limit points. Suppose $x \notin \bar{A}$. Then there is an open set O containing x which contains no point of A. But if O contains no point of A, then it cannot contain a limit point of A either. (If an open set contains a limit point of A, then it must contain a point of A, by the definition of limit point.) Thus O contains no point of $\bar{A}$, so x is not a limit point of $\bar{A}$. This means that all limit points of $\bar{A}$ must necessarily be in $\bar{A}$. By Theorem 3.7, this is equivalent to saying that $\bar{A}$ is a closed set.*

Suppose now that F is a closed subset of X for which $A \subset F$. Then $\bar{A} \subset \bar{F}$ (as the reader will prove in Problem 5 of Exercise 3.3) and, since F contains all its limit points, then $\bar{F} = F \cup F' = F$. Thus $\bar{A} \subset F$ for every closed set F containing A.   □

Since $\bar{A}$ is a closed set which is a subset of every closed set containing A, we may justifiably say that $\bar{A}$ is the smallest closed set which contains A. Equivalently, $\bar{A}$ is the intersection of all closed sets containing A. Note the duality between $\bar{A}$, the smallest closed set containing A, and int A, the largest open set contained in A. This duality is further illustrated by the next theorem, whose proof is left to the reader.

Theorem 3.11:   *Let A be a subset of a metric space X.*

(a)   *A is open if and only if $A = $ int A.*
(b)   *A is closed if and only if $A = \bar{A}$.*

Definition: *Let A be a subset of a metric space X. A point $x \in X$ is a **boundary point** of A provided that x belongs to $\bar{A}$ and to $\overline{(X\backslash A)}$. The set of boundary points of A is called the **boundary** of A and is denoted by bdy A.*

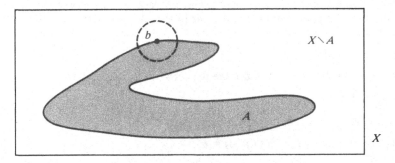

FIGURE 3.7   **The point *b* is a boundary point of *A*.**

It follows immediately from the definition that a set and its complement have the same boundary. The readers should test their knowledge of the definitions in this chapter by explaining why the following statements are equivalent for a subset A and point x in a metric space X:

(1) $x \in$ bdy A.
(2) $x \in (\bar{A} \backslash \text{int } A)$.
(3)   Every open set containing x contains a point of A and a point of $X\backslash A$.
(4)   Every neighborhood of x contains a point of A and a point of $X\backslash A$.
(5) $d(x, A) = d(x, X\backslash A) = 0$.
(6) $x \in \bar{A} \cap \overline{(X\backslash A)}$.

Example 3.3.5

(a)   The boundary of any interval in $\mathbb{R}$ with endpoints a and b is $\{a, b\}$.
(b)   In $\mathbb{R}^n$,

$$\text{bdy } B(a, r) = \text{bdy } B[a, r] = \{x \in \mathbb{R}^n : d(a, x) = r\}.$$

(c)   The boundary of the set of all points of $\mathbb{R}^n$ having only rational coordinates is $\mathbb{R}^n$.
(d)   In any metric space X,

$$\text{bdy } \varnothing = \text{bdy } X = \varnothing.$$

EXERCISE 3.3

1. Show that Example 3.3.2 generalizes to $\mathbb{R}^n$.

2. In $\mathbb{R}^n$, let R denote the set of points having only rational coordinates and I its complement, the set of points having at least one irrational coordinate.
   Prove that

   (a) int R = int $I = \varnothing$.

   (b) $R' = I' = \mathbb{R}^n$, where R' and I' are the derived sets of R and I.

   (c) bdy R = bdy $I = \mathbb{R}^n$.

3. For a subset A of a metric space X, prove that

   (a) $\bar{A} = X$ if and only if int $(X \backslash A) = \varnothing$.

   (b) $\overline{(X \backslash A)} = X \backslash \text{int } A$.

4. For a subset A of a metric space (X, d), prove that

   (a) $x \in \bar{A}$ if and only if $d(x, A) = 0$.

   (b) $x \in \text{int } A$ if and only if $d(x, X \backslash A) > 0$. (Assume $A \neq X$.)

5. Let A, B be subsets of a metric space with $A \subset B$.
   Prove that

   (a) int $A \subset$ int B;

   (b) $\bar{A} \subset \bar{B}$;

   (c) $A' \subset B'$.

   Give an example for which $A \subset B$ but neither bdy A nor bdy B is a subset of the other.

6. Prove Theorem 3.11.

7. Show that in any metric space,

   (a) $\bar{\bar{A}} = \bar{A}$, (b) int (int A) = int A.

8. Prove that the boundary of a subset A of a metric space X is always a closed set.

9. Let X be a metric space and A a subset of X.
   Prove:

   (a) A is open if and only if bdy $A \subset X \backslash A$.

   (b) A is closed if and only if bdy $A \subset A$.

   (c) A is both open and closed if and only if bdy $A = \varnothing$.

10. Let X be a space with the discrete metric. Show that every subset of X has empty boundary.

11. Let A, B be subsets of a metric space. Show that $\overline{A \cup B} = \bar{A} \cup \bar{B}$ and that $\overline{A \cap B} \subset \bar{A} \cap \bar{B}$. Give an example to show that $\overline{A \cap B}$ and $\bar{A} \cap \bar{B}$ may not be equal.

3.4 CONTINUOUS FUNCTIONS

This section introduces continuity for functions from one metric space to another. The definition is a natural generalization of continuity for a function $f: \mathbb{R} \to \mathbb{R}$. The primary purpose of the section is to show that continuity for functions on metric spaces can be described in terms of the topologies of the domain and range spaces.

The reader is probably familiar with the definition of continuity for functions on $\mathbb{R}$:

Definition: *Let $f: A \to \mathbb{R}$ be a function from a subset A of $\mathbb{R}$ to $\mathbb{R}$ and let $a \in A$. Then f is **continuous at** a if for each positive number ϵ there is a positive number δ such that if $x \in A$ and $|x - a| < \delta$, then $|f(x) - f(a)| < \epsilon$. If f is continuous for each $a \in A$, then it is said simply that f is **continuous.***

The definition of continuity is extended to functions on arbitrary metric spaces as follows:

Definition: *Let (X, d) and (Y, d') be metric spaces and $f: X \to Y$ a function. Then f is **continuous at the point** a in X if for each positive number ϵ there is a positive number δ such that if $x \in X$ and $d(x, a) < \delta$, then $d'(f(x), f(a)) < \epsilon$. A function is said to be **continuous** provided that it is continuous at each point of its domain.*

The definition of continuity can be restated in terms of open balls as follows: f is **continuous at** $a \in X$ means that for each open ball $B_{d'}(f(a), \epsilon)$ centered at $f(a)$, there is an open ball $B_d(a, \delta)$ such that the image $f(B_d(a, \delta))$ is a subset of $B_{d'}(f(a), \epsilon)$.

The next theorem is a direct analogue of a theorem for real functions.

Theorem 3.12:   *Let $f: X \to Y$ be a function from metric space (X, d) to metric space (Y, d') and let $a \in X$. Then f is continuous at a if and only if for each sequence $\{x_n\}_{n=1}^{\infty}$ in X converging to a, the sequence $\{f(x_n)\}_{n=1}^{\infty}$ converges to $f(a)$.*

Proof: *Suppose first that f is continuous at a and let $\{x_n\}_{n=1}^{\infty}$ be a sequence in X converging to a. It must be shown that $\{f(x_n)\}_{n=1}^{\infty}$ converges to $f(a)$. With this in mind, let ϵ be a positive number. Since f is continuous at a, there is a positive number δ such that if $x \in X$ and $d(x, a) < \delta$, then $d'(f(x), f(a)) < \epsilon$. Since $\{x_n\}_{n=1}^{\infty}$ converges to a, there is a positive integer N such that if $n \geq N$, then $d(x_n, a) < \delta$. The choice of δ now insures that $d'(f(x_n), f(a)) < \epsilon$ for $n \geq N$, so $\{f(x_n)\}_{n=1}^{\infty}$ converges to $f(a)$.*

The reverse implication will be proved in contrapositive form: If f is not con-

tinuous at a, then there is a sequence $\{x_n\}_{n=1}^{\infty}$ in X converging to a for which $\{f(x_n)\}_{n=1}^{\infty}$ does not converge to $f(a)$. If f is not continuous at a, then there is a positive number ϵ with the property that if $\delta > 0$ then there is an x (depending on δ) in X such that $d(x, a) < \delta$ but $d'(f(x), f(a)) \geq \epsilon$. In particular, there is such a point x_n for the reciprocal $1/n$ of each positive integer n:

$$d(x_n, a) < 1/n, \quad \text{but } d'(f(x_n), f(a)) \geq \epsilon.$$

The preceding line shows that $\{x_n\}_{n=1}^{\infty}$ converges to a and that $\{f(x_n)\}_{n=1}^{\infty}$ does not converge to $f(a)$.   □

Theorem 3.13:   The following statements are equivalent for a function f from metric space (X, d) to metric space (Y, d'):

(1) f is continuous.

(2)   For each sequence $\{x_n\}_{n=1}^{\infty}$ converging to a point a in X, the sequence $\{f(x_n)\}_{n=1}^{\infty}$ converges to $f(a)$.

(3)   For each open set O in Y, $f^{-1}(O)$ is open in X.

(4)   For each closed set C in Y, $f^{-1}(C)$ is closed in X.

Proof: The equivalence of (1) and (2) is established by applying Theorem 3.12 at each point $a \in X$.

The equivalence of (3) and (4) follows from the duality between open sets and closed sets. Suppose $f^{-1}(O)$ is open in X for each open set O in Y and let C be a closed subset of Y. Then $Y \backslash C$ is open so $f^{-1}(Y \backslash C)$ is open in X, and $X \backslash f^{-1}(Y \backslash C))$ is closed in X. But

$$X \backslash f^{-1}(Y \backslash C) = X \backslash (X \backslash f^{-1}(C)) = f^{-1}(C),$$

so $f^{-1}(C)$ is closed in X for each closed subset C of Y. The analogous argument that (4) implies (3) is left as an exercise.

It now remains to be proved that (1) and (3) are equivalent. Suppose first that f is continuous and let O be open in Y. It must be proved that $f^{-1}(O)$ is open in X. Let $a \in f^{-1}(O)$. Then $f(a)$ belongs to the open set O so there is an open ball $B_{d'}(f(a), r)$, $r > 0$, in Y centered at $f(a)$ and contained in O. Since f is continuous at a, there is a positive number δ such that if $x \in X$ and $d(x, a) < \delta$, then $d'(f(x), f(a)) < r$. This means that

$$f(B_d(a, \delta)) \subset B_{d'}(f(a), r) \subset O$$

so

$$B_d(a, \delta) \subset f^{-1}(O).$$

Since $f^{-1}(O)$ contains such an open ball centered at each of its points, then $f^{-1}(O)$ is open.

Suppose now that $f^{-1}(O)$ is open in X for each open subset O of Y. For $a \in X$ and $\epsilon > 0$, $B_{d'}(f(a), \epsilon)$ is an open set in Y, so $f^{-1}(B_{d'}(f(a), \epsilon))$ must be open in X. Since a belongs to $f^{-1}(B_{d'}(f(a), \epsilon))$, there is an open ball $B_d(a, \delta)$ of positive radius δ contained in $f^{-1}(B_{d'}(f(a), \epsilon))$. Then

$$f(B_d(a, \delta)) \subset B_{d'}(f(a), \epsilon).$$

But this simply means that if $x \in X$ and $d(x, a) < \delta$, then $d'(f(x), f(a)) < \epsilon$. Thus f is continuous at each point a in X, so f is continuous.   □

Statement (3) of Theorem 3.13 describes continuity of functions on metric spaces in terms of the topologies of the domain and range spaces. In the next chapter, where we shall deal with collections of open sets not necessarily determined by metrics, this property will be used as an alternate definition of continuity.

EXERCISE 3.4

1. Let X and Y be metric spaces and let a be a point of X which is not a limit point of X. Show that every function $f: X \to Y$ is continuous at a. Illustrate this phenomenon with a function $f: A \to \mathbb{R}$ from a subset A of $\mathbb{R}$ to $\mathbb{R}$.

2. Prove that (4) implies (3) in Theorem 3.13.

3. Let $f: X \to Y$ be a function on the indicated metric spaces and let a be a point of X. Prove that f is continuous at a if and only if for each open set O containing $f(a)$, $f^{-1}(O)$ is a neighborhood of a.

4. Let $f: X \to Y$ be a function on the indicated metric spaces. Prove that the following statements are equivalent:

    (a) f is continuous.

    (b) For each subset A of X, $f(\bar{A}) \subset \overline{f(A)}$.

    (c) For each subset B of Y, $f^{-1}(\text{int } B) \subset \text{int } f^{-1}(B)$.

5. Show that every function $f: X \to Y$ for which the domain X has the discrete metric is continuous.

6. Let X be a metric space with metric d and A a non-empty subset of X. Define $f: X \to \mathbb{R}$ by

$$f(x) = d(x, A), \quad x \in X.$$

Show that f is continuous.

7. Suppose that $f: X \to Y$ and $g: Y \to Z$ are continuous functions on the indicated metric spaces. Prove that the composite function $g \cdot f: X \to Z$ is continuous.

3.5 EQUIVALENCE OF METRIC SPACES

What should it mean to say that two metric spaces (X, d) and (Y, d') are equivalent? The present section answers this question in two ways, both of which require a one-to-one correspondence between X and Y. The first form of equivalence, called *metric equivalence*, requires that for every pair of points $a, b \in X$, the distance from a to b must be the same as the distance between the corresponding points of Y. The second definition of equivalence, called *topological equivalence*, is a condition on the topologies for X and Y determined by their respective metrics.

Definition: *Metric spaces (X, d) and (Y, d') are **metrically equivalent** or **isometric** if there is a one-to-one function $f: X \to Y$ from X onto Y such that for all $a, b \in X$,*

$$d(a, b) = d'(f(a), f(b)).$$

*The function f is called an **isometry**.*

The following observations show that metric equivalence is an equivalence relation:

(a)   The identity function on any metric space is an isometry, so metric equivalence is a reflexive relation.

(b)   If $f: X \to Y$ is an isometry from X onto Y, then the inverse function f^{-1}: $Y \to X$ is an isometry from Y onto X. Thus the relation is symmetric.

(c)   The composition of two isometries is an isometry, so metric equivalence is also a transitive relation.

Definition: *Metric spaces (X, d) and (Y, d') are **topologically equivalent** or **homeomorphic** if there is a one-to-one function $f: X \to Y$ from X onto Y for which f and the inverse function f^{-1} are both continuous. The function f is called a **homeomorphism**.*

Recall from Theorem 3.13 that continuity of $f: X \to Y$ can be expressed by saying that $f^{-1}(O)$ is open in X for each open subset O of Y. Similarly, $f^{-1}: Y \to X$ is continuous provided that $(f^{-1})^{-1}(U) = f(U)$ is open in Y for such open subset U of X. Thus a one-to-one function f from X onto Y is a homeomorphism provided that a subset O of Y is open if and only if $f^{-1}(O)$ is open in X.

It is left as an easy exercise to show that topological equivalence is an equivalence relation.

Since each isometry is a continuous map, it follows that topological equivalence is weaker than metric equivalence. In other words, if (X, d) and (Y, d') are metrically equivalent, then they must be topologically equivalent also.

Example 3.5.1

Consider the metric spaces $X = (0, 1)$ and $Y = (0, 2)$ with metrics determined by the usual metric on the real line. The function $f: X \rightarrow Y$ defined by

$$f(x) = 2x, \quad x \in (0, 1),$$

is a homeomorphism but not an isometry. Since X has diameter 1 and Y has diameter 2, X and Y cannot be isometric.

Example 3.5.2

The open intervals (a, b), $a < b$, and $(0, 1)$ are topologically equivalent when considered as metric spaces with metrics given by the usual method of measuring distances in $\mathbb{R}$. This follows from the fact that the function $f: (0, 1) \rightarrow (a, b)$ defined by

$$f(x) = (b - a)x + a, \quad x \in (0, 1),$$

is a homeomorphism.

Example 3.5.3

Recall from calculus that the function $g: (- \pi/2, \pi/2) \rightarrow \mathbb{R}$ defined by

$$g(x) = \tan x, \quad x \in (-\pi/2, \pi/2),$$

is a one-to-one correspondence, is continuous, and has as inverse function the principal arctangent function, which is also continuous. Thus $(-\pi/2, \pi/2)$ is topologically equivalent to $\mathbb{R}$. It is left as an exercise to show that unbounded open intervals $(-\infty, a)$ and (a, ∞) are topologically equivalent to $\mathbb{R}$. Since topological equivalence is an equivalence relation, this example shows that all open intervals on $\mathbb{R}$ are topologically equivalent to each other and to the entire real line.

The next theorem gives conditions under which two different metrics for a set X determine the same family of open sets. A lemma will be needed.

Lemma: *Let d_1 and d_2 be two metrics for the set X and suppose that there is a positive number c such that $d_1(x, y) \leq cd_2(x, y)$ for all $x, y \in X$. Then the identity function $i: (X, d_2) \rightarrow (X, d_1)$ is continuous.*

Proof: *Let $a \in X$ and let ϵ be a positive number. Then if $\delta = \epsilon/c$ and x is a member of X for which $d_2(x, a) < \delta$, then*

$$d_1(i(x), i(a)) = d_1(x, a) \le cd_2(x, a) < c\delta = \epsilon.$$

Thus $d_1(i(x), i(a)) < \epsilon$ whenever $d_2(x, a) < \delta$. Thus $i: (X, d_2) \rightarrow (X, d_1)$ is continuous. □

Theorem 3.14: *Let d_1 and d_2 be two metrics for the set X and suppose there are positive numbers c and c' such that*

$$d_1(x, y) \le cd_2(x, y), \quad d_2(x, y) \le c'd_1(x, y)$$

for all x, $y \in X$. Then the identity function on X is a homeomorphism between (X, d_1) and (X, d_2).

Proof: *The identity map is clearly a one-to-one correspondence from X onto itself. Continuity in both directions is guaranteed by the preceding lemma.* □

Definition: *Metrics d_1 and d_2 for a set X which determine the same topology are called* **equivalent metrics.**

For metrics d_1, d_2 on a set X satisfying the hypotheses of Theorem 3.14, the description of continuity for the identity function in terms of open sets shows that d_1 and d_2 are equivalent: Since $i: (X, d_1) \rightarrow (X, d_2)$ is continuous, then for each d_2-open set O, $i^{-1}(O) = O$ is also d_1-open. Since $i^{-1} = i: (X, d_2) \rightarrow (X, d_1)$ is continuous, then for each d_1-open set U, $i^{-1}(U) = U$ is also d_2-open. Hence d_1 and d_2 determine precisely the same open sets.

Example 3.5.4

Consider the usual metric d and the taxicab metric d' for $\mathbb{R}^n$:

$$d(x, y) = \left(\sum_{i=1}^{n} (x_i - y_i)^2 \right)^{1/2}, \quad d'(x, y) = \sum_{i=1}^{n} |x_i - y_i|$$

for $x = (x_1, \ldots, x_n)$ and $y = (y_1, \ldots, y_n)$ in $\mathbb{R}^n$. Since $\sqrt{u^2 + v^2} \le u + v$ for all non-negative real numbers u and v, it follows that $d(x, y) \le d'(x, y)$. It is also a simple matter to observe that $d'(x, y) \le nd(x, y)$. Hence, by Theorem 3.13, the metrics d and d' are equivalent.

The equivalence of d and d' is visualized in $\mathbb{R}^2$ as follows: An open ball $B_d(a, r)$ is an "open disk" centered at a with radius r excluding the bounding circle. An open ball $B_{d'}(a, r)$ is an "open diamond" centered at a and excluding the four bounding segments. Note that every open disk can be expressed as a union of open diamonds, and that every open diamond can be expressed as a union of open disks. This says simply that d and d' determine identical open sets for $\mathbb{R}^2$.

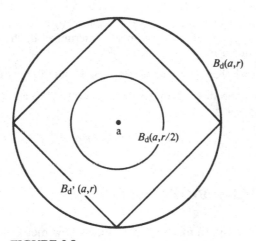

FIGURE 3.8

By considerations like those of Example 3.5.4, the reader can show that the topology for $\mathbb{R}^n$ determined by the max metric is identical with that determined by the usual and taxicab metrics.

EXERCISE 3.5

1. Show that any two non-degenerate closed and bounded intervals are topologically equivalent

2. Show that two metric spaces with discrete metrics are isometric if and only if they have the same cardinal number.

3. Prove that topological equivalence is an equivalence relation for metric spaces.

4. Suppose that d_1 and d_2 are metrics for X and c is a positive number for which $d_1(x, y) \leq cd_2(x, y)$. Prove that $B_{d_2}(x, r/c) \subset B_{d_1}(x, r)$.

5. For the usual metric d and the max metric d'' on $\mathbb{R}^n$, prove that, for all x, y in $\mathbb{R}^n$,

$$d''(x, y) \le d(x, y) \le \sqrt{n}d''(x, y).$$

Conclude that d and d'' are equivalent metrics.

6. Show that every metric space (X, d) is topologically equivalent to a bounded metric space. (*Hint:* Let $d'(x, y) = \min \{1, d(x, y)\}$, the minimum of 1 and $d(x, y)$. The metric $d''(x, y) = d(x, y)/(1 + d(x, y))$ is somewhat more complicated but illustrates the same phenomenon.)

7. The *open unit n-cube* J^n is the subset of $\mathbb{R}^n$ defined by $J^n = \{x = (x_1, \ldots, x_n) \in \mathbb{R}^n: 0 < x_i < 1$ for $i = 1, \ldots, n\}$ with the metric determined by the usual metric d of $\mathbb{R}^n$. Prove that J^n is topologically equivalent to $\mathbb{R}^n$.

8. For a, $b \in \mathbb{R}^n$, prove that there is an isometry of $\mathbb{R}^n$ onto itself which maps a to b.

9. Let $f: X \to Y$ be an isometry between metric spaces (X, d) and (Y, d'). Show that for each $a \in X$ and $r > 0$,

$$f(B_d(a, r)) = B_{d'}(f(a), r).$$

3.6 NEW SPACES FROM OLD

There are two standard methods of building new metric spaces from those at hand. The first method, which produces *subspaces,* involves simply taking a subset Y of a given metric space X and measuring distances in Y with the metric of X. We have already considered intervals as subspaces of $\mathbb{R}$ in the preceding section. The second method of building new spaces assigns a metric to the Cartesian product of metric spaces.

Definition: *Let (X, d) be a metric space and Y a subset of X. The metric space (Y, d'), where d' is the restriction of d to $Y \times Y$, is called a **subspace** of (X, d).*

Example 3.6.1

The following are commonly used subspaces of Euclidean spaces.

(a) The *unit n-cube* is the set

$$I^n = \{x = (x_1, \ldots, x_n) \in \mathbb{R}^n: 0 \le x_i \le 1 \quad \text{for } i = 1, \ldots, n\}$$

with the subspace metric induced by the usual metric d:

$$d'(x, y) = \left(\sum_{i=1}^{n} (x_i - y_i)^2 \right)^{1/2}.$$

(b) The *n-dimensional unit sphere* S^n is the set

$$S^n = \left\{ x = (x_1, \ldots, x_{n+1}) \in \mathbb{R}^{n+1} : \sum_{i=1}^{n+1} x_i^2 = 1 \right\}$$

with the metric induced by d. Thus S^1 is the unit circle in $\mathbb{R}^2$ and S^2 is the unit sphere in $\mathbb{R}^3$. Note that S^1 is a curve, so it is a one-dimensional object even though it is a subspace of $\mathbb{R}^2$. Also, S^2 is a surface, so it is a two-dimensional object even though it is a subspace of $\mathbb{R}^3$. In general, S^n is an n-dimensional object in $\mathbb{R}^{n+1}$.

(c) The *n-dimensional unit ball* B^n is the set

$$B^n = \left\{ x = (x_1, \ldots, x_n) \in \mathbb{R}^n : \sum_{i=1}^{n} x_i^2 \le 1 \right\}$$

with metric induced by d. The boundary of B^n is S^{n-1}.

(d) For $n \ge 2$, consider the set

$$A^{n-1} = \{ x = (x_1, \ldots, x_n) \in \mathbb{R}^n : x_n = 0 \}$$

with metric induced by d. Then A^{n-1} is a subspace of $\mathbb{R}^n$ and is isometric to $\mathbb{R}^{n-1}$ under the correspondence

$$(x_1, \ldots, x_{n-1}, 0) \leftrightarrow (x_1, \ldots, x_{n-1}).$$

(The only distinction between $\mathbb{R}^{n-1}$ and A^{n-1} is the extra 0 for points in A^{n-1}.) For this reason, it is often said that $\mathbb{R}^{n-1}$ is a subspace of $\mathbb{R}^n$. Thus we consider the real line to be a subspace of the plane, the plane to be a subspace of three-dimensional space, and so on. To be absolutely correct, it should be said that $\mathbb{R}^{n-1}$ is metrically equivalent to a subspace of $\mathbb{R}^n$ or that $\mathbb{R}^{n-1}$ is *isometrically embedded* in $\mathbb{R}^n$. For all practical purposes, no distinction is made between a metric space and an isometric copy of it.

Definition: *Let $\{(X_i, d_i)\}_{i=1}^n$ be a finite collection of metric spaces. The* **product metric space** *(X, d) is the Cartesian product*

$$X = \prod_{i=1}^{n} X_i$$

of the sets $X_1, \ldots, X_n$ with the **product metric** *d defined by*

$$d(x, y) = \left(\sum_{i=1}^{n} (d_i(x_i, y_i))^2 \right)^{1/2}$$

for $x = (x_1, \ldots, x_n)$ *and* $y = (y_1, \ldots, y_n)$ *in X. The spaces* (X_i, d_i) *are called the* **coordinate spaces** *or the* **factors** *of the product space* (X, d).

It is not altogether obvious that what has been called the product metric is actually a metric. This is proved in the next theorem.

Theorem 3.15: *If* $\{(X_i, d_i)\}_{i=1}^{n}$ *is a sequence of metric spaces, then the product metric is a metric for the product set* $X = \prod_{i=1}^{n} X_i$.

Proof: *Consider points* $x = (x_1, \ldots, x_n)$, $y = (y_1, \ldots, y_n)$, *and* $z = (z_1, \ldots, z_n)$ *in X. Since each* d_i *is a metric, the properties* $d(x, y) \geq 0$, $d(x, y) = 0$ *only when* $x = y$, *and* $d(x, y) = d(y, x)$ *follow from the corresponding properties in the coordinate spaces. As usual, it is the Triangle Inequality that requires more attention. Note that*

$$d(x, z)^2 = \sum_{i=1}^{n} (d_i(x_i, z_i))^2 \leq \sum_{i=1}^{n} [d_i(x_i, y_i) + d_i(y_i, z_i)]^2$$

$$= \sum_{i=1}^{n} (d_i(x_i, y_i))^2 + 2 \sum_{i=1}^{n} d_i(x_i, y_i)d_i(y_i, z_i) + \sum_{i=1}^{n} (d_i(y_i, z_i))^2$$

$$\leq (d(x, y))^2 + 2d(x, y)d(y, z) + (d(y, z))^2 = [d(x, y) + d(y, z)]^2.$$

The last inequality follows from the Cauchy-Schwarz Inequality (Theorem 3.1). Thus we conclude that $d(x, z) \leq d(x, y) + d(y, z)$ *and that d is actually a metric.* $\square$

Example 3.6.2

Euclidean n-dimensional space $\mathbb{R}^n$, with its usual metric d, is the product of the real line $\mathbb{R}$ taken as coordinate space n times:

$$\mathbb{R}^n = \prod_{i=1}^{n} X_i$$

where $(X_i, d_i) = (\mathbb{R}, \text{usual metric})$ for $i = 1, \ldots, n$.

Example 3.6.3

Hilbert space H consists of all infinite sequences $x = (x_1, \ldots, x_n, \ldots)$ for which each coordinate x_i is a real number and for which $\sum_{i=1}^{\infty} x_i^2$ converges to a finite limit. The number

$$\|x\| = \left(\sum_{i=1}^{\infty} x_i^2\right)^{1/2}$$

is called the *norm* of x. For $x = (x_1, \ldots, x_n, \ldots)$ and $y = (y_1, \ldots, y_n, \ldots)$ in H, the *distance* from x to y is defined by

$$d(x, y) = \left(\sum_{i=1}^{\infty} (x_i - y_i)^2 \right)^{1/2}.$$

In order to show that $d(x, y)$ is actually well-defined, it must be proved that $\sum_{i=1}^{\infty} (x_i - y_i)^2$ is a convergent series. To this end, consider a finite sum $\sum_{i=1}^{n} (x_i - y_i)^2$:

$$\sum_{i=1}^{n} (x_i - y_i)^2 = \sum_{i=1}^{n} x_i^2 - 2 \sum_{i=1}^{n} x_i y_i + \sum_{i=1}^{n} y_i^2$$

$$\leq \sum_{i=1}^{n} x_i^2 + 2 \left(\sum_{i=1}^{n} x_i^2 \right)^{1/2} \left(\sum_{i=1}^{n} y_i^2 \right)^{1/2} + \sum_{i=1}^{n} y_i^2,$$

the last inequality following from the Cauchy-Schwarz Inequality. Since $\sum_{i=1}^{n} x_i^2 \leq \|x\|^2$ and $\sum_{i=1}^{n} y_i^2 \leq \|y\|^2$ regardless of the value of n, then

$$\sum_{i=1}^{n} (x_i - y_i)^2 \leq \|x\|^2 + 2\|x\|\|y\| + \|y\|^2 = (\|x\| + \|y\|)^2,$$

so $\sum_{i=1}^{\infty} (x_i - y_i)^2$ is bounded for all n by $(\|x\| + \|y\|)^2$. Since a bounded series of non-negative real numbers is convergent, then $\sum_{i=1}^{\infty} (x_i - y_i)^2$ converges. The proof that d is a metric is similar to the proof that the usual distance function on $\mathbb{R}^n$ is a metric and is left to the reader.

Note that the correspondence

$$(x_1, \ldots, x_n) \leftrightarrow (x_1, \ldots, x_n, 0, 0, 0, \ldots),$$

between points of $\mathbb{R}^n$ and points of H which have non-zero coordinates in at most the first n terms, is an isometry between $\mathbb{R}^n$ and a subspace of H. Thus $\mathbb{R}^n$ is isometrically embedded in H, and we may consider $\mathbb{R}^n$ as a subspace of H.

Hilbert space, with its infinite number of coordinates, suggests the possibility of infinite products of metric spaces. We shall consider such products in Chapters 7 and 8.

EXERCISE 3.6

1. Let (X, d) be a metric space and (Y, d') a subspace. Prove that the inclusion map $i: Y \to X$ defined by

$$i(y) = y, \quad y \in Y,$$

is continuous.

2. Suppose that metric space (X_i, d_i) is topologically equivalent to (Y_i, d'_i) for $i = 1, \ldots,$ n. Show that the product metric spaces $X = \prod_{i=1}^{n} X_i$ and $Y = \prod_{i=1}^{n} Y_i$ are topologically equivalent.

3. For $i = 1, \ldots, n$, let $[a_i, b_i]$ be a non-degenerate closed, bounded interval. Prove that the product metric space $K = \prod_{i=1}^{n} [a_i, b_i]$ is homeomorphic to the unit n-cube I^n.

4. Prove that each open ball $B(a, r)$, $a \in \mathbb{R}^n$, $r > 0$, considered as a subspace of $\mathbb{R}^n$, is homeomorphic to $\mathbb{R}^n$. (*Hint:* Show first that the unit open ball $B(\theta, 1)$ with center at the origin and radius 1 is homeomorphic to $\mathbb{R}^n$.)

5. Let $\{(X_i, d_i)\}_{i=1}^{n}$ be a sequence of metric spaces and let $X = \prod_{i=1}^{n} X_i$ be the Cartesian product set. Define metrics d' and d'' on X as follows: For $x = (x_1, \ldots, x_n)$ and $y = (y_1, \ldots, y_n)$ in X,

$$d'(x, y) = \sum_{i=1}^{n} d_i(x_i, y_i)$$

$$d''(x, y) = \max \{d_i(x_i, y_i)\}_{i=1}^{n}.$$

Show that d' and d'' are metrics and that both are equivalent to the usual product metric d for X.

6. (a) Prove the Cauchy-Schwarz Inequality for Hilbert space H: For $x = (x_1, \ldots, x_n, \ldots)$ and $y = (y_1, \ldots, y_n, \ldots)$ in H

$$\left| \sum_{i=1}^{\infty} x_i y_i \right| \leq \left(\sum_{i=1}^{\infty} x_i^2 \right)^{1/2} \left(\sum_{i=1}^{\infty} y_i^2 \right)^{1/2}.$$

(b) Prove the Minkowski Inequality for H:

$$\left(\sum_{i=1}^{\infty} (x_i + y_i)^2 \right)^{1/2} \leq \left(\sum_{i=1}^{\infty} x_i^2 \right)^{1/2} + \left(\sum_{i=1}^{\infty} y_i^2 \right)^{1/2}.$$

(c) Define dot product $x \cdot y$ and vector addition $x + y$ for H so that the above inequalities can be restated

(a) $|x \cdot y| \leq \|x\| \|y\|$.

(b) $\|x + y\| \leq \|x\| + \|y\|$.

(d) Prove that the distance function d for H defined in Example 3.6.3 is a metric.

7. Let (X, d) be a metric space, (Y, d') a bounded metric space, and $C(X, Y)$ the set of all continuous functions $f: X \to Y$. Show that the function ρ defined for f, g in $C(X, Y)$ by

$$\rho(f, g) = \sup \{d'(f(x), g(x)): x \in X\}$$

is a metric for $C(X, Y)$.

3.7 COMPLETE METRIC SPACES

Convergence of sequences was discussed in Section 3.2. In this section, that discussion continues in the context of a property of metric spaces which insures the convergence of certain sequences. The property of interest is *completeness*. Intuitively speaking, this property is characteristic of those spaces in which every convergent sequence converges to a point *in the space*. For example, the open unit interval (0, 1) is not complete since the sequence $\{1/n\}_{n=1}^{\infty}$ converges to a point not in (0, 1). This idea is made precise in the definitions that follow.

Definition: *Let (X, d) be a metric space. A sequence $\{x_n\}_{n=1}^{\infty}$ of points of X is a* **Cauchy sequence** *provided that for each positive number ϵ there is a positive integer N such that if m and n are integers greater than or equal to N, then $d(x_n, x_m) < \epsilon$.*

A comparison of definitions will reveal that every convergent sequence is Cauchy.

Definition: *A metric space (X, d) is* **complete** *if every Cauchy sequence in X converges to a point in X.*

Example 3.7.1

(a) The completeness of the real line $\mathbb{R}$ is a fact of elementary analysis. A proof can also be made using Cauchy's Nested Intervals Theorem (Theorem 2.11). The details of this process are left as an exercise for the reader.

(b) Completeness of $\mathbb{R}^n$ follows from that of $\mathbb{R}$. To see this, consider a Cauchy sequence $\{x_k\}_{k=1}^{\infty}$ in $\mathbb{R}^n$, $n \geq 2$. For $1 \leq i \leq n$, the sequence of ith coordinates of the points x_k is a Cauchy sequence in $\mathbb{R}$ and hence converges to a real number z_i. It follows easily that $\{x_k\}_{k=1}^{\infty}$ converges to $z = (z_1, z_2, \ldots, z_n)$.

(c) Each closed interval $[a, b]$ is complete. To prove this, consider a Cauchy sequence $\{x_k\}_{k=1}^{\infty}$ in $[a, b]$. Since $\mathbb{R}$ is complete, this sequence converges to a real number x in $\mathbb{R}$. Since $[a, b]$ is closed, it follows easily that x belongs to $[a, b]$.

(d) Open intervals and half-open, half-closed intervals are not complete. For example, $\{1/n\}_{n=1}^{\infty}$ is a Cauchy sequence in (0, 1) which does not converge to a point of (0, 1). Analogous examples show the incompleteness of (a, b), $(a, b]$, and $[a, b)$ for all real numbers $a < b$.

(e) Hilbert space (Example 3.6.3) is complete. The proof of this fact is left as an exercise.

The preceding example shows that a subspace of a complete space may fail to be complete. The next theorem characterizes those subspaces that do inherit the completeness property.

Theorem 3.16:   *Let (X, d) be a complete metric space. A subspace A of X is complete if and only if it is closed.*

Proof: *Suppose first that A is a complete subspace. It will be proved that A is closed by showing that A contains all its limit points. If x is a limit point of A, then by Theorem 3.9 there is a sequence of distinct points of A which converges to x. Since each convergent sequence is Cauchy and A is complete, the limit of this sequence, namely x, must be in A. Thus A is closed.*

Suppose now that A is a closed subspace of a complete metric space X. To demonstrate that A is complete, consider a Cauchy sequence $\{x_n\}_{n=1}^{\infty}$ of points of A. Since X is complete, this sequence converges to a point x belonging to X. By Theorem 3.7(b), the fact that A is closed insures that the limit x belongs to A. Thus each Cauchy sequence of points of A converges to a point of A, and we conclude that A is a complete subspace. $\square$

Example 3.7.2

The property of completeness is not preserved by topological equivalence. In other words, there are pairs of metric spaces which are topologically equivalent with one space complete and the other not complete. The real line $\mathbb{R}$ and the open interval (0, 1) illustrate this phenomenon.

Definition: *A subset A of a metric space X is **nowhere dense** in X if $\bar{A}$ has empty interior.*

Example 3.7.3

  (a)   As subsets of the real line $\mathbb{R}$, each of the following is nowhere dense:
    (i)   any finite set
    (ii)   the range of the sequence $\{1/n\}_{n=1}^{\infty}$
    (iii)   the set $\mathbb{Z}$ of integers
  (b)   As subsets of the plane, each of the following is nowhere dense:
    (i)   any finite set
    (ii)   the points whose coordinates are integers
    (iii)   any finite collection of lines
    (iv)   a circle

The property of being nowhere dense is designed to describe those sets which are "very thinly distributed" in their containing space. The thinness of the distribution is reflected in the fact that the closure of a nowhere dense set does not contain any open ball of positive radius. A set A which fails to be nowhere dense, int $\bar{A} \neq \emptyset$, is thought of as being "densely distributed" near the interior points of $\bar{A}$ and is often called *somewhere dense*. Although we shall not have reason to use this term, it does seem preferable to the double negative "not nowhere dense."

Definition: *A metric space or subspace that is the union of a countable family of nowhere dense sets is said to be of the first category. A metric space which is not of the first category is said to be of the second category.*

Example 3.7.4

(a) As a subspace of $\mathbb{R}$, the set R of rational numbers is of the first category. It is the union of a countable collection of nowhere dense singleton sets, each containing one rational number. Similarly, the set of points in $\mathbb{R}^n$ having all coordinates rational is also of the first category.

(b) The next theorem, the Baire Category Theorem, shows that every complete metric space is of the second category. It will justify the present assertion that $\mathbb{R}^n$ is of the second category for each positive integer n.

From an intuitive viewpoint, the category concept describes the thinness or thickness of the distribution of the points of a set relative to the containing space. As we have noted, a nowhere dense set, whose closure contains no ball of positive radius, is thought of as very thinly distributed. A set of the first category is the union of a countable family of such thin sets, and a set of the second category is not.

The following lemma provides a characterization of nowhere dense sets. Its proof is left as an exercise.

Lemma: *A subset A of a metric space X is nowhere dense in X if and only if each non-empty open set in X contains an open ball whose closure is disjoint from A.*

Theorem 3.17: The Baire Category Theorem *Every complete metric space, considered as a subspace of itself, is of the second category.*

Proof: *The proof is by contradiction. Supposing that the theorem is false, let X be a complete metric space which is not of the second category. Then there is a sequence $\{A_n\}_{n=1}^{\infty}$ of nowhere dense sets whose union is X.*

By the lemma, the open set X must contain an open ball B_1 whose closure $\bar{B}_1$ is disjoint from A_1. We choose such an open ball B_1 of radius less than 1. Since B_1 is a non-empty open set, it must contain an open ball B_2 of radius less than 1/2 for which $\bar{B}_2$ is disjoint from A_2. Proceeding inductively, we define a nested sequence $\{B_n\}_{n=1}^{\infty}$ of open balls for which the radius of B_n is less than $1/n$ and $\bar{B}_n$ is disjoint from A_n. The sequence $\{x_n\}_{n=1}^{\infty}$ of centers of these open balls is a Cauchy sequence which, by completeness, converges to a point x in X. Since x belongs to each of the sets $\bar{B}_n$, then x belongs to none of the sets A_n. Thus the union of the sets A_n cannot equal X, contrary to the assumption that X is of the first category. Thus complete metric spaces can be only of the second category. □

Definition: *Let (X, d) be a metric space and $f: X \to X$ a function. Then f is* **contractive with respect to the metric** *d provided that there is a positive number $\alpha < 1$ such that, for all x, y in X,*

$$d(f(x), f(y)) \le \alpha d(x, y).$$

It is an easy exercise to show that a contractive function is always continuous. In fact, any function that does not increase distance is continuous.

Theorem 3.18:   The Contraction Lemma   *Let (X, d) be a complete metric space and $f: X \to X$ a contractive function. Then there is exactly one point x in X for which $f(x) = x$.*

Proof: *To show the existence of such a point, choose a point x_1 in X and define*

$$x_2 = f(x_1), \quad x_3 = f(x_2), \ldots, x_n = f(x_{n-1}), \quad n \ge 2.$$

The fact that f is a contractive function insures that $\{x_n\}_{n=1}^{\infty}$ is a Cauchy sequence. By completeness, this sequence has a limit x in X. Since $f: X \to X$ is continuous, then the sequence $\{f(x_n)\}_{n=1}^{\infty}$ converges to $f(x)$. But $f(x_n) = x_{n+1}$, $n \ge 1$, so $\{f(x_n)\}_{n=1}^{\infty}$ is simply $\{x_n\}_{n=2}^{\infty}$, whose limit is x. Thus $f(x) = x$.

To show the required uniqueness property, suppose that y is a second point satisfying $f(y) = y$. Then

$$d(x, y) = d(f(x), f(y)) \le \alpha d(x, y).$$

Since $\alpha < 1$, this relation cannot hold unless $d(x, y) = 0$ and $x = y$. □

The point x for which $f(x) = x$ in the proof of Theorem 3.18 is called the *fixed point* of the function *f*, and theorems of this type are called *fixed point theorems*. This particular example, the Contraction Lemma, is extremely useful for solving equations in function spaces and illustrates the significance of the completeness property. We shall see fixed point theorems again in a more general context in Chapters 5 and 9.

The final result of this chapter shows that every metric space can be considered a subspace of a complete metric space, called the *completion* of the given space. Several preliminary definitions are needed.

Definition: *Let (X, d) and (Y, d') be metric spaces. A distance preserving function f: X → Y from X into Y is called an **isometric embedding**.*

Definition: *A subspace A of a metric space X is **dense** in X provided that $\bar{A} = X$.*

Theorem 3.19: *Let (X, d) be a metric space. Then there is a complete metric space (Y, d') and an isometric embedding e: X → Y for which e(X) is a dense subspace of Y. The space (Y, d') is unique up to metric equivalence.*

Proof: *The lengthy proof of this theorem, which defines the completion of a metric space (X, d), is presented in outline form. Some details of the proof are left for the reader as exercises.*

Let $\mathcal{C}$ be the family of all Cauchy sequences $\{x_n\}_{n=1}^{\infty}$ in X. For brevity, let us denote a typical Cauchy sequence by $\langle x_n \rangle$. Define an equivalence relation $\sim$ on $\mathcal{C}$ as follows: Cauchy sequences $\langle x_n \rangle$ and $\langle y_n \rangle$ are to be considered equivalent, $\langle x_n \rangle \sim \langle y_n \rangle$, provided that the sequence $\{d(x_n, y_n)\}_{n=1}^{\infty}$ of real numbers has limit 0. It is left as an exercise for the reader to verify that $\sim$ is an equivalence relation. Let $Y = \mathcal{C}/\sim$ denote the family of equivalence classes, where $[\langle x_n \rangle]$ denotes the equivalence class of $\langle x_n \rangle$. For $[\langle x_n \rangle]$, $[\langle y_n \rangle]$ in Y, define

$$d'([\langle x_n \rangle], [\langle y_n \rangle]) = \text{limit } d(x_n, y_n)$$

to be the limit of the sequence $\{d(x_n, y_n)\}_{n=1}^{\infty}$ of real numbers.

Since d' is defined for pairs of equivalence classes, it is necessary to show that $d'([\langle x_n \rangle], [\langle y_n \rangle])$ is independent of the choice of the representatives $\langle x_n \rangle$ and $\langle y_n \rangle$ in their respective classes. For Cauchy sequences $\langle x'_n \rangle$ and $\langle y'_n \rangle$ equivalent respectively to $\langle x_n \rangle$ and $\langle y_n \rangle$, it must be demonstrated that

$$\text{limit } d(x_n, y_n) = \text{limit } d(x'_n, y'_n).$$

This follows from properties of the distance function d and the fact that

$$\text{limit } d(x_n, x'_n) = \text{limit } d(y_n, y'_n) = 0$$

and is left as an exercise. The proof that d' is a metric is also left as an exercise.

For x in X, the constant sequence $\langle x \rangle$ whose only value is x is clearly Cauchy and determines a member $[\langle x \rangle]$ of Y. Define e: X → Y by

$$e(x) = [\langle x \rangle], \quad x \in X.$$

For x, y in X,

$$d'(e(x), e(y)) = d'([\langle x \rangle], [\langle y \rangle]) = limit \; d(x, y) = d(x, y),$$

so e is an isometric mapping from X into Y.

To see that the closure of e(X) is Y, let $[\langle x_n \rangle]$ be a point of Y and ϵ a positive number. The desired conclusion will follow if it can be shown that there is some member of e(X) whose distance from $[\langle x_n \rangle]$ is less than ϵ. Since $\langle x_n \rangle$ is Cauchy, there is a positive integer N for which m, n $\geq$ N imply $d(x_n, x_m) < \epsilon/2$. Let $z = x_N$, and consider the member $[\langle z \rangle]$ of e(X) determined by z. Then

$$d'([\langle x_n \rangle], [\langle z \rangle]) = limit \; d(x_n, z) = limit \; d(x_n, x_N) \leq \epsilon/2 < \epsilon,$$

so $[\langle z \rangle]$ is within the prescribed distance of $[\langle x_n \rangle]$.

The completeness of (Y, d') can be proved as follows. The details are left as an exercise. First note that each Cauchy sequence in the dense subspace e(X) of Y converges. To see this, let $\{[\langle z^k \rangle]\}_{k=1}^{\infty}$ be a Cauchy sequence in e(X), where each sequence $\langle z^k \rangle$ has constant value z^k, a point in X, k = 1, 2, 3, ... Then $\{z^k\}_{k=1}^{\infty}$ is a Cauchy sequence in X and determines a member of Y to which the given Cauchy sequence in e(X) converges. Now let $\{[\langle x_n^k \rangle]\}_{k=1}^{\infty}$ be a Cauchy sequence in Y. Since e(X) is dense in Y, there is for each positive integer k a member $[\langle z^k \rangle]$ in e(X) whose distance from $[\langle x_n^k \rangle]$ is less than $1/k$. It follows that $\{[\langle z^k \rangle]\}_{k=1}^{\infty}$ is a Cauchy sequence which, as just demonstrated, converges to some member of Y. By the way the members $[\langle z^k \rangle]$ were chosen, it follows that $\{[\langle x_n^k \rangle]\}_{k=1}^{\infty}$ converges to the same member of Y. Thus Y is complete.

It remains to be shown that (Y, d') is unique up to metric equivalence. To establish this, let (Z, d'') be another complete metric space and $f: (X, d) \to (Z, d'')$ an isometric embedding of X in Z for which f(X) is dense in Z. It must be proved that (Y, d') is metrically equivalent to (Z, d''). Define $F: e(X) \to f(X)$ by

$$F(e(x)) = f(x), \quad x \in X.$$

This function is extended to an isometry $\hat{F}: Y \to Z$ from Y onto Z as follows: Let y be a member of $Y \backslash e(X)$. Since e(X) is dense in Y, there is a sequence $\{y_n\}_{n=1}^{\infty}$ of members of e(X) which converges to y. Then $\{F(y_n)\}_{n=1}^{\infty}$ is a Cauchy sequence in Z which, by completeness, converges to a point $\hat{F}(y)$ in Z. It is left as an exercise to show that $\hat{F}$ is an isometry from (Y, d') onto (Z, d''). □

Definition: *Let (X, d) be a metric space. The space (Y, d') defined by Theorem 3.19 is called the **completion** of (X, d').*

Since we do not distinguish between isometric metric spaces, Theorem 3.19 guarantees the existence of exactly one completion for each metric space. In fact, the uniqueness condition is often more useful in determining the completion of metric space than the rather complicated construction described by the theorem. This is illustrated by the following example.

Example 3.7.5

(a) The completion of the space of rational numbers is the real line $\mathbb{R}$. The completion construction is, in fact, often used to define the real numbers from the rational numbers. Assuming, however, that the real numbers have already been defined, the assertion is established as follows: The real line is a complete metric space which contains the metric space of rational numbers as a dense subspace. The uniqueness of the completion shows that $\mathbb{R}$ is the desired completion.

(b) The completion of $(0, 1)$ is $[0, 1]$. This follows easily from the uniqueness of the completion: $[0, 1]$ is a complete metric space which contains $(0, 1)$ as a dense subspace.

Example 3.7.6   The Space $C(X, \mathbb{R})$

This example generalizes the space $\mathcal{C}[a, b]$ of continuous, real-valued functions defined on a closed interval $[a, b]$. For a given metric space (X, d), $C(X, \mathbb{R})$ denotes the family of continuous, bounded, real-valued functions with domain X. (A *bounded function* $f: X \to \mathbb{R}$ is a function whose image $f(X)$ is bounded.) For f, g in $C(X, \mathbb{R})$, define

$$\rho(f, g) = \text{lub } \{ |f(x) - g(x)| : x \in X\}.$$

The proof that ρ is a metric is left as an exercise. The metric ρ is called the *supremum metric* or *uniform metric* for $C(X, \mathbb{R})$.

The metric space $(C(X, \mathbb{R}), \rho)$ is a complete metric space. To see this, let $\{f_n\}_{n=1}^{\infty}$ be a Cauchy sequence in $C(X, \mathbb{R})$. Then for x in X, the sequence $\{f_n(x)\}_{n=1}^{\infty}$ is a Cauchy sequence of real numbers which converges to a real number $f(x)$. This defines a function $f: X \to \mathbb{R}$, called the *limit* of the sequence $\{f_n\}_{n=1}^{\infty}$. The sequence $\{f_n\}_{n=1}^{\infty}$ converges to the limit function in the following rather strong sense: Given $\epsilon > 0$, there is a positive integer N such that if $n \geq N$ and $x \in X$, then

$$|f_n(x) - f(x)| < \epsilon.$$

The essential feature here is the fact that the integer N is dependent only upon ϵ and not on the choice of x. The same integer N will suffice for each point x in X. For this reason, the sequence $\{f_n\}_{n=1}^{\infty}$ is said to *converge uniformly* to f. The completeness of $C(X, \mathbb{R})$ will be established by showing that the limit function f is bounded and continuous.

The fact that f is bounded follows easily: Let M be a positive integer for which $n \geq M$ implies $\rho(f_n, f) < 1$. Then since f_M is bounded and $f(x)$ and $f_M(x)$ differ by no more than 1 for all x in X, f must be bounded also.

The continuity of f is established as follows. Let x_0 be a point of X and ϵ a positive number. Let N be a positive integer such that if $n \geq N$ then $\rho(f_n, f) < \epsilon/3$. Since f_N is continuous at x_0, there is a positive number δ

such that if $d(x, x_0) < \delta$, then $d(f_N(x), f_N(x_0)) < \epsilon/3$. Thus for x in X satisfying $d(x, x_0) < \delta$,

$$d(f(x), f(x_0)) \le d(f(x), f_N(x)) + d(f_N(x), f_N(x_0)) + d(f_N(x_0), f(x_0))$$
$$< \epsilon/3 + \epsilon/3 + \epsilon/3 = \epsilon.$$

Thus the limit function f is continuous, and $C(X, \mathbb{R})$ is a complete metric space.

The space $C(X, \mathbb{R})$ has additional structure which, although interesting, is not central to the purpose of this text. It has an algebraic structure of addition and multiplication by real numbers defined as follows:

$$(f + g)(x) = f(x) + g(x), \quad f, g \in C(X, \mathbb{R}),$$
$$(\alpha f)(x) = \alpha f(x), \quad f \in C(X, \mathbb{R}), \ \alpha \in \mathbb{R}.$$

With these operations, $C(X, \mathbb{R})$ is a vector space. A norm for $C(X, \mathbb{R})$, having properties analogous to those for the norm in $\mathbb{R}^n$, is defined by $\|f\| = \text{lub}\,\{\,|f(x)| : x \in X\}$, $f \in C(X, \mathbb{R})$. This norm defines the metric for $C(X, \mathbb{R})$ as follows:

$$\rho(f, g) = \|f - g\|, \quad f, g \in C(X, \mathbb{R}).$$

Thus $C(X, \mathbb{R})$ is a vector space with a norm, and it is complete in the metric defined by that norm. Such a space is called a *Banach space*. Banach spaces are extremely important in topology and analysis. Hilbert space is a Banach space; the proof of this is left as an exercise. Additional information on the topic of Banach spaces can be found in the supplementary reading list at the end of the chapter.

The convergence of sequences in $C(X, \mathbb{R})$ is defined in a more general context in the next definition.

Definition: *Let (X, d) and (Y, d') be metric spaces and $\{f_n : X \to Y\}_{n=1}^{\infty}$ a sequence of functions from X to Y. This sequence **converges uniformly** to a function $f: X \to Y$ provided that for each positive number ϵ there is a positive integer N such that if $n \ge N$ and x is a point of X, then $d'(f_n(x), f(x)) < \epsilon$.*

The proof of the following theorem, which can be based on Example 3.7.6, is left as an exercise.

Theorem 3.20:   *Let (X, d) and $Y, d')$ be metric spaces and $\{f_n\}_{n=1}^{\infty}$ a sequence of continuous functions from X to Y which converges uniformly to a function $f: X \to Y$. Then f is continuous.*

The major theorems of this section have wide applicability to other areas of mathematics. For example, the Contraction Lemma is used to prove the Inverse Function Theorem of analysis and the Picard Existence Theorem of differential equations. The Baire Category Theorem has a multitude of applications, including the Uniform Boundedness Principle of functional analysis and theorems on approximation of functions; it can be used, for example, to show that there is a continuous function $f: [0, 1] \to \mathbb{R}$ which is nowhere differentiable. The applicability of Baire's theorem rests largely on the fact that $C(X, \mathbb{R})$, Hilbert space, and many related spaces are complete metric spaces. References for investigation of these applications are given at the end of the chapter.

EXERCISE 3.7

1. Show the incompleteness of (a, b), $(a, b]$, and $[a, b)$ by exhibiting Cauchy sequences that do not converge.

2. Prove that every convergent sequence is Cauchy.

3. (a) Prove the following:
    Cantor's Intersection Theorem: *Let (X, d) be a complete metric space and $\{A_n\}_{n=1}^{\infty}$ a nested sequence of non-empty closed sets whose diameters $D(A_n)$ have limit 0. Then $\bigcap_{n=1}^{\infty} A_n$ has exactly one member.*

    (b) Show that, in part (a), $\bigcap_{n=1}^{\infty} A_n$ may be empty if the requirement that the diameters approach 0 is deleted.

4. Show that Hilbert space (Example 3.6.3) is complete and conclude that it is a Banach space.

5. Prove that the following statements are equivalent for a subset A of a metric space X:

    (a) A is nowhere dense.

    (b) Each non-empty open set in X has a non-empty open subset disjoint from $\bar{A}$.

    (c) Each non-empty open set in X contains an open ball whose closure is disjoint from A.

6. Let (X, d) be a metric space, M a positive number, and $f: X \to X$ a continuous function for which

$$d(f(x), f(y)) \le Md(x, y)$$

for all x, y in X. Prove that f is continuous. Use this to conclude that every contractive function is continuous.

7. Give an example of two metric spaces (X_1, d_1) and (X_2, d_2) which are topologically equivalent and for which (X_1, d_1) is complete and (X_2, d_2) is not.

8. Give an example of a set X with two equivalent metrics d and d' for which (X, d) is complete and (X, d') is not.

9. Let (X, d) be a complete metric space. Show that the completion process of Theorem 3.19 defines a metric space (Y, d') which is metrically equivalent to (X, d).

10. Complete the details in the proof of Theorem 3.19.

11. Prove Theorem 3.20.

12. Give an example of a sequence $\{f_n\}_{n=1}^\infty$ of continuous functions $f_n: I \to I$ such that, for each x in $I = [0, 1]$, $\{f_n(x)\}_{n=1}^\infty$ converges to a real number $f(x)$, but the limit function f is not continuous.

13. Let (X, d) be a metric space and A a dense subspace such that every Cauchy sequence in A converges in X. Prove that X is complete. Identify the use of this result in the proof of Theorem 3.19.

14. **Definition:** *Let (X, d) be a metric space. A point p in X is an **isolated point** if the singleton set $\{p\}$ is an open set.*

    (a) Let (X, d) be a metric space and p a point of X. Assume $X \neq \{p\}$.
    Prove that the following conditions are equivalent:

    (i) p is an isolated point.

    (ii) $d(p, X\backslash\{p\}) > 0$.

    (iii) $p \notin \overline{(X\backslash\{p\})}$.

    (b) Prove that a complete metric space without isolated points must be uncountable. (*Hint:* Use the Baire Category Theorem.)

15. Let (X, d) be a complete metric space and $\{U_n\}_{n=1}^\infty$ a sequence of open dense subsets of X. Prove that $\bigcap_{n=1}^\infty U_n$ is dense in X.

16. The Baire Category Theorem (Theorem 3.17) shows that for metric spaces, completeness implies the property of being of the second category. Show that these properties are not equivalent by giving an example of a metric space which is of the second category but is not complete. (*Hint:* Any open interval (a, b) in $\mathbb{R}$ is topologically equivalent to $\mathbb{R}$.)

17. Let $f: \mathbb{R} \to \mathbb{R}$ be a continuous, unbounded function. Show that there is a number t_0 for which $\{f(nt_0): n \text{ an integer}\}$ is an unbounded set.

SUGGESTIONS FOR FURTHER READING

For additional reading on metric spaces, *Set Theory and Metric Spaces* by Kaplansky and *Introduction to Topology and Modern Analysis* by Simmons are recommended. Simmons' text is particularly recommended for Banach and Hilbert spaces.

Picard's Theorem and related applications of the Contraction Lemma can be found in textbooks on differential equations and real analysis. See, for example, *Differential and Difference Equations* by Brand. Some applications of the Baire Category Theorem are given in Willard's *General Topology* and Eisenberg's *Topology*. For applications to functional analysis, *Introduction to Topology and Modern Analysis* by Simmons is a good place to start.

HISTORICAL NOTES FOR CHAPTER 3

The extension of topological considerations beyond the realm of Euclidean space was first made by Maurice Fréchet. Fréchet introduced metric spaces in 1906 in a very general context that allowed the "points" under consideration to be abstract objects, not just real numbers or n-tuples of real numbers. This revolutionary idea brought under one theory the work of an essentially topological nature done on sets of curves by G. Ascoli, the study of sets of lines and planes by Emile Borel, and the function space studies of C. Arzela, V. Volterra, David Hilbert, I. Fredholm, and others.

The supremum metric for functions in $\mathcal{C}[a, b]$ and $C(X, \mathbb{R})$ is usually attributed to Fréchet, but it was used as early as 1885 by Weierstrass in his work on uniform convergence. The systematic study of continuous functions and homeomorphisms on abstract spaces was initiated by Fréchet, although the idea of homeomorphism had been used in a less general context by Henri Poincaré in 1895.

Hilbert space H was the invention of David Hilbert (1862–1943) in 1906. The Cauchy-Schwarz Inequality in the form used in this chapter is due to Cauchy. The Minkowski Inequality was proved by Hermann Minkowski (1864–1909) in 1909.

The idea of a completion of a metric space can be traced to Cauchy, who attempted in 1821 to define the irrational numbers as limits of Cauchy sequences of rational numbers, thus effecting a completion for the space of rational numbers. Cauchy's method depended to a considerable extent on intuition and was revised and put on a logically sound basis by Charles Méray in 1869. Méray referred to the completion technique as a definition of "fictitious numbers." A similar completion for the set of rational numbers was defined by Cantor. The general concept of complete metric space was defined by Fréchet, and the general completion construction was presented by Hausdorff in 1914.

The Contraction Lemma is due to Stefan Banach (1892–1945). The concept of a general normed space is due to Banach and others, notably Hans Hahn (1879–1934), Eduard Helly (1884–1943), and Norbert Wiener (1894–1964). Banach spaces were introduced by Banach in 1923.

The Baire Category Theorem was proved by the French mathematician René Baire (1874–1932) for the real line in 1889. The general theorem for complete metric spaces first appeared in *Grundzüge der Mengenlehre* in 1914 and is attributed to Hausdorff.

4 Topological Spaces

It was shown in Chapter 3 that such properties as limit point, interior, closure, boundary, and continuity of functions on metric spaces can be expressed in terms of the open sets or topologies on the spaces involved. In addition, the latter part of the chapter demonstrated that the same topology may be determined by several different metrics. Thus, for the study of ideas like continuity, it appears that the families of open sets for the spaces involved are more basic than their metrics. For this reason, we turn in this chapter to a general definition of the term "topology for a set," defining it in a manner consistent with the topology determined by a metric. This will allow us to extend the ideas of Chapter 3 to situations in which metrics are not available.

4.1 THE DEFINITION AND SOME EXAMPLES

Theorem 3.4 lists the primary properties of the topology generated by a metric which were used in our study of metric spaces. The term "topology for a set" is extended to the nonmetric case by using the conditions of Theorem 3.4 as the defining axioms.

Definition: *Let X be a set and let T be a family of subsets of X satisfying the following conditions:*

   (a)   The set X and the empty set $\emptyset$ belong to T.
   (b)   The union of any family of members of T is a member of T.
   (c)   The intersection of any finite family of members of T is a member of T.

*Then T is called a **topology** for X and the members of T are called **open sets**. The ordered pair (X, T) is called a **topological space** or simply a **space**.*

Using the term "open set" instead of "member of T," the definition of a topology may be restated as follows: *A family of subsets of X is a topology for X means that:*

   (a)   Both X and $\emptyset$ are open sets.
   (b)   The union of any family of open sets is an open set.
   (c)   The intersection of any finite family of open sets is an open set.

When the topology T is understood, it is common practice to refer to topological space X instead of (X, T), omitting mention of the topology.

Example 4.1.1

(a) The *usual topology* for the real line $\mathbb{R}$ is the topology generated by its usual metric. We shall refer to the real line with the usual topology as simply "the real line," or $\mathbb{R}$, understanding that the usual topology is to be applied unless a different topology is specified.

(b) The *usual topology* for $\mathbb{R}^n$ is the topology generated by the usual metric. By Example 3.5.4, the taxicab metric d' and the max metric d'' also determine the usual topology for $\mathbb{R}^n$. We shall refer to $\mathbb{R}^n$ with the usual topology as *Euclidean n-space,* or simply $\mathbb{R}^n$, understanding that the usual topology is to be used unless a different one is specified.

(c) For a set X, the topology generated by the discrete metric is the *discrete topology*. In the discrete topology, every subset of X is open. A set with the discrete topology is called a *discrete space*. Note that the discrete topology is the largest possible collection of open subsets of X.

(d) At the opposite extreme, the *trivial topology* for X is the family $\mathcal{T} = \{\varnothing, X\}$ whose only members are $\varnothing$ and X. (Refer to the definition to see that this is a topology.) A set with its trivial topology is called a *trivial space*. As one might surmise, neither discrete nor trivial spaces are of much interest as topological spaces. However, discrete spaces can be combined by various constructions to produce very interesting spaces which are not discrete. Several such examples will appear in later chapters.

(e) As another example, consider a set X with topology $\mathcal{T}'$ consisting of the empty set $\varnothing$ and all subsets O of X for which $X \backslash O$ is a finite set. Then $\mathcal{T}'$ is a topology for X and is called the *finite complement topology*. This topology is of interest only when X is an infinite set; if X is finite, it coincides with the discrete topology in which every subset is open.

Definition: *A subset C of a topological space X is **closed** provided that its complement $X \backslash C$ is an open set.*

Theorem 4.1: *The closed sets of a topological space X have the following properties:*

(a) *X and $\varnothing$ are closed sets.*

(b) *The intersection of any family of closed sets is a closed set.*

(c) *The union of any finite family of closed sets is a closed set.*

The proof of Theorem 4.1 is left as an exercise.

Definition: *Let (X, T) be a topological space and A a subset of X. A point x in X is a **limit point, cluster point,** or **accumulation point** of A if every open set containing x contains a point of A distinct from x. The set of limit points of A is called the **derived set** of A.*

Example 4.1.2

Consider the topological space $(\mathbb{R}, T')$, where T' is the finite complement topology of Example 4.1.1(e). For any infinite subset A of $\mathbb{R}$, the derived set A' is $\mathbb{R}$ itself. To see this, consider any point x in $\mathbb{R}$ and an open set O containing x. Since $\mathbb{R} \backslash O$ is finite, then O must contain all but a finite number of members of A. In particular, O must contain at least one point of A distinct from x. Hence $x \in A'$ so $A' = \mathbb{R}$.

A finite subset B of $\mathbb{R}$ has no limit points with respect to the finite complement topology. If x does not belong to B, then $\mathbb{R} \backslash B$ is an open set containing x which contains no point of B; if x does belong to B, then $\{x\} \cup (\mathbb{R} \backslash B)$ is an open set containing x which contains no point of B different from x.

The next theorem should come as no surprise. Its proof is left as an exercise.

Theorem 4.2:   *A subset A of a topological space X is closed if and only if A contains all its limit points.*

Definition: *Let X be a topological space and $\{x_n\}_{n=1}^{\infty}$ a sequence of points of X. Then $\{x_n\}_{n=1}^{\infty}$ **converges** to the point $x \in X$, or x is a **limit** of the sequence, if for each open set O containing x there is a positive integer N such that $x_n \in O$ for all $n \geq N$.*

The following examples show that Theorems 3.8 and 3.9 for sequences in metric spaces do not carry over to general topological spaces. These examples suggest that sequences will not play in general topological spaces the fundamental role that they play in metric spaces.

Example 4.1.3

Consider the set $\mathbb{R}$ of real numbers with the finite complement topology. Every sequence $\{x_n\}_{n=1}^{\infty}$ of distinct points converges to every point. To see this, let $x \in \mathbb{R}$ and let O be an open set containing x. Then O has finite complement and hence must contain all but a finite number of terms of the sequence. In particular, there is a positive integer N such that if $n \geq N$, then $x_n \in O$. We conclude that

$\{x_n\}_{n=1}^{\infty}$ converges to every member of $\mathbb{R}$ and, therefore, that a sequence may have more than one limit.

Example 4.1.4

Consider a trivial topological space $X = \{a, b\}$ consisting of only two points a and b. Then a is a limit point of $\{b\}$ since X is the only open set containing a. However, the singleton set $\{b\}$ contains no sequence of distinct points converging to a.

EXERCISE 4.1

1. Show that the trivial topology $\mathcal{T}_1$ and the discrete topology $\mathcal{T}_2$ are, respectively, the smallest and largest topologies for any set X.

2. Let X be a set with at least two members. Show that there is no metric for X which generates the trivial topology.

3. Show that the finite complement topology is actually a topology for any set X.

4. Show that a space $(X, \mathcal{T})$ is discrete if and only if each set consisting of only one point is open.

5. Prove Theorems 4.1 and 4.2.

6. Let A, B be subsets of a space X with $A \subset B$, and let A', B' denote the derived sets of A and B, respectively. Show that $A' \subset B'$. Show by an example that A' may equal B' even though A is a proper subset of B.

7. Let X be a space, A a subset of X, and x a member of X. Prove that if there is a sequence of distinct points of A converging to x, then x is a limit point of A.

8. Let X be a set and $\mathcal{T}'$ the finite complement topology for X.

   (a) Show that $(X, \mathcal{T}')$ is discrete if and only if X is a finite set.

   (b) Show that if A is an infinite subset of X, then every point of X is a limit point of A.

9. Let X be a set. The *countable complement topology* $\mathcal{T}''$ for X consists of $\varnothing$ and all subsets O of X for which $X \backslash O$ is a countable set.

   (a) Show that $\mathcal{T}''$ is actually a topology for X.

   (b) For the space $(X, \mathcal{T}'')$, show that a countable subset A of X has derived set $A' = \varnothing$ and that an uncountable set B has $B' = X$.

   (c) Show that the intersection of any countable family of members of $\mathcal{T}''$ is a member of $\mathcal{T}''$.

10. Let $S = \{a, b\}$ be a two-element set and let $\mathcal{T} = \{\varnothing, \{a\}, \{a, b\}\}$. Show that $\mathcal{T}$ is a topology and identify the limit points of each subset of S. (The space S is called *Sierpiński space*.)

11. How many different topologies are there for a set with three members?

4.2 INTERIOR, CLOSURE, AND BOUNDARY

The interior, closure, and boundary for subsets of a topological space are defined in complete analogy with their counterparts for subsets of metric spaces.

Definition: *Let A be a subset of a topological space X. A point x in A is an **interior point** of A if there is an open set O containing x and contained in A. Equivalently, A is called a **neighborhood** of x. The **interior** of A, denoted int A, is the set of all interior points of A.*

*The **closure** $\bar{A}$ of A is the union of A with its set of limit points:*

$$\bar{A} = A \cup A'$$

where A' is the derived set of A.

*A point x in X is a **boundary point** of A if x belongs to both $\bar{A}$ and $\overline{(X \backslash A)}$. The set of boundary points of A is called the **boundary** of A and is denoted bdy A.*

Theorem 4.3: *For any subsets A, B of a topological space X:*

(1) The interior of A is the union of all open sets contained in A and is therefore the largest open set contained in A.

(2) A is open if and only if A = int A.

(3) If $A \subset B$, then int $A \subset$ int B.

(4) int $(A \cap B) =$ int $A \cap$ int B.

Proof: *Statements (1) and (2) carry over from Chapter 3, and (3) is an immediate consequence of the definition of interior. To prove (4), note first that since $A \cap B$ is a subset of both A and B, then int $(A \cap B)$ is a subset of int $A \cap$ int B by (3). For the reverse inclusion, note that int $A \cap$ int B is an open set and is a subset of $A \cap B$. Since int $(A \cap B)$ is the largest open set contained in $A \cap B$, then*

$$\text{int } A \cap \text{int } B \subset \text{int } (A \cap B). \qquad \square$$

Example 4.2.1

It is not true in general that int $(A \cup B)$ equals int $A \cup$ int B. As a counterexample, consider the real line with $A = [0, 1]$ and $B = [1, 2]$. Then

$$\text{int } (A \cup B) = \text{int } [0, 2] = (0, 2)$$

while

$$\text{int } A \cup \text{int } B = (0, 1) \cup (1, 2)$$

so int $(A \cup B)$ contains 1 while int $A \cup$ int B does not. The reader is asked in one of the exercises to prove that the inclusion

$$\text{int } A \cup \text{int } B \subset \text{int } (A \cup B)$$

is always valid.

Theorem 4.4:   *For any subsets A, B of a space X:*

(1)   The closure of A is the intersection of all closed sets containing A and is therefore the smallest closed set containing A.
(2)   A is closed if and only if $A = \bar{A}$.
(3)   If $A \subset B$, then $\bar{A} \subset \bar{B}$.
(4) $\overline{A \cup B} = \bar{A} \cup \bar{B}$.

Proof: *Again statements (1) and (2) carry over from Chapter 3 (Theorems 3.10 and 3.11). For (3), note that if $A \subset B$, then the definition of limit point guarantees that $A' \subset B'$. Then*

$$\bar{A} = A \cup A' \subset B \cup B' = \bar{B}.$$

To prove (4), note first that $\bar{A} \cup \bar{B}$ is a closed set which contains $A \cup B$. Since $\overline{A \cup B}$ is the smallest closed set containing $A \cup B$, then

$$\overline{A \cup B} \subset \bar{A} \cup \bar{B}.$$

For the reverse inclusion, use (3) and the fact that both A and B are subsets of $A \cup B$.   □

Example 4.2.2

It is not true in general that $\overline{A \cap B}$ equals $\bar{A} \cap \bar{B}$. For example, let $A = (0, 1)$ and $B = (1, 2)$ on the real line. Then

$$\overline{A \cap B} = \bar{\varnothing} = \varnothing$$

but

$$\bar{A} \cap \bar{B} = [0, 1] \cap [1, 2] = \{1\}.$$

The reader is left the easy exercise of showing that the inclusion

$$\overline{A \cap B} \subset \bar{A} \cap \bar{B}$$

is always valid.

Theorem 4.5: *Let A be a subset of a topological space X.*

(1) *bdy $A = \bar{A} \cap \overline{(X \backslash A)} = bdy\ (X \backslash A)$.*

(2) *bdy A, int A, and int $(X \backslash A)$ are pairwise disjoint sets whose union is X.*

(3) *bdy A is a closed set.*

(4) *$\bar{A} = int\ A \cup bdy\ A$.*

(5) *A is open if and only if bdy $A \subset (X \backslash A)$.*

(6) *A is closed if and only if bdy $A \subset A$.*

(7) *A is open and closed if and only if bdy $A = \varnothing$.*

Proof: *Properties (1) through (4) follow immediately from the definitions. To prove (5), note that if A is open, then $A = int\ A$ by Theorem 4.3, part (2). Since int A and bdy A are disjoint by (2), then A and bdy A are disjoint, so bdy A must be a subset of $X \backslash A$. For the reverse implication, suppose bdy $A \subset X \backslash A$. Then no point of A is a boundary point of A, so every point of A is an interior point. Thus $A = int\ A$, so A is open.*

Statement (6) follows from the duality between open sets and closed sets: A is closed if and only if $X \backslash A$ is open. By (5), this is equivalent to saying that

$$bdy\ (X \backslash A) \subset X \backslash (X \backslash A)$$

or

$$bdy\ A \subset A.$$

Statement (7) is proved by combining (5) and (6): A is both open and closed if and only if bdy A is contained in both A and $X \backslash A$. Since A and $X \backslash A$ are disjoint, this occurs if and only if bdy $A = \varnothing$. ☐

According to Theorem 4.5, the points of a subset A of a space X may be of two types, interior points and boundary points. The set A may have additional boundary points outside A, however; the union of all interior points and boundary points of A is $\bar{A}$. The points of X are of three non-overlapping types: (1) interior points of A, (2) interior points of $X \backslash A$, and (3) boundary points of A, which are identical with the boundary points of $X \backslash A$. (Of course, any of these three sets may be empty.)

The following examples are an attempt to spare the reader some of the common misconceptions about boundaries and closures in metric spaces.

Example 4.2.3

For an open ball $B(a, r)$ in a metric space (X, d), $\overline{B(a, r)}$ may *not* be the closed ball $B[a, r]$, and bdy $B(a, r)$ may *not* be $\{x \in X: d(x, a) = r\}$.

(a) Consider first the case of a discrete metric space (X, d) and an open ball $B(a, 1)$ of radius 1:

$$\overline{B(a, 1)} = \{a\}, \quad B[a, 1] = X.$$

Note also that

$$\text{bdy } B(a, 1) = \varnothing, \quad \{x \in X: d(x, a) = 1\} = X \backslash \{a\}.$$

(b) These phenomena are not restricted to discrete spaces. Let Y be the subspace of $\mathbb{R}^2$ shaded in Figure 4.1: $Y = \{\theta\} \cup \{x \in \mathbb{R}^2: \|x\| \geq 1\}$.

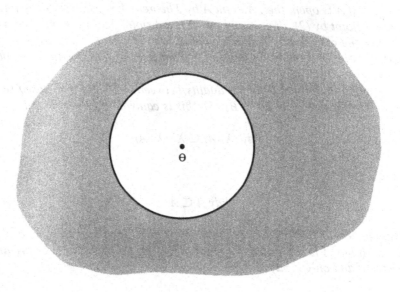

FIGURE 4.1

In Y, $\overline{B(\theta, 1)} = \{\bar{\theta}\} = \{\theta\}$ while $B[\theta, 1]$ is the union of $\{\theta\}$ with the unit circle. Also, bdy $B(\theta, 1) = \varnothing$ and $\{x \in Y: d(x, \theta) = 1\}$ is the unit circle.

Definition: *A subset A of a space X is **dense** in X provided that $\bar{A} = X$. If X has a countable dense subset, then X is a **separable** space.*

It is a simple consequence of the definitions of closure and dense set that a subset A of X is dense in X if and only if every non-empty open set in X contains at least one point of A.

Example 4.2.4

(a) The real line $\mathbb{R}$ is separable. The set of rational numbers is countable and dense in $\mathbb{R}$.

(b) Euclidean *n*-space $\mathbb{R}^n$ is separable. The set R of points of $\mathbb{R}^n$ having only rational coordinates is dense in $\mathbb{R}^n$ by Example 3.3.4(c). This set is countable since it is the product of the set of rational numbers (a countable set) taken as a factor *n* times.

(c) Hilbert space H is separable. Let C denote the set of all points $x = (x_1, \ldots, x_n, \ldots)$ all of whose coordinates are rational and for which only finitely many coordinates x_i are non-zero. In other words,

$$C = \bigcup_{n=1}^{\infty} C_n$$

where $C_n = \{x = (x_1, \ldots, x_n, 0, 0, \ldots) \in H : x_i$ is rational for $i = 1, \ldots, n$ and $x_i = 0$ for $i > n\}$. Since each set C_n is countable, then C is the union of a countable family of countable sets and is hence countable. To see that C is dense in H, consider a non-empty open set O. Let $B(a, r)$ be a ball with center $a = (a_1, \ldots, a_n, \ldots)$ and positive radius r contained in O. Since $\sum_{n=1}^{\infty} a_n^2$ converges, there is a positive integer N such that

$$\sum_{n=N+1}^{\infty} a_n^2 < r^2/2.$$

For $i = 1, \ldots, N$ there is a rational number x_i between $a_i - r/\sqrt{2N}$ and $a_i + r/\sqrt{2N}$. Then $x = (x_1, \ldots, x_N, 0, 0, \ldots)$ belongs to C and

$$d(a, x) = \left(\sum_{n=1}^{\infty} (a_i - x_i)^2 \right)^{1/2} = \left(\sum_{n=1}^{N} (a_i - x_i)^2 + \sum_{n=N+1}^{\infty} a_i^2 \right)^{1/2}$$

$$< \left(\sum_{n=1}^{N} (r/\sqrt{2N})^2 + r^2/2 \right)^{1/2} = (Nr^2/2N + r^2/2)^{1/2} = r$$

so

$$x \in B(a, r) \subset O.$$

Thus $\bar{C} = H$ and H is separable.

(d) The real line with the finite complement topology is separable since, by Example 4.1.2, every countably infinite subset is dense.

Definition: *A subset B of a space X is **nowhere dense** provided that int $\bar{B}$ = Ø.*

The relations between dense sets and nowhere dense sets are explored in the exercises for this section.

EXERCISE 4.2

1. Let A and B be subsets of a space X. Show that
   (a) int $A \cup$ int $B \subset$ int $(A \cup B)$.
   (b) $\overline{A \cap B} \subset \bar{A} \cap \bar{B}$.
   (c) int (int A) = int A.
   (d) $\bar{\bar{A}} = \bar{A}$.

2. Prove statements (1) through (4) of Theorem 4.5.

3. Let (X, d) be a metric space, a a point of X, and r a positive number. Prove that
   (a) $\overline{B(a, r)} \subset B[a, r]$.
   (b) bdy $B(a, r) \subset \{x \in X: d(x, a) = r\}$.

4. Identify int A, bdy A, int $(X \backslash A)$, $\bar{A}$, and the derived set A' in each of the following cases:
   (a) $A = \{x = (x_1, x_2) \in \mathbb{R}^2: x_2 > 0\}$ in $\mathbb{R}^2$;
   (b) $A = [0, 1]$, as a subset of $\mathbb{R}$ with the finite complement topology;
   (c) $A = \{a\}$ where $X = \{a, b\}$ with the discrete topology;
   (d) $A = \{a\}$ where $X = \{a, b\}$ with the trivial topology.

5. (a) If (X_1, d_1) and (X_2, d_2) are separable metric spaces, prove that the product metric space $X_1 \times X_2$ is separable.
   (b) Use (a) to prove that $\mathbb{R}^n$ is separable for each positive integer n.

6. Let A be a subset of a space X. Prove that A is dense in X if and only if int $(X \backslash A)$ = Ø.

7. Let B be a subset of a space X. Prove that the following statements are equivalent.
   (a) B is nowhere dense.
   (b) $X \backslash \bar{B}$ is dense in X.
   (c) $X \backslash \overline{(X \backslash \bar{B})}$ = Ø.
   (d) $B \subset \overline{(X \backslash \bar{B})}$.

8. **Definition:**   *For a subset A of a space X, the **exterior** of A is the set ext $A = X \backslash \bar{A}$.*
   Prove that ext A = int $(X \backslash A)$.

9. Prove:
   (a) Every finite subset of $\mathbb{R}^n$ is nowhere dense.
   (b) The set of points of $\mathbb{R}^n$ all of whose coordinates are integers is nowhere dense.
   (c) $\mathbb{R}^{n-1}$ is nowhere dense when considered as a subset of $\mathbb{R}^n$.

10. The purpose of this problem is to show that the concept of topology for a set X can be defined in terms of the closure operation.

    Definition: *Let X be a set. A **closure operator** on X is a function c which associates with each subset A of X a subset $c(A)$ of X satisfying the following properties:*

    (1) $c(\emptyset) = \emptyset$,
    (2) $A \subset c(A)$,
    (3) $c(c(A)) = c(A)$,
    (4) $c(A \cup B) = c(A) \cup c(B)$,

    for all subsets A, B of X.

    *A subset A of X is **c-closed** provided that $c(A) = A$, and a subset B of X is **c-open** provided that $X\backslash B$ is c-closed.*

    Assume that c is a closure operator for a given set X. Prove that:

    (a) The family $\mathcal{T}$ of c-open sets is a topology for X.

    (b) For each subset A of X, $c(A) = \bar{A}$, where $\bar{A}$ is the closure of A in the topology $\mathcal{T}$.

4.3 BASIS AND SUBBASIS

A topology for a set X can be a very large and complicated family of subsets. Often it simplifies matters to deal with a smaller collection which generates the topology by taking unions. Such a subcollection is called a *basis;* the precise definition follows.

Definition: *Let $(X, \mathcal{T})$ be a topological space. A **base** or **basis** $\mathcal{B}$ for $\mathcal{T}$ is a subcollection of $\mathcal{T}$ with the property that each member of $\mathcal{T}$ is a union of members of $\mathcal{B}$. Reference to the topology is sometimes omitted, and we speak of basis for X rather than a basis for the topology of X. The members of $\mathcal{B}$ are called **basic open sets**, and $\mathcal{T}$ is the topology **generated** by $\mathcal{B}$.*

Example 4.3.1

(a) The collection $\mathcal{B}$ of all open intervals is a basis for the usual topology of $\mathbb{R}$.

(b) For any metric space (X, d), the collection $\mathcal{B}$ of all open balls $B(a, r)$, $a \in X, r > 0$, is a basis for the topology generated by d.

(c) For any set X, the collection of all singleton sets $\{x\}$, $x \in X$, is a basis for the discrete topology.

(d) For any space $(X, \mathcal{T})$, the topology $\mathcal{T}$ is a basis for itself. This fact is of little use because the point of defining a basis is to produce a smaller collection of open sets with which to work.

Definition: *Let (X, T) be a space and let a be a member of X. A **local base** or **local basis** at a is a subcollection $\mathcal{B}_a$ of T such that*

    (1)   a belongs to each member of $\mathcal{B}_a$, and
    (2)   each open set containing a contains a member of $\mathcal{B}_a$.

Example 4.3.2

(a)   For $a \in \mathbb{R}$, the collection $\mathcal{B}_a$ of all open intervals of the form $(a - \epsilon, a + \epsilon)$ $\epsilon > 0$, is a local basis at a.

(b)   For any metric space (X, d) and $a \in X$, the collection $\mathcal{B}_a$ of all open balls centered at a is a local base at a.

(c)   For a discrete space X, the singleton set $\{a\}$ forms a local basis at a. (The local basis is the collection whose only member is $\{a\}$.)

(d)   For any space (X, T) and $a \in X$, the collection of all open sets containing a is a local basis at a.

(e)   If $\mathcal{B}$ is a basis for a space X, then the collection of all members of $\mathcal{B}$ which contain a is a local basis at a. Conversely, if for each $a \in X$, $\mathcal{B}_a$ is a local basis at a, then $\bigcup_{a \in X} \mathcal{B}_a$ is a basis for the topology of X.

Definition: *A space X is **first countable** or **satisfies the first axiom of countability** provided that there is a countable local basis at each point of X. The space X is **second countable** or **satisfies the second axiom of countability** provided that the topology of X has a countable basis.*

If a space X has a countable basis, that is, a basis $\mathcal{B}$ consisting of a countable family of open sets, then the members of $\mathcal{B}$ which contain a particular point a form a countable local basis at a. Thus each second countable space is first countable.

Theorem 4.6:   *Every second countable space is separable.*

Proof: *Let X be a second countable space with countable basis $\mathcal{B}$. Let A be the countable set formed by choosing a member from each basic open set. (If $\mathcal{B}$ happens to have $\varnothing$ as a member, then choose one member from each non-empty member of $\mathcal{B}$.) It follows from the definition of basis that A is dense in X.* $\square$

Theorem 4.7:

    (a)   Every metric space is first countable.
    (b)   Every separable metric space is second countable.

Proof:

(a)   The reader is left the exercise of showing that the collection of open balls $\{B(a, 1/n)\}_{n=1}^{\infty}$ is a countable local base at a, for each point a in a metric space X.

(b)   Let (X, d) be a separable metric space with countable dense set A. Then for each $a \in A$, $\{B(a, 1/n)\}_{n=1}^{\infty}$ is a countable collection and hence $\mathcal{B} = \{B(a, 1/n): a \in A, n \text{ a positive integer}\}$ is the union of a countable family of countable sets and is therefore countable. The proof will be completed by showing that $\mathcal{B}$ is a basis for the metric topology generated by d. To this end, let O be an open set and let $x \in O$. There is an open ball B(x, r) of positive radius r centered at x and contained in O. Let n be a positive integer for which $1/n < r/2$. Since A is dense in X, then there is some member a of A in B(x, 1/n). Then $x \in B(a, 1/n)$, and B(a, 1/n) is a member of $\mathcal{B}$. For any $y \in B(a, 1/n)$,

$$d(x, y) \le d(x, a) + d(a, y) < 1/n + 1/n = 2/n < 2r/2 = r,$$

so $y \in B(x, r)$. Thus

$$x \in B(a, 1/n) \subset B(x, r) \subset O.$$

Hence, for each x in O, there is a member of $\mathcal{B}$ which contains x and is contained in O. Then O is a union of members of $\mathcal{B}$, so $\mathcal{B}$ is a countable basis for X.   □

Corollary: Euclidean n-space $\mathbb{R}^n$ is second countable for each positive integer n.

Proof: By Example 4.2.4, the collection of points of $\mathbb{R}^n$ having only rational coordinates is a countable dense set. Thus $\mathbb{R}^n$ is a separable metric space and is second countable by Theorem 4.8.   □

Corollary: Hilbert space H is second countable.

Proof: Example 4.2.4(c) shows that H is separable.   □

Example 4.3.3

Consider the space $(\mathbb{R}, \mathcal{T}')$ of real numbers with the finite complement topology. This space does not have a countable local basis at any point. To see this, let

$a \in \mathbb{R}$ and suppose that $\mathcal{T}'$ has a countable local basis $\mathcal{B}_a = \{B_n\}_{n=1}^{\infty}$ at a. Then for each positive integer n, $\mathbb{R} \backslash B_n$ is finite so

$$\bigcup_{n=1}^{\infty} (\mathbb{R} \backslash B_n) = \mathbb{R} \backslash \bigcap_{n=1}^{\infty} B_n$$

is countable. Since $\mathbb{R}$ is uncountable, then $\bigcap_{n=1}^{\infty} B_n$ must be uncountable and must contain at least one point (actually, uncountably many points) b distinct from a. Then

$$O = X \backslash \{b\}$$

is an open set containing a, but O does not contain any member of the collection $\mathcal{B}_a = \{B_n\}_{n=1}^{\infty}$. Thus $\mathcal{B}_a$ is not a local basis at a, and $(\mathbb{R}, \mathcal{T}')$ has no countable local basis at any point.

There are two viewpoints from which to consider bases. Thus far, we have started with a topology and considered the problem of finding a basis for it. On the other hand, one might start with a basis and generate a topology from it by forming unions. But not every family of subsets of X is a basis for a topology. The next theorem gives necessary and sufficient conditions that a family of subsets generate a topology.

Theorem 4.8:   *A family $\mathcal{B}$ of subsets of a set X is a basis for some topology for X if and only if both of the following conditions hold:*

(a)   *The union of the members of $\mathcal{B}$ is X.*

(b)   *For each B_1, B_2 in $\mathcal{B}$ and $x \in B_1 \cap B_2$, there is a member B_x of $\mathcal{B}$ such that*

$$x \in B_x \subset B_1 \cap B_2.$$

Proof: *Suppose first that $\mathcal{B}$ is a basis for a topology for X. Then (a) follows from the fact that X is an open set and must be a union of members of $\mathcal{B}$. Since each member of $\mathcal{B}$ is a subset of X, then*

$$X = \bigcup_{B \in \mathcal{B}} B.$$

For (b), let B_1, B_2 be members of $\mathcal{B}$ and x a member of $B_1 \cap B_2$. Then $B_1 \cap B_2$ is an open set and is therefore some union of members of $\mathcal{B}$. Thus there is some $B_x \in \mathcal{B}$ such that

$$x \in B_x \subset B_1 \cap B_2.$$

   Now suppose that $\mathcal{B}$ satisfies properties (a) and (b) and consider the collection $\mathcal{T}$ of all unions of members of $\mathcal{B}$. It must be shown that $\mathcal{T}$ is a topology for X. Note that $\emptyset$, the union of the empty collection of members of $\mathcal{B}$, is in $\mathcal{T}$; X is in $\mathcal{T}$ by (a).

   Since $\mathcal{T}$ consists of all unions of members of $\mathcal{B}$, then the union of any family of members of $\mathcal{T}$ is also in $\mathcal{T}$. In other words, the union of any family of open sets is open.

   It remains to be proved that the intersection of any finite collection of open sets is open. This follows easily by induction provided it is first shown that for O_1, O_2 in $\mathcal{T}$, $O_1 \cap O_2$ belongs to $\mathcal{T}$. For $x \in O_1 \cap O_2$, there must be members B_1, B_2 of $\mathcal{B}$ such that

$$x \in B_1 \subset O_1, \quad x \in B_2 \subset O_2.$$

Thus

$$x \in B_1 \cap B_2 \subset O_1 \cap O_2.$$

By (b), there is a member B_x of $\mathcal{B}$ with

$$x \in B_x \subset O_1 \cap O_2.$$

Then $O_1 \cap O_2$ is the union of such members B_x of $\mathcal{B}$ and is therefore a member of $\mathcal{T}$. Thus $\mathcal{T}$ is a topology for X.   □

Example 4.3.4

Let $\mathcal{B}$ be the family of all intervals in $\mathbb{R}$ of the form $[a, b)$, $a < b$. It is easily observed that $\mathcal{B}$ satisfies the conditions of Theorem 4.8 and is a basis for a topology, called the *half-open interval topology* $\mathcal{T}''$ for $\mathbb{R}$. It is left as an exercise for the reader to show that $(\mathbb{R}, \mathcal{T}'')$ is first countable and separable but not second countable. (The real line with the half-open interval topology is sometimes called the *Sorgenfrey line*.)

Definition: *Let $\mathcal{B}$ and $\mathcal{B}'$ be bases for topologies $\mathcal{T}$ and $\mathcal{T}'$ for a set X. Then $\mathcal{B}$ and $\mathcal{B}'$ are **equivalent bases** provided that the topologies $\mathcal{T}$ and $\mathcal{T}'$ are identical.*

The proof of the following theorem is left as an exercise.

Theorem 4.9:   *Bases $\mathcal{B}$ and $\mathcal{B}'$ for topologies on a set X are equivalent if and only if both of the following conditions hold:*

(a) For each $B \in \mathcal{B}$ and $x \in B$, there is a member $B' \in \mathcal{B}'$ such that $x \in B' \subset B$.

(b) For each $B' \in \mathcal{B}'$ and $x \in B'$, there is a member $B \in \mathcal{B}$ such that $x \in B \subset B'$.

In some instances it is advantageous to have a smaller collection of sets which generates a basis by the process of forming finite intersections. Such a family, called a *subbasis*, is defined as follows:

Definition: *Let (X, T) be a space. A subcollection $\mathcal{S}$ of T is a **subbasis** or **subbase** for T if the family $\mathcal{B}$ of all finite intersections of members of $\mathcal{S}$ is a basis for T.*

Example 4.3.5

The collection $\mathcal{S}$ of all open intervals of the form (a, ∞) and $(-\infty, b)$, $a, b \in \mathbb{R}$, is a subbasis for the usual topology for $\mathbb{R}$.

EXERCISE 4.3

1. Let (X, T) be a space and $\mathcal{B}$ a subcollection of T. Suppose that for each a in X, the set $\mathcal{B}_a$ of members of $\mathcal{B}$ which contain a is a local base at a. Show that $\mathcal{B}$ is a basis for T.

2. Prove part (a) of Theorem 4.7.

3. Give an example different from Example 4.3.4 of a space X that is first countable but not second countable.

4. Let X be a first countable space and x a limit point of a subset A of X. Show that there is a sequence of points of $A \backslash \{x\}$ which converges to x.

5. Describe the bases $\mathcal{B}'$ and $\mathcal{B}''$ for $\mathbb{R}^n$ determined by the open balls of the taxicab metric d' and the max metric d'', respectively. Show that $\mathcal{B}'$ and $\mathcal{B}''$ are both equivalent to the basis $\mathcal{B}$ of open balls in the usual metric d.

6. Let X be a first countable space and x a member of X. Prove that there is a local nested basis $\{S_n\}_{n=1}^{\infty}$ at a (i.e., a local basis such that $S_{n+1} \subset S_n$ for each positive integer n).

7. Let $(\mathbb{R}, T'')$ be the real line with the half-open interval topology of Example 4.3.4.

   (a) Find the closure, interior, and boundary of the set $A = [0, 2]$ and the set $B = (0, 2)$.

   (b) Prove that $(\mathbb{R}, T'')$ is separable and first countable but not second countable.

8. Let d_1 and d_2 be metrics for a set X, and let $\mathcal{B}_1$ and $\mathcal{B}_2$ denote, respectively, the families of all open balls generated by d_1 and d_2. Show that d_1 and d_2 are equivalent metrics if and only if $\mathcal{B}_1$ and $\mathcal{B}_2$ are equivalent bases.

9. Prove Theorem 4.9.

10. Let X be a set and $\mathscr{S}$ any family of subsets of X whose union equals X. Show that $\mathscr{S}$ is a subbasis for a topology for X.

4.4 CONTINUITY AND TOPOLOGICAL EQUIVALENCE

Since distance between points is not defined for general topological spaces, how should one define continuity for a function from one topological space to another? Keep in mind that the definition, when applied to metric spaces, should agree with the definition of continuity given in Chapter 3. Recall that a function $f: X \to Y$ from metric space (X, d) to metric space (Y, d') is continuous at a point a in X if and only if for each open ball $B_{d'}(f(a), \epsilon)$ in Y centered at $f(a)$, there is an open ball $B_d(a, \delta)$ in X centered at a such that $f(B_d(a, \delta))$ is contained in $B_{d'}(f(a), \epsilon)$. Continuity at a for a function $f: X \to Y$ on topological spaces X and Y is defined by simply replacing the open balls centered at $f(a)$ and a by open sets containing $f(a)$ and a, respectively.

Definition: *Let $(X, \mathcal{T})$ and $(Y, \mathcal{T}')$ be topological spaces, $f: X \to Y$ a function, and a a point of X. Then f is **continuous at** a provided that for each open set V in Y containing $f(a)$ there is an open set U in X containing a such that $f(U) \subset V$. The function f is **continuous** if it is continuous at each point of its domain.*

We shall deal most often with continuity of a function on its entire domain rather than at each point separately, so we examine the definition a bit more closely. The function f is continuous simultaneously for all a in X means that for every open set V in Y and every point a with $f(a)$ in V, there is an open set U_a in X with a in U_a and $f(U_a) \subset V$. Since $f(U_a) \subset V$ is equivalent to $U_a \subset f^{-1}(V)$, this means that $f^{-1}(V)$ contains an open set about each of its members; in other words, $f^{-1}(V)$ is open in X for each open set V in Y.

Alternate Definition: *A function $f: (X, \mathcal{T}) \to (Y, \mathcal{T}')$ is **continuous** means that for each open set V in Y, $f^{-1}(V)$ is an open set in X.*

The following theorem restates the definition of continuity in several equivalent forms.

Theorem 4.10:   *Let $f: X \to Y$ be a function on the indicated topological spaces and let $a \in X$. The following statements are equivalent:*

(1) f is continuous at a.

(2)   For each open set V in Y containing $f(a)$, there is an open set U in X such that $a \in U$ and $U \subset f^{-1}(V)$.

(3)   For each neighborhood V of $f(a)$, $f^{-1}(V)$ is a neighborhood of a.

(4)   For each subset V of Y with $f(a) \in int\ V$, a belongs to $int\ f^{-1}(V)$.

A proof of Theorem 4.10 can be formulated directly from the definitions of the terms involved; this is left as an exercise for the reader.

Theorem 4.11:   *Let $f: X \to Y$ be a function on the indicated topological spaces. The following statements are equivalent.*

(1)   *f is continuous.*

(2)   *For each closed subset C of Y, $f^{-1}(C)$ is closed in X.*

(3)   *For each subset A of X, $f(\bar{A}) \subset \overline{f(A)}$.*

(4)   *There is a basis $\mathcal{B}$ for the topology of Y such that $f^{-1}(B)$ is open in X for each basic open set B in $\mathcal{B}$.*

(5)   *There is a subbasis $\mathcal{S}$ for the topology of Y such that $f^{-1}(S)$ is open in X for each subbasic open set S in $\mathcal{S}$.*

Proof: *We use the open set formulation to describe continuity: f is continuous if and only if for each open set V in Y, $f^{-1}(V)$ is open in X. The equivalence of (1) and (2) follows from the duality between open sets and closed sets, precisely as in the proof of Theorem 3.13.*

[(2) $\Rightarrow$ (3)]: Suppose that (2) holds, and let A be a subset of X. Then $\overline{f(A)}$ is a closed subset of Y, so its inverse image $f^{-1}(\overline{f(A)})$ is closed in X. Since

$$A \subset f^{-1}(\overline{f(A)})$$

and the latter set is closed, then

$$\bar{A} \subset f^{-1}(\overline{f(A)})$$

so

$$f(\bar{A}) \subset \overline{f(A)}$$

and (3) holds.

[(3) $\Rightarrow$ (2)]: Assume (3) and let C be a closed subset of Y. Then

$$f(\overline{f^{-1}(C)}) \subset \overline{ff^{-1}(C)} \subset \bar{C} = C$$

so

$$\overline{f^{-1}(C)} \subset f^{-1}(C)$$

and $f^{-1}(C)$ must be a closed set. Thus (3) implies (2).

We have completed the proof that (1), (2), and (3) are equivalent. Since a basis $\mathcal{B}$ and subbasis $\mathcal{S}$ for Y consist of open sets, it should be clear that (1) implies both (4) and (5). Similarly, since a basis is a subbasis, (4) implies (5). The proof will be completed by showing that (5) implies (4) and (4) implies (1).

[(5) $\Rightarrow$ (4)]: Suppose (5) holds and consider the basis $\mathcal{B}$ generated from $\mathcal{S}$ by taking finite intersections. For any basic open set $B \in \mathcal{B}$,

$$B = \bigcap_{i=1}^{n} S_i$$

for some finite collection of members $S_1, \ldots, S_n$ of $\mathcal{S}$. Then

$$f^{-1}(B) = f^{-1}\left(\bigcap_{i=1}^{n} S_i\right) = \bigcap_{i=1}^{n} f^{-1}(S_i)$$

by Theorem 1.7. Since each set $f^{-1}(S_i)$ is open in X and the intersection of any finite collection of open sets is open, then $f^{-1}(B)$ is open in X. Thus (5) implies (4).

[(4) $\Rightarrow$ (1)]: Assuming (4), let O be an open set in Y. By the definition of basis,

$$O = \bigcup_{\alpha \in I} B_\alpha$$

for some subcollection $\{B_\alpha : \alpha \in I\}$ of the basis $\mathcal{B}$. Then

$$f^{-1}(O) = f^{-1}\left(\bigcup_{\alpha \in I} B_\alpha\right) = \bigcup_{\alpha \in I} f^{-1}(B_\alpha).$$

Since each set $f^{-1}(B_\alpha)$ is open in X and the union of any family of open sets is open, then $f^{-1}(O)$ is open in X and f is continuous. $\square$

Conditions (4) and (5) of Theorem 4.11 will be useful when we want to deal with a basis or subbasis rather than with the entire topology of the range space. Condition (3) is the description of continuity which seems to fit best with the intuitive idea of closeness. We visualize continuity of f at x by thinking that whenever x is "close" to a set A, then $f(x)$ is "close" to $f(A)$. In topological language, x is "close" to A means $x \in \bar{A}$, and $f(x)$ is "close" to $f(A)$ means $f(x) \in \overline{f(A)}$.

Theorem 4.12: *If $f: X \to Y$ and $g: Y \to Z$ are continuous functions on the indicated spaces, then the composite function $g \circ f: X \to Z$ is continuous.*

The proof of Theorem 4.12 is left as an exercise.

Example 4.4.1

It should be emphasized that continuity for a function $f: X \to Y$ is expressed in terms of *inverse images* of open sets in Y: f is continuous if and only if for each open set O in Y, $f^{-1}(O)$ is open in X. This is not to be confused with f mapping open sets in X to open sets in Y, which is quite a different property. Consider, for example, the function $f: \mathbb{R} \to Y$ from the real line $\mathbb{R}$ to a discrete two-point space $Y = \{a, b\}$ defined by

$$f(x) = \begin{cases} a & \text{if } x \le 0 \\ b & \text{if } x > 0. \end{cases}$$

This function does map open sets in X to open sets in Y, because every subset of Y is open. But f is not continuous; $\{a\}$ is open in Y but

$$f^{-1}(a) = (-\infty, 0]$$

is not open in $\mathbb{R}$.

Definition: *Let $f: X \to Y$ be a function on the indicated spaces. Then f is an **open function** or **open mapping** if for each open set O in X, $f(O)$ is open in Y. The function f is a **closed function** or **closed mapping** if for each closed set C in X, $f(C)$ is closed in Y.*

Example 4.4.1 shows that an open mapping may fail to be continuous. The reader is asked in the exercises to find examples to illustrate the following:

(a) A closed mapping may not be continuous.
(b) An open mapping may not be closed, and conversely.
(c) A continuous function may be neither open nor closed.
(d) A mapping that is both open and closed may not be continuous.

Definition: *Topological spaces X and Y are **topologically equivalent** or **homeomorphic** if there is a one-to-one function $f: X \to Y$ from X onto Y for which both f and the inverse function f^{-1} are continuous. The function f is called a **homeomorphism**.*

Topological equivalence is an equivalence relation for topological spaces.

A homeomorphism $f: X \to Y$ between spaces X and Y is a one-to-one function from X onto Y for which both f and f^{-1} are continuous. Since $(f^{-1})^{-1} = f$ for a bijection, continuity of f^{-1} can be expressed by the fact that for each open set O

in X, $f(O)$ is open in Y. In other words, a homeomorphism is just a bijection which is also an open, continuous function. Since continuity of f^{-1} can also be expressed by the fact that for each closed set C in X, $f(C)$ is closed in Y, then f is a homeomorphism if and only if it is a closed, continuous bijection.

Definition: *A property P of topological spaces is a **topological property** or **topological invariant** provided that if space X has property P, then so does every space Y which is topologically equivalent to X.*

The next three theorems give examples of topological properties.

Theorem 4.13:   *Separability is a topological property.*

Proof: *Let X be a separable space with countable dense subset A and Y a space homeomorphic to X. Let f: X → Y be a homeomorphism. The obvious candidate for a countable dense subset of Y is f(A). To see that f(A) is dense in Y, let O be a non-empty open set in Y. Then $f^{-1}(O)$ is a non-empty open set in X. Since A is dense in X, $f^{-1}(O)$ contains some member a of A. Then O contains the member f(a) of f(A), so every non-empty open set in Y contains at least one member of f(A). Thus $\overline{f(A)} = Y$ and Y is separable.*   □

Proofs of the next two theorems are left as exercises.

Theorem 4.14:   *First countability and second countability are topological properties.*

Definition: *A topological space X is **metrizable** provided that the topology of X is generated by a metric.*

Theorem 4.15:   *The property of being a metrizable space is a topological property.*

Hint: Let (X, d) be a metric space, Y a topological space homeomorphic to X, and $f: X \to Y$ a homeomorphism. Define d' on $Y \times Y$ by

$$d'(y_1, y_2) = d(f^{-1}(y_1), f^{-1}(y_2)), \quad y_1, y_2 \in Y.$$

Show that d' is a metric and that d' generates the topology of Y.

Since $\mathbb{R}$ and $(0, 1)$ are homeomorphic, we note that the property of being a bounded metric space is not a topological property.

Until this point, it has not been possible to give a precise definition of the branch of mathematics known as topology. *Topology* is the branch of mathematics which deals with topological properties. One objective of topology is to give criteria in terms of topological invariants which allow one to determine whether or not two given spaces are homeomorphic. Ideally, one would want a list of topological properties which are easy to check and for which two spaces are homeomorphic if and only if they share the same properties from the list. Theorems of this type are called *classification theorems* because they divide topological spaces into classes, with two members of the same class being homeomorphic. Mathematicians have experienced only limited success in classifying topological spaces. There is no known list of topological properties which completely classifies topological spaces.

A second objective of topology is to establish relations among various topological properties and to show which combinations of topological properties are equivalent. Theorems of this type are called *characterization theorems* since they completely describe or characterize spaces of a certain type. We shall see one of the most famous characterization theorems, the Urysohn characterization, which gives necessary and sufficient conditions for a space to be homeomorphic to a separable metric space (Theorem 8.18), in Chapter 8.

EXERCISE 4.4

1. Let $f: X \to Y$ be a function and let $a \in X$. Prove that f is continuous at a if and only if for each subset A of X with $a \in \bar{A}$, $f(a) \in \overline{f(A)}$.

2. Let $f: X \to Y$ be a function and let $a \in X$. Prove that f is continuous at a if and only if there is a local basis $\mathcal{B}_{f(a)}$ at $f(a)$ such that for each $B \in \mathcal{B}_{f(a)}$, $f^{-1}(B)$ is a neighborhood of a.

3. Find an example of each of the following:

    (a) A closed mapping that is not continuous

    (b) An open mapping that is not closed and a closed mapping that is not open

    (c) A continuous function that is neither open nor closed

    (d) A function that is both open and closed but not continuous

4. Prove that the composition of continuous functions is continuous (Theorem 4.12).

5. Prove that topological equivalence is an equivalence relation.

6. Let $f: X \to Y$ be a one-to-one correspondence from space X onto space Y. Prove that the following statements are equivalent:

    (a) f is a homeomorphism.

    (b) f and f^{-1} are both open mappings.

    (c) f and f^{-1} are both closed mappings.

7. Prove Theorem 4.14.

8. Prove Theorem 4.15.

9. Let $f: X \to Y$ be a one-to-one function from space X onto space Y.

   (a) Show that the following statements are equivalent:

      (1) f^{-1} is continuous.

      (2) f is an open mapping.

      (3) f is a closed mapping.

   (b) Show that the following statements are equivalent:

      (1) f is a homeomorphism.

      (2) f is an open, continuous mapping.

      (3) f is a closed, continuous mapping.

10. Let X be a separable space, Y a space, and $f: X \to Y$ a continuous function from X onto Y. Prove that Y is separable.

11. **Definition:** *Let X be a space and $f: X \to \mathbb{R}$ a real-valued function on X. Then f is **upper semicontinuous** if $f^{-1}(-\infty, a)$ is open for each a in $\mathbb{R}$; f is **lower semicontinuous** if $f^{-1}(a, \infty)$ is open for each a in $\mathbb{R}$.*

   (a) Prove that a function $f: X \to \mathbb{R}$ is continuous if and only if it is both upper and lower semicontinuous.

   (b) Give an example of an upper semicontinuous function that is not continuous.

   (c) Repeat (b) for a lower semicontinuous function.

12. (a) Prove that a function $f: X \to \mathbb{R}$ from a space X into $\mathbb{R}$ is upper semicontinuous if and only if $\{x \in X: f(x) \geq a\}$ is closed in X for each a in $\mathbb{R}$.

   (b) State and prove the condition analogous to (a) for lower semicontinuous functions.

13. **Definition:** *Let A be a subset of a given space X. The **characteristic function** of A is the function $f_A: X \to \mathbb{R}$ having value 1 at each point of A and value 0 at each point of $X \backslash A$.*

      Let X be a space with subspace A. Prove:

   (a) The characteristic function of A is lower semicontinuous if and only if A is open.

   (b) The characteristic function of A is upper semicontinuous if and only if A is closed.

   (c) The characteristic function of A is continuous if and only if A is both open and closed.

14. (a) Let X be a space and $\{f_\alpha\}_{\alpha \in A}$ a family of lower semicontinuous functions $f_\alpha: X \to \mathbb{R}$ for which $\{f_\alpha(x): \alpha \in A\}$ has an upper bound for each x in X. Prove that the function $g: X \to \mathbb{R}$ defined by

$$g(x) = \text{lub } \{f_\alpha(x): \alpha \in A\}, \quad x \in X,$$

   is lower semicontinuous.

   (b) State and prove the corresponding result for upper semicontinuous functions.

4.5 SUBSPACES

A subspace of a metric space (X, d) is simply a subset A of X with the metric of X used to measure distances between points of A. In other words, the metric of X is essentially "cut down" to A. The main definition of this section defines subspaces of a general topological space X in an analogous manner by "cutting down" the open sets of X.

Definition: *Let $(X, \mathcal{T})$ be a topological space and A a subset of X. The **relative topology** or **subspace topology** $\mathcal{T}'$ for A determined by $\mathcal{T}$ consists of all sets of the form $O \cap A$ for which O is an open set of $\mathcal{T}$:*

$$\mathcal{T}' = \{O \cap A : O \in \mathcal{T}\}.$$

*The members of $\mathcal{T}'$ are called **relatively open sets** or simply **open sets** in A, and $(A, \mathcal{T}')$ is called a **subspace** of $(X, \mathcal{T})$.*

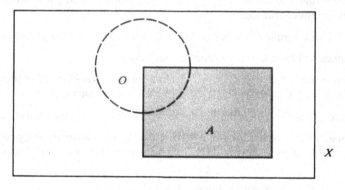

FIGURE 4.2   Open sets in A are of the form $O \cap A$, where O is open in X.

When no confusion is likely, it is common practice to refer to A as a subspace of X, omitting mention of the topologies $\mathcal{T}$ and $\mathcal{T}'$.

It is a simple matter to show that what we have called the subspace topology $\mathcal{T}'$ for a subset A of a space $(X, \mathcal{T})$ is actually a topology for A:

(a) $\varnothing$ and A are relatively open sets since

$$\varnothing = \varnothing \cap A, \quad A = X \cap A$$

and both $\varnothing$ and X are open in X.

(b) For any family $\{O_\alpha \cap A\}$ of relatively open sets, where each O_α is open in X,

$$\bigcup_\alpha (O_\alpha \cap A) = \left(\bigcup_\alpha O_\alpha\right) \cap A$$

is relatively open because the union of any family of open sets in X is open.

(c) For any finite family $\{O_i \cap A\}_{i=1}^n$ of relatively open sets, where each O_i is open in X,

$$\bigcap_{i=1}^n (O_i \cap A) = \left(\bigcap_{i=1}^n O_i\right) \cap A$$

is relatively open because the intersection of any finite family of open sets in X is open.

A subset D of A is *relatively closed* if it is a closed set in the subspace topology for A: D is relatively closed if and only if

$$A \backslash D = O \cap A$$

for some open set O in X. The next theorem shows that a relatively closed set could be defined equivalently as the intersection with A of a closed set in X.

Theorem 4.16:   *Let (A, T') be a subspace of a topological space (X, T). A subset D of A is closed in the subspace topology for A if and only if $D = C \cap A$ for some closed subset C of X.*

Proof: *Suppose first that D is a relatively closed set. Then*

$$A \backslash D = O \cap A$$

for some open set O in X, and

$$D = A \backslash (A \backslash D) = A \backslash (O \cap A) = (X \backslash O) \cap A,$$

so D is the intersection of A with the closed set $C = X \backslash O$ in X.

For the reverse implication, suppose that $D = C \cap A$ for some closed set C in X. Then $O = X \backslash C$ is open in X and

$$A \backslash D = A \backslash (C \cap A) = (X \backslash C) \cap A = O \cap A,$$

so $A \backslash D$ is open in the subspace topology of A, and D is a relatively closed set. $\square$

Proofs of the next two theorems are left as exercises.

Theorem 4.17:   *Let (A, T') be a subspace of a space (X, T), a a member of A, and N a subset of A. Then N is a neighborhood of a with respect to the subspace topology for A if and only if N = U ∩ A where U is a neighborhood of a with respect to the topology T on X.*

Theorem 4.18:   *Let (X, d) be a metric space and (A, d') a metric subspace. Let T be the topology for X generated by d, T' the subspace topology for A determined by T, and T'' the metric topology for A determined by d'. Then T' = T''.*

Definition: *A property P of topological spaces is **hereditary** provided that if X has property P, then every subspace of X has property P.*

Example 4.5.1

First countability and second countability are hereditary properties. If X has a countable local basis $\{B_n\}_{n=1}^{\infty}$ at point $a \in X$ and A is a subset of X containing a, then $\{B_n \cap A\}_{n=1}^{\infty}$ is a local base at a in the subspace topology for A. If X is second countable, the same method of proof shows that every subspace is second countable.

Example 4.5.2

Separability is not hereditary. Consider, for example, the subset X of $\mathbb{R}^2$ consisting of the real axis $\mathbb{R}$ and the one additional point $a = (0, 1)$. Define a topology T on X to consist of the empty set $\varnothing$ and all subsets of X which contain a. Then (X, T) is separable since the singleton set $\{a\}$ is dense. However, the subspace topology T' for $\mathbb{R}$ as a subspace of X is the discrete topology, so $(\mathbb{R}, T')$ is not separable.

Definition: *A topological space X is a **Hausdorff space** if for each pair a, b of distinct points of X there exist disjoint open sets U and V such that a ∈ U and b ∈ V.*

Example 4.5.3

Every metric space (X, d) is Hausdorff. To see this, note that if a, b are distinct points of X, then $r = d(a, b)$ is a positive number. Thus $U = B(a, r/2)$ and $V = B(b, r/2)$ are disjoint open sets containing a and b, respectively. Thus $\mathbb{R}^n$, Hilbert space, and discrete spaces are examples of Hausdorff spaces.

The space $(X, \mathcal{T})$ of Example 4.5.2 is not Hausdorff since every non-empty open set in X contains the point $(0, 1)$. The real line with the finite complement topology and trivial spaces with more than one point are also not Hausdorff.

Example 4.5.4   The Zariski Topology

Let n be a positive integer and consider the family $\mathcal{P}$ of all polynomials in n real variables $x_1, x_2, \ldots, x_n$. For such a polynomial P, let $Z(P)$ denote its solution set in $\mathbb{R}^n$:

$$Z(P) = \{(x_1, x_2, \ldots, x_n) \in \mathbb{R}^n : P(x_1, x_2, \ldots, x_n) = 0\}.$$

It is left as an exercise for the reader to show that the set $\mathcal{B}$ of all complements of the sets $Z(P)$, $P \in \mathcal{P}$, is a basis for a topology for $\mathbb{R}^n$. This topology is called the *Zariski topology* for $\mathbb{R}^n$.

For $n = 1$, the Zariski topology equals the finite complement topology on $\mathbb{R}$. The reason is that the finite subsets of $\mathbb{R}$ coincide precisely with the solution sets of polynomials in one real variable. To see this, note that if $A = \{a_1, \ldots, a_n\}$ is a finite subset of $\mathbb{R}$, then

$$P(x) = (x - a_1) \cdot (x - a_2) \cdot \cdots \cdot (x - a_n)$$

is a polynomial for which $Z(P) = A$. Furthermore, the solution set of a polynomial in one real variable is always a finite subset of $\mathbb{R}$.

For $n > 1$, the Zariski topology does not coincide with the finite complement topology. The reason is that the finite subsets of $\mathbb{R}^n$ do not coincide with the solution sets of polynomials in n real variables, $n > 1$. For example, the line $y = 1$ in $\mathbb{R}^2$ is the solution set of the polynomial

$$P(x, y) = y - 1,$$

but this solution set is not finite.

It is left as an exercise for the reader to show that $\mathbb{R}^n$ with the Zariski topology is not Hausdorff.

Theorem 4.19:

(1) *The property of being a Hausdorff space is a topological and hereditary property.*

(2) *A sequence $\{x_n\}_{n=1}^\infty$ in a Hausdorff space cannot converge to more than one point.*

Proof:

(1) *Suppose that X is Hausdorff and Y is homeomorphic to X by homeomorphism $f: X \to Y$. For distinct points a and b in Y, $f^{-1}(a)$ and $f^{-1}(b)$ are distinct points of X, so there are disjoint open sets U and V in X such that*

$$f^{-1}(a) \in U, \quad f^{-1}(b) \in V.$$

Then $f(U)$ and $f(V)$ are disjoint open sets in Y containing a and b, respectively.

   The proof that the Hausdorff property is hereditary is even easier. If A is a subspace of X and a, b are distinct points of A, then there are disjoint open sets U and V in X containing a and b, respectively. Then $U \cap A$ and $V \cap A$ are disjoint relatively open sets in A containing a and b, respectively, so A is Hausdorff.

(2) *Suppose that $\{x_n\}_{n=1}^{\infty}$ converges to two distinct limits a and b in a Hausdorff space X. Then there are disjoint open sets U and V containing a and b. But by the definition of convergence, there are positive integers N_1 and N_2 such that if $n \geq N_1$, then $x_n \in U$ and if $n \geq N_2$, then $x_n \in V$. If n is greater than or equal to the larger of N_1 and N_2, then x_n belongs to the empty set $U \cap V$. This contradiction shows that $\{x_n\}_{n=1}^{\infty}$ cannot converge to two distinct limits in a Hausdorff space.* □

Note that the preceding proof for the uniqueness of the limit of a convergent sequence in a Hausdorff space is essentially the same as the proof of the corresponding property for metric spaces (Theorem 3.8).

The Hausdorff property is sometimes called a *separation property* since it states that any two distinct points can be "separated" by disjoint open sets. Additional separation properties will be studied in Chapters 6 and 8.

Definition: *If X is a space which is homeomorphic to a subspace A of a space Y, then X is said to be **embedded** in Y. The homeomorphism $f: X \to A$ is called an **embedding** of X in Y.*

The isometric embeddings of $\mathbb{R}^{n-1}$ in $\mathbb{R}^n$ and of $\mathbb{R}^n$ in Hilbert space discussed in Examples 3.6.1 and 3.6.3 are topological embeddings.

Following the outline of our earlier work on metric spaces, it would be natural to introduce the product of topological spaces, the second method of forming new spaces from old ones, at this point. This is postponed until Chapter 7, however, in order to study the most important topological properties, connectedness and compactness, first. Any reader who cannot wait to learn about product spaces may read that chapter now and then return to Chapter 5.

EXERCISE 4.5

1. (a) Give an example to show that if A is a subspace of a space X, then a relatively open set in A may fail to be an open subset of X. Prove that if A is an open subset of X, then every relatively open set in A is open in X.

    (b) Repeat (a) for closed sets.

2. Give an example of a subspace A of a topological space X and subsets B and C of X for which

    (a) The closure of $B \cap A$ in the subspace topology for A does not equal $\bar{B} \cap A$.

    (b) The interior of $C \cap A$ in the subspace topology for A does not equal (int C) $\cap A$.

3. Let A be a subspace of a space X and let B be a subset of A.
    Prove that:

    (a) A point x in A is a limit point of B in the subspace topology for A if and only if x is a limit point of B in the topology for X.

    (b) The closure of B in the subspace topology for A equals $\bar{B} \cap A$.

4. Let $f: X \to Y$ be a continuous function on the indicated spaces and A a subspace of X. Prove that the restriction $f|_A : A \to Y$ of f to A is continuous.

5. Prove Theorem 4.17.

6. Prove Theorem 4.18.

7. Let $(X, \mathcal{T})$ be a space, Y a subset of X and Z a subset of Y. Then Y has a subspace topology $\mathcal{T}'$ and Z can be assigned a subspace topology in two ways: Z has a subspace topology $\mathcal{T}_1$ as a subspace of $(X, \mathcal{T})$, and a subspace topology $\mathcal{T}_2$ as a subspace of $(Y, \mathcal{T}')$. Prove that $\mathcal{T}_1 = \mathcal{T}_2$.

8. Give an example of a separable Hausdorff space which has a non-separable subspace.

9. Prove that a finite subset A of a Hausdorff space X has no limit points. Conclude that A must be closed.

10. Let X be a Hausdorff space, A a subset of X, and x a limit point of A. Prove that every open set containing x contains infinitely many members of A.

11. Prove: If there is an embedding of X in Y and an embedding of Y in Z, then there is an embedding of X in Z.

12. Give an example of spaces A and B for which A can be embedded in B and B can be embedded in A, but A and B are not homeomorphic. (*Hint:* Simple examples can be found in $\mathbb{R}$.)

13. (a) Let X be a space and $\{X_n\}_{n=1}^{\infty}$ a sequence of separable subspaces of X for which $\bigcup_{n=1}^{\infty} X_n$ is dense in X. Prove that X is separable.

    (b) Use (a) to prove that Hilbert space H is separable.

14. (a) Show that the family of sets $\mathcal{B}$ of Example 4.5.4 is a basis for a topology for $\mathbb{R}^n$.

    (b) Show that the Zariski topology for $\mathbb{R}^n$ is not Hausdorff.

    (c) For $\mathbb{R}^n$ with the Zariski topology, show that each finite set is closed.

15. **Definition:** *A space X is* **locally Euclidean of dimension** *n provided that each point in X belongs to an open set homeomorphic to n-dimensional Euclidean space $\mathbb{R}^n$.*

This exercise shows that a locally Euclidean space may not be Hausdorff. Let X be the subset of $\mathbb{R}^2$ defined by

$$X = (\mathbb{R} \times \{0\}) \cup ([0, \infty) \times \{1\}).$$

Let $\mathcal{B}$ consist of all subsets of X of the following types:

$$(a, b) \times \{0\} \quad \text{for } a < b,$$

$$(a, b) \times \{1\} \quad \text{for } 0 \le a < b,$$

$$((a, 0) \times \{0\}) \cup ([0, b) \times \{1\}) \quad \text{for } a < 0 < b.$$

(a)   Show that $\mathcal{B}$ is a basis for a topology for X.

(b)   Show that the subspace $\mathbb{R} \times \{0\}$ of X is homeomorphic to the real line.

(c)   Show that the subspace $((-\infty, 0) \times \{0\}) \cup ([0, \infty) \times \{1\})$ of X is homeomorphic to the real line.

(d)   Show that X is locally Euclidean of dimension 1.

(e)   Show that X is not a Hausdorff space.

SUGGESTIONS FOR FURTHER READING

For additional reading on general topological spaces, several textbooks are listed in the bibliography. *Introduction to Topology* by Gamelin and Green, *Basic Topology* by Armstrong, and *Topology* by Hocking and Young are recommended for a general review of the subject. Willard's *General Topology* and Kelley's *General Topology* have accessible accounts of nets, which are a generalization of sequences; these texts also explain the theory of convergence based on filters, another approach to convergence. A readable introduction to filters also appears in *General Topology* by Bourbaki.

Counterexamples in Topology by Steen and Seebach is an excellent resource for locating spaces of various special types.

HISTORICAL NOTES FOR CHAPTER 4

Point-set topology emerged as a coherent discipline with the publication in 1914 of the classic treatise *Grundzüge der Mengenlehre* by Felix Hausdorff. Hausdorff defined topological space in terms of neighborhoods of the points of the underlying set. His definition is equivalent to the definition in this chapter with the additional requirement that a space satisfy what is now called the Hausdorff property. Each pair of distinct points must have disjoint neighborhoods in Hausdorff's original definition.

Hausdorff's axiom system was the culmination of many attempts to formulate defining axioms for abstract spaces. The study of abstract spaces beyond the Euclidean spaces had been foreseen by Riemann in the 1860s. The first definition of abstract space, based on axioms for limits of sequences, was given by Maurice Fréchet in 1906. This formulation, although suitable for metric spaces, was restricted by the denumerability of the terms of a sequence and did not prove adequate for spaces in general. Frigyes Riesz (1880–1956) in 1908 proposed a set of axioms for abstract spaces based on limit points, but his definition was too complicated and a theory based on his ideas was never developed in detail. The major drawback in the Riesz approach was that it did not require the closure of a set to be a closed set. An axiom system for the plane, from which Hausdorff drew some ideas, was proposed by David Hilbert in 1902. A set of axioms for Riemann surfaces, based on the neighborhood concept, was formulated by Hermann Weyl (1885–1955) in 1913. Axioms for a topological space based on a closure operator, listed in Problem 10 of Exercise 4.2 and called the Kuratowski Closure Properties, came later than the Hausdorff axioms. The closure axioms were proposed by K. Kuratowski in 1922.

In *Grundzüge der Mengenlehre,* Hausdorff gave his definition of topological space and brought the topological research of the late nineteenth and early twentieth centuries into a logical and unified framework. Hausdorff went well beyond what was known at that time, however, and introduced many new properties of interest in topology. These included the first and second axioms of countability, the subspace topology, a systematic treatment of continuity and homeomorphism, and other properties which will be mentioned as they are studied in this text.

Separability, in the context of metric spaces, was introduced by Fréchet in 1906. Nowhere dense sets were introduced by du Bois-Reymond in his research on Euclidean spaces.

There is now a description of topological spaces along the lines envisioned by Fréchet, but for which his sequence approach was inadequate. The new approach involves nets, which are a generalization of sequences introduced by E. H. Moore and H. L. Smith in 1922. Additional information on Moore-Smith convergence can be found in the suggested reading list for this chapter.

5

Connectedness

The property of connectedness, that is, the property of being "unbroken" or "all in one piece," is one of the most important topological properties. It is a relatively simple property for the more common topological spaces, and it seems to fit well with our intuition in those spaces. For example, the real line ℝ is connected, and its connected subspaces are precisely the intervals. The plane is also connected but, as we shall see, its connected subspaces are considerably more complicated. Connectedness is one of the oldest topological properties, having been defined in essentially the form used today by Camille Jordan in 1892.

Connectedness is one of the more useful topological invariants for other branches of mathematics. As was hinted in Chapter 1 and will be proved in this chapter, the Intermediate Value Theorem of calculus depends upon the fact that ℝ is connected.

The properties of being "connected near a point" and "connected by paths" are also introduced in this chapter.

5.1 CONNECTED AND DISCONNECTED SPACES

It is easier to say what it means for a space to be disconnected, so that term is defined first.

Definition: *A topological space X is **disconnected** or **separated** if it is the union of two disjoint, non-empty open sets. Such a pair A, B of subsets of X is called a **separation** of X. A space X is **connected** provided that it is not disconnected. In other words, X is connected if there do not exist open subsets A and B of X such that*

$$A \neq \varnothing, \quad B \neq \varnothing, \quad A \cap B = \varnothing, \quad A \cup B = X.$$

*A subspace Y of X is **connected** provided that it is a connected space when assigned the subspace topology. The terms **connected set** and **connected subset** are sometimes used to mean connected space and connected subspace, respectively.*

Example 5.1.1

(a) A discrete space with more than one point is disconnected.

(b) Any trivial space is connected since there fail to exist two non-empty open sets.

(c)   Let Y be the set of non-zero real numbers with the subspace topology of $\mathbb{R}$. Then Y is disconnected since $(-\infty, 0)$ and $(0, \infty)$ form a separation.

(d)   Let $Z = \mathbb{R}^2 \backslash \mathbb{R}$ denote the plane minus the real axis, with the subspace topology. Then Z is disconnected since the upper half plane $U = \{(x_1, x_2) \in \mathbb{R}^2: x_2 > 0\}$ and lower half plane $V = \{(x_1, x_2) \in \mathbb{R}^2: x_2 < 0\}$ form a separation.

(e)   Let $X = [0, 1] \cup [2, 3]$ with the subspace topology of the real line. Then $A = [0, 1]$ and $B = [2, 3]$ are disjoint, non-empty open subsets of X for which $X = A \cup B$, so X is disconnected. (Note that A and B are relatively open since $A = (-\infty, 3/2) \cap X$ and $B = (3/2, \infty) \cap X$.)

(f)   Let X denote the set of real numbers with the one additional point $a = (0, 1)$, and let the topology $\mathcal{T}$ for X consist of $\varnothing$ and all subsets of X which contain a. Then there do not exist two disjoint non-empty open sets, so X is connected. Note that as a subspace of $(X, \mathcal{T})$, $\mathbb{R}$ is assigned the discrete topology and is therefore disconnected. This example (as well as examples (c), (d), and (e) above) demonstrate that the property of being connected is definitely not hereditary.

Example 5.1.2

The real line $\mathbb{R}$ with the usual topology is connected. To see this, suppose to the contrary that $\mathbb{R}$ is disconnected. Then

$$\mathbb{R} = A \cup B$$

for some disjoint, non-empty open sets A and B of $\mathbb{R}$. Since

$$A = \mathbb{R}\backslash B, \quad B = \mathbb{R}\backslash A,$$

then A and B are closed as well as open. Consider two points a and b with $a \in A$ and $b \in B$. Without loss of generality we may assume $a < b$.

Let

$$A' = A \cap [a, b].$$

Now A' is a closed and bounded subset of $\mathbb{R}$ and consequently contains its least upper bound c. Note that $c \neq b$ since A and B have no point in common. Thus $c < b$. Since A contains no point of $(c, b]$, then

$$(c, b] \subset B$$

and hence $c \in \bar{B}$. But B is closed, so $c \in B$. Thus c belongs to both A and B, contradicting the assumption that A and B are disjoint. This contradiction shows that $\mathbb{R}$ is connected.

It will be shown in Section 5.3 that the connected subspaces of ℝ are precisely the intervals.

EXERCISE 5.1

1. Prove that connectedness is a topological property.
2. Show that the set of rational numbers, with the subspace topology of ℝ, is disconnected.
3. Modify the proof of Example 5.1.2 to show that every closed interval is connected.
4. Let (ℝ, $\mathcal{T}'$) be the space of real numbers with the finite complement topology. Is (ℝ, $\mathcal{T}'$) connected or is it disconnected? Prove your answer.

5.2 THEOREMS ON CONNECTEDNESS

The argument of Example 5.1.2, a proof by contradiction, is typical of arguments involving connectedness. To show that a space is connected, one must demonstrate that two sets having certain properties cannot exist. Proceeding in the contrapositive form by assuming disconnectedness gives a pair of sets with which to work. Several examples of such arguments are given in this section to prove the fundamental theorems about connectedness.

Connectedness has been described in many apparently different but equivalent ways. The major ones are presented in the Corollary to Theorem 5.1.

Definition: *Non-empty subsets A and B of a space X are **separated sets** if $\bar{A} \cap B$ and $A \cap \bar{B}$ are both empty.*

Theorem 5.1: *The following statements are equivalent for a topological space X:*

(1) X is disconnected.
(2) X is the union of two disjoint, non-empty closed sets.
(3) X is the union of two separated sets.
(4) There is a continuous function from X onto a discrete two-point space {a, b}.
(5) X has a proper subset A which is both open and closed.
(6) X has a proper subset A such that

$$\bar{A} \cap \overline{(X \backslash A)} = \varnothing.$$

Proof: *It will be shown that (1) implies each of the other statements and that each statement implies (1). Assume first that X is disconnected and let A, B be disjoint, non-empty open sets whose union is X.*

[(1) ⟹ (2)]: $B = X\backslash A$ and $A = X\backslash B$ are disjoint, non-empty closed sets whose union is X.

[(1) ⟹ (3)]: Since A and B are closed as well as open, then

$$\bar{A} \cap B = A \cap B = \varnothing, \quad A \cap \bar{B} = A \cap B = \varnothing,$$

so X is the union of the separated sets A and B.

[(1) ⟹ (4)]: The function $f\colon X \to \{a, b\}$ defined by

$$f(x) = \begin{cases} a & if \ x \in A \\ b & if \ x \in B \end{cases}$$

is continuous and maps X onto the discrete space $\{a, b\}$.

[(1) ⟹ (5)]: $A \neq \varnothing$, and

$$A = X\backslash B \neq X$$

since $B \neq \varnothing$. Thus A is the required set. (B will do equally well.)

[(1) ⟹ (6)]: Either A or B can be used as the required set.

[(2) ⟹ (1)]: If $X = C \cup D$ where C and D are disjoint, non-empty closed sets, then

$$D = X\backslash C, \quad C = X\backslash D$$

are open as well as closed.

[(3) ⟹ (1)]: If X is the union of separated sets C and D, then C and D are both non-empty, by definition. Since $X = C \cup D$ and $\bar{C} \cap D = \varnothing$, then $\bar{C} \subset C$, so C is closed. The same argument shows that D is also closed, and it follows as before that C and D must be open as well.

[(4) ⟹ (1)]: If $f\colon X \to \{a, b\}$ is continuous, then $f^{-1}(a)$ and $f^{-1}(b)$ are disjoint open subsets of X whose union is X. Since f is required to have both a and b as images, both $f^{-1}(a)$ and $f^{-1}(b)$ are non-empty.

[(5) ⟹ (1)]: Suppose X has a proper subset A which is both open and closed. Then $B = X\backslash A$ is a non-empty open set disjoint from A for which $X = A \cup B$.

[(6) ⟹ (1)]: Suppose X has a proper subset A for which

$$\bar{A} \cap \overline{(X\backslash A)} = \varnothing.$$

Then $\bar{A}$ and $\overline{(X\backslash A)}$ are disjoint, non-empty closed sets whose union is X, and it follows as before that $\bar{A}$ and $\overline{(X\backslash A)}$ are also open.   □

Corollary: *The following statements are equivalent for a topological space X:*

(1)   X is connected.

(2)   X is not the union of two disjoint, non-empty closed sets.

(3)   X is not the union of two separated sets.

(4)   There is no continuous function from X onto a discrete two-point space {a, b}.

(5)   The only subsets of X which are both open and closed are X and $\varnothing$.

(6)   X has no proper subset A for which

$$\bar{A} \cap \overline{(X \backslash A)} = \varnothing.$$

Theorem 5.2:   *Let X be a connected space and f: X $\to$ Y a continuous function from X onto a space Y. Then Y is connected.*

Proof: *Using the contrapositive form, assume that Y is disconnected. Then there are disjoint, non-empty open sets A and B in Y such that Y = A $\cup$ B. Then the sets $f^{-1}(A)$ and $f^{-1}(B)$*

(a)   are open sets because f is continuous;

(b)   are disjoint because f is a function;

(c)   are non-empty because f is surjective;

(d)   have union X because

$$X = f^{-1}(Y) = f^{-1}(A \cup B) = f^{-1}(A) \cup f^{-1}(B).$$

Thus, if Y is disconnected, then X is also disconnected and the proof is complete. $\square$

Corollary: *If f: X $\to$ Y is a continuous function on the indicated spaces and X is connected, then the image f(X) is a connected subspace of Y.*

Theorem 5.2 is sometimes rephrased by saying that connectedness is preserved by continuous functions. Properties preserved by continuous functions are called *continuous invariants*.

Example 5.2.1

Each open interval on the real line, being homeomorphic to $\mathbb{R}$, is connected. (A later example will show that the connected subsets of $\mathbb{R}$ are precisely the intervals.)

The next theorem is a useful criterion for determining whether or not a subspace is connected.

Theorem 5.3:   *A subspace Y of a space X is disconnected if and only if there exist open sets U and V in X such that*

$$U \cap Y \neq \emptyset, \quad V \cap Y \neq \emptyset, \quad U \cap V \cap Y = \emptyset, \quad Y \subset U \cup V.$$

Proof: *Suppose first that Y is disconnected. Then there are disjoint, non-empty open sets A and B in the subspace topology for Y such that Y = A ∪ B. By the definition of relatively open sets, there must be open sets U and V in X such that*

$$A = U \cap Y, \quad B = V \cap Y.$$

It is a simple matter to check that U and V have the required properties.

For the reverse implication, suppose that U and V are open subsets of X such that

$$U \cap Y \neq \emptyset, \quad V \cap Y \neq \emptyset, \quad U \cap V \cap Y = \emptyset, \quad Y \subset U \cup V.$$

Then

$$A = U \cap Y, \quad B = V \cap Y$$

are non-empty, disjoint relatively open sets whose union is Y, so Y is disconnected.   □

The analogue of Theorem 5.3 for closed sets is left as an exercise.

Theorem 5.4:   *If Y is a connected subspace of a space X, then $\bar{Y}$ is connected.*

Proof: *Suppose Y is connected. For a change of pace, the connectedness of $\bar{Y}$ will be shown by proving that there is no continuous function from $\bar{Y}$ onto a discrete two-point space.*

Consider a continuous function $f: \bar{Y} \rightarrow \{a, b\}$ from $\bar{Y}$ into such a discrete space. We must show that f is not surjective. The restriction $f|_Y$ cannot be surjective. This means that f maps Y to only one point of $\{a, b\}$, say a:

$$f(Y) = \{a\}.$$

Since f is continuous, Theorem 4.11 guarantees that

$$f(\bar{Y}) \subset \overline{f(Y)} = \{\bar{a}\} = \{a\},$$

so f is not surjective. Thus, by Theorem 5.1, $\bar{Y}$ is connected.   □

An examination of the preceding proof will reveal that Theorem 5.4 can be strengthened as follows:

Corollary: *Let Y be a connected subspace of a space X and Z a subspace of X such that $Y \subset Z \subset \bar{Y}$. Then Z is connected.*

Example 5.2.2

Every interval on the real line is connected. This can be proved as follows: It has already been shown in Example 5.2.1 that each open interval is connected. A non-degenerate closed interval is the closure of an open interval, so Theorem 5.4 shows that every non-degenerate closed interval is connected. A degenerate closed interval is connected since it has only one member. Any other non-empty interval is contained between an open interval and its closure, so the Corollary to Theorem 5.4 establishes connectedness. The empty set $\varnothing$, which is an interval, is connected since it has no non-empty subsets.

Example 5.2.3   The Topologist's Sine Curve

Let $A = \{(0, y) \in \mathbb{R}^2 : -1 \leq y \leq 1\}$, $B = \{(x, y) \in \mathbb{R}^2 : 0 < x \leq 1, y = \sin(\pi/x)\}$. The subspace $T = A \cup B$ of $\mathbb{R}^2$, shown in Figure 5.1, is called the

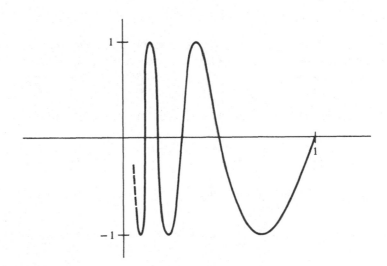

FIGURE 5.1   The topologist's sine curve.

topologist's sine curve. Note that B is connected since it is the image of $(0, 1]$ under a continuous function. Since $T = \bar{B}$, it follows from Theorem 5.4 that T is connected.

Obviously, it is not true in general that the union of connected sets is always connected. The reader should also be able to give an example to show that the intersection of two connected sets may fail to be connected. It does seem reasonable, however, that if a family of connected sets all have a point in common, then their union is connected. This is true and is proved in the next theorem.

Theorem 5.5:   *Let X be a space and $\{A_\alpha: \alpha \in I\}$ a family of connected subsets of X for which $\cap_{\alpha \in I} A_\alpha$ is not empty. Then $\cup_{\alpha \in I} A_\alpha$ is connected.*

Proof: *Theorem 5.3 will be used to show that $Y = \cup_{\alpha \in I} A_\alpha$ is connected. Suppose that U and V are open sets in X for which*

$$U \cap Y \ne \varnothing, \quad U \cap V \cap Y = \varnothing, \quad and \quad Y \subset U \cup V.$$

It will be shown that $V \cap Y = \varnothing$, thus proving that Y is connected. Now $U \cap Y \ne \varnothing$, so U contains some point in $A_{\alpha'}$, for some $\alpha' \in I$. Since $A_{\alpha'}$ is connected, then $A_{\alpha'} \subset U$. If $b \in \cap_{\alpha \in I} A_\alpha$, then b must be in $A_{\alpha'}$, so $b \in U$. Thus U contains a point b in each A_α, $\alpha \in I$. Since A_α is connected, then $A_\alpha \subset U$ for each $\alpha \in I$. Thus

$$Y = \bigcup_{\alpha \in I} A_\alpha \subset U$$

so $V \cap Y = \varnothing$.   □

Corollary: *Let X be a space, $\{A_\alpha: \alpha \in I\}$ a family of connected subsets of X, and B a connected subset of X such that, for each $\alpha \in I$, $A_\alpha \cap B \ne \varnothing$. Then $B \cup (\cup_{\alpha \in I} A_\alpha)$ is connected.*

Proof: *By Theorem 5.5, each set $B \cup A_\alpha$, $\alpha \in I$, is connected and the intersection*

$$\bigcap_{\alpha \in I} (B \cup A_\alpha) \ne \varnothing$$

since it contains B. Thus, by Theorem 5.5,

$$B \cup \left(\bigcup_{\alpha \in I} A_\alpha\right) = \bigcup_{\alpha \in I} (B \cup A_\alpha)$$

is connected.   □

The proof of the following variation on the union of connected sets is left as an exercise.

Theorem 5.6:   *Let $\{A_n\}_{n=1}^{\infty}$ be a sequence of connected subsets of a space X such that for each integer $n \geq 1$, A_n has at least one point in common with one of the preceding sets $A_1, \ldots, A_{n-1}$. Then $\bigcup_{n=1}^{\infty} A_n$ is connected.*

Definition: *A **component** of a topological space X is a connected subset C of X which is not a proper subset of any connected subset of X.*

The following properties of the components of a space X should be noted:

(1)   Each point $a \in X$ belongs to exactly one component. The component C_a containing a is the union of all the connected subsets of X which contain a and thus may be thought of as the largest connected subset of X which contains a.

(2)   For points a, b in X, the components C_a and C_b are either identical or disjoint.

(3)   Every connected subset of X is contained in a component.

(4)   Each component of X is a closed set.

(5) X is connected if and only if it has only one component.

(6)   If C is a component of X and A, B form a separation of X, then C is a subset of A or a subset of B.

Example 5.2.4

(a)   For the subspace $X = (0, 1) \cup (2, 3)$ of $\mathbb{R}$, there are two components, $(0, 1)$ and $(2, 3)$. Note that both components are closed sets with respect to the subspace topology for X determined by $\mathbb{R}$.

(b)   In a discrete space, each component contains only one point.

(c)   For the set R of rational numbers with the subspace topology determined by the real line, each component contains only one point. Note, however, that the topology in this case is not discrete.

Example 5.2.5

Property (6) of components states that if two points a and b belong to the same component of X, then they must belong to the same member of any separation of X. This example shows that the converse is false: It is possible for points a, b to be always in the same member of any separation A, B of X yet to belong to different components.

Consider the subspace X of $\mathbb{R}^2$ in Figure 5.2 consisting of a sequence of line segments converging to a line segment whose midpoint c has been deleted. Then $[a, c)$ is the component of X which contains a and $(c, b]$ is the component

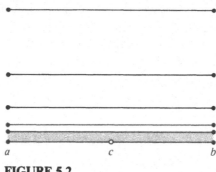

FIGURE 5.2

which contains b, so a and b belong to different components. However, for any separation of X into disjoint non-empty open sets A and B whose union is X, both a and b belong to A or both a and b belong to B.

Definition: *A space X is **totally disconnected** provided that each component of X consists of a single point.*

According to Example 5.2.4(b) and (c), discrete spaces of more than one point and the set of rational numbers are totally disconnected spaces. The property of total disconnectedness results from an attempt to classify the "degree of disconnectedness" of a space. Most people would agree, for example, that the set of rational numbers, in which the components consist of single points, is disconnected to a greater extent than is the space $X = [0, 1] \cup [2, 3]$, which has only two components.

EXERCISE 5.2

1. (a) Prove that the union of two disjoint closed intervals is always a disconnected subspace of $\mathbb{R}$.

    (b) Repeat (a) for disjoint open intervals.

    (c) Is the union of two disjoint intervals in $\mathbb{R}$ always disconnected? Give reasons for your answer.

2. Prove the analogue of Theorem 5.3 for closed sets.

3. Prove Theorem 5.4 and its corollary using a separation into disjoint, non-empty open sets.

4. Prove Theorem 5.6.

5. Determine whether each of the following subspaces of $\mathbb{R}^2$ is connected or disconnected. Give a reason for each answer.

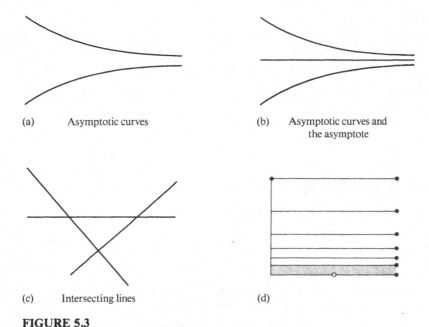

(a)   Asymptotic curves   (b)   Asymptotic curves and
                                                  the asymptote

(c)   Intersecting lines   (d)

FIGURE 5.3

6. Prove that every countable subset of $\mathbb{R}$ is totally disconnected.

7. Prove properties (1)–(6) for components listed in the text.

8. Give examples of subsets A and B of $\mathbb{R}^2$ to illustrate each of the following:

   (a) A and B are connected, but $A \cap B$ is disconnected.

   (b) A and B are connected, but $A \backslash B$ is disconnected.

   (c) A and B are disconnected, but $A \cup B$ is connected.

   (d) A and B are connected and $\bar{A} \cap \bar{B} \neq \varnothing$, but $A \cup B$ is disconnected.

9. (a) Suppose that A and B are open subsets of a space X for which $A \cup B$ and $A \cap B$ are connected. Prove that A and B are connected.

   (b) Repeat part (a) for closed sets.

10. Prove that a homeomorphism $h: X \to Y$ between spaces X and Y induces a one-to-one correspondence between components of X and components of Y.

11. Give an example of spaces X and Y for which there is a one-to-one correspondence between components of X and components of Y with corresponding components homeomorphic, but X is not homeomorphic to Y. (*Hint:* Consider the rational numbers and the integers.)

12. **Definition:** *A Hausdorff space X is **0-dimensional** if X has a basis $\mathcal{B}$ of sets which are simultaneously open and closed.*

    Prove that every 0-dimensional space is totally disconnected.

13. Prove:

(a) The property of being totally disconnected is a topological invariant but not a continuous invariant.

(b) The property of total disconnectedness is hereditary.

14. **Definition:** *A point x is a **cut point** of a connected space X provided that X\\{x} is a disconnected subspace.*

Prove:

(a) Every point of $\mathbb{R}$ is a cut point.

(b) If $h: X \rightarrow Y$ is a homeomorphism between the indicated spaces and x is a cut point of X, then $h(x)$ is a cut point of Y.

(c) Every point of the subspace

$$((-1, 0] \times \{0\}) \cup \{(x, y) \in \mathbb{R}^2 : 0 < x < 1, \quad y = \sin (1/x)\}$$

of $\mathbb{R}^2$ is a cut point.

15. Let X be a space such that for each pair a, b of points of X, $\{a, b\}$ is a subset of a connected set. Prove that X is connected.

16. In $\mathbb{R}^2$, let A_n be the edges of the rectangle of height $2 - 2/n$ and length n centered at the origin, $n = 2, 3, \ldots$. Let L_1 and L_2 be the horizontal lines $y = 1$ and $y = -1$, respectively, and let

$$X = L_1 \cup L_2 \cup \left(\bigcup_{n=2}^{\infty} A_n \right)$$

with the subspace topology of $\mathbb{R}^2$. Prove that the components of X are the sets L_1, L_2, and A_n, $n \geq 2$, but that there is no separation $X = A \cup B$ of X with $L_1 \subset A$ and $L_2 \subset B$.

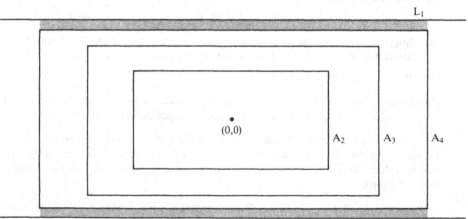

FIGURE 5.4

5.3 CONNECTED SUBSETS OF THE REAL LINE

Examples 5.1.2, 5.2.1, and 5.2.2 have shown that the real line and all intervals on the real line are connected. This section shows that there are no other connected subsets of $\mathbb{R}$.

Lemma: *A non-empty subset A of $\mathbb{R}$ is an interval if and only if for each pair c, d of members of A, every real number between c and d is in A.*

Proof: *It is evident from the definition of intervals in Chapter 2 that every interval has the stated property.*

Consider, then, a subset A of $\mathbb{R}$ which contains every real number between any two of its members. The proof breaks into several cases depending upon whether or not A has a least upper bound or greatest lower bound and whether or not these bounds, if they exist, belong to A.

Suppose, for example, that A has neither a least upper bound nor a greatest lower bound. Then for $x \in \mathbb{R}$ there are members c and d of A for which $c < x$ and $d > x$. Then $x \in A$, so it follows that $A = \mathbb{R}$ and A is the interval $(-\infty, \infty)$.

Suppose A has greatest lower bound a which does not belong to A and A has no least upper bound. Then A contains no real number $x \le a$. If $y > a$ then y is not an upper bound for A so there is a member d of A with $d > y$. Similarly, y is not the greatest lower bound for A, so there is a member c of A with $c < y$. Then $c < y < d$ so $y \in A$. Thus $A = (a, \infty)$ and is an interval.

The remaining cases are left to the reader.   □

Theorem 5.7: *The connected subsets of $\mathbb{R}$ are precisely the intervals.*

Proof: *Since we know that every interval is connected, it remains only to be proved that a subset B of $\mathbb{R}$ which is not an interval must be disconnected. Let B be a subset of $\mathbb{R}$ that is not an interval. Then, by the lemma, there are members c and d in B and a real number y with $c < y < d$ for which $y \notin B$. Then the open sets*

$$U = (-\infty, y), \quad V = (y, \infty)$$

have the following properties:

  (a) $c \in U \cap B$, so $U \cap B \ne \emptyset$;
       $d \in V \cap B$, so $V \cap B \ne \emptyset$;
  (b) $U \cap V = \emptyset$, so $U \cap V \cap B = \emptyset$;
  (c) $B \subset U \cup V$.

By Theorem 5.3, B is disconnected. Hence every connected subset of $\mathbb{R}$ must be an interval.   □

EXERCISE 5.3

1. Give another proof of the fact that every interval on $\mathbb{R}$ is connected by showing that every interval is the image of $\mathbb{R}$ under a continuous function. (*Hint:* The function whose graph appears below maps $\mathbb{R}$ onto $[a, b]$.)

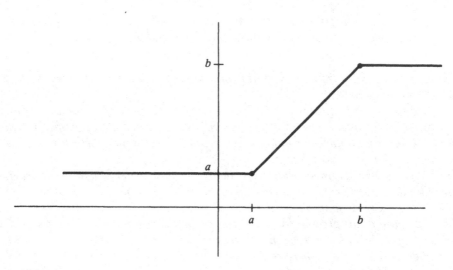

FIGURE 5.5

2. Let A be an interval on $\mathbb{R}$ and $f: A \rightarrow \mathbb{R}$ a continuous function. Prove that $f(A)$ is an interval.

3. Let $f: [a, b] \rightarrow [c, d]$ be a homeomorphism on the indicated intervals. Prove that f maps endpoints to endpoints.

4. Prove that an open interval (a, b) and a closed interval $[c, d]$ are not homeomorphic.

5.4 APPLICATIONS OF CONNECTEDNESS

In Chapter 1 a suggestive but incomplete argument was given for a special case of the Intermediate Value Theorem (Theorem 1.1). The Intermediate Value Theorem will be shown in the present section to be a simple consequence of our work on connectedness. The subject of fixed-point theorems and their importance will also be discussed in this section.

Theorem 5.8:   The Intermediate Value Theorem   *Let $f: [a, b] \rightarrow \mathbb{R}$ be a continuous function from a closed interval $[a, b]$ into $\mathbb{R}$ and y_0 a real number between $f(a)$ and $f(b)$. Then there is a number $c \in [a, b]$ for which $f(c) = y_0$.*

Proof: *The interval [a, b] is connected. Since f is continuous, then f([a, b]) is a connected subset of ℝ and, by Theorem 5.7, must be an interval. Thus any number y_0 between f(a) and f(b) must be in the image f([a, b]). This simply means that y_0 = f(c) for some real number c between a and b.*   □

Corollary: *Let f: [a, b] → ℝ be a continuous function for which one of f(a) and f(b) is positive and the other is negative. Then the equation f(x) = 0 has a root between a and b.*

Theorem 5.9:   *Let f: [a, b] → [a, b] be a continuous function from a closed interval [a, b] into itself. Then there is a member c ∈ [a, b] such that f(c) = c.*

Proof: *If f(a) = a or f(b) = b, then f has the required property, so we assume that f(a) ≠ a and f(b) ≠ b. Thus a < f(a) and f(b) < b, since f(a) and f(b) must be in [a, b]. Define g: [a, b] → ℝ by*

$$g(x) = x - f(x), \quad x \in [a, b].$$

Then g is continuous and

$$g(a) = a - f(a) < 0, \quad g(b) = b - f(b) > 0.$$

The Intermediate Value Theorem (Theorem 5.8) applies to show the existence of a number of c ∈ [a, b] for which g(c) = 0. Then f(c) = c for this number c.   □

Recall from Chapter 3 that a *fixed point* of a function $f: X \to X$ is a point x for which $f(x) = x$. A topological space X has the *fixed-point property* if every continuous function from X into itself has at least one fixed point.

Theorem 5.9 can now be restated as follows: *Every closed and bounded interval has the fixed-point property.* In Chapter 9 we shall investigate the Brouwer Fixed Point Theorem, which shows that every closed ball in $ℝ^n$ has the fixed-point property.

The next theorem shows that if X is homeomorphic to a space with the fixed-point property, then X has the fixed-point property as well. Combining this result with the Brouwer Fixed Point Theorem will allow us to conclude, for example, that any subset of $ℝ^3$ which is homeomorphic to a closed disk has the fixed-point property. Thus squares, triangles, and other closed figures homeomorphic to a closed disk have the fixed-point property. The $n = 2$ case of the Brouwer Fixed Point Theorem will be proved in Chapter 9.

Theorem 5.10:   *The fixed-point property is a topological invariant.*

Proof: *Let X be a space which has the fixed-point property, Y a space homeomorphic to X, and h: X → Y a homeomorphism. Let f: Y → Y be a continuous* *function. Since the composite function $h^{-1}fh: X \to X$ is a continuous function on X, it has at least one fixed point x_0:*

$$h^{-1}fh(x_0) = x_0.$$

Then

$$f(h(x_0)) = hh^{-1}fh(x_0) = h(x_0),$$

so the point $h(x_0)$ is a fixed point for f. Thus Y has the fixed-point property.   □

Example 5.4.1

The real line does not have the fixed-point property since, for example, the function

$$f(x) = x + 1, \quad x \in \mathbb{R},$$

has no fixed point. Since each open interval is homeomorphic to $\mathbb{R}$, Theorem 5.10 shows that no open interval has the fixed-point property. It is left as an exercise for the reader to show that intervals of the form $[a, b)$, $(a, b]$, $(-\infty, a]$, and $[a, \infty)$ do not have the fixed-point property.

Example 5.4.2

The n-sphere S^n, $n \geq 1$, does not have the fixed-point property since the function

$$g(x) = -x, \quad x \in S^n,$$

has no fixed point.

EXERCISE 5.4

1. Prove that every polynomial having real coefficients and odd degree has a real root.

2. Prove that intervals of the form $(a, b]$, $[a, b)$, $(-\infty, a]$, and $[a, \infty)$ do not have the fixed-point property.

3. Which of the following subsets of the plane have the fixed-point property? Give reasons for your answers:

    (a) $B(\theta, 1) = \{(x_1, x_2) \in \mathbb{R}^2: x_1^2 + x_2^2 < 1\}$

    (b) $R = \{(x_1, x_2) \in \mathbb{R}^2: x_1 \text{ and } x_2 \text{ are both rational}\}$

(c) $A = \{(x_1, x_2) \in \mathbb{R}^2 : x_1 = 0 \text{ or } x_2 = 0\}$ (the axes).

(d) $C = \{(x_1, x_2) \in \mathbb{R}^2 : 0 \le x_1 \le 1 \text{ and } x_2 = 0, \text{ or } x_1 = 0 \text{ and } 0 \le x_2 \le 1\}$.

4.   (a)   Prove that $\mathbb{R}^n$ does not have the fixed-point property.

(b)   Prove that no open ball $B(a, r)$, $a \in \mathbb{R}^n$, $r > 0$, in $\mathbb{R}^n$ has the fixed-point property.

5.   Does the topologist's sine curve of Example 5.2.3 have the fixed-point property? Justify your answer.

5.5 PATH CONNECTED SPACES

Path connectedness is a topological property stronger than connectedness which is useful in many applications. Intuitively speaking, a space X is path connected provided that each pair of points in X can be joined by a continuous curve in X.

Definition: *A **path** in a space X is a continuous function $p: [0, 1] \to X$. The points $p(0)$ and $p(1)$ are the **endpoints** of the path. The path is said to **join** the **initial point** $p(0)$ and the **terminal point** $p(1)$. The path p is also called a **path from** $p(0)$ **to** $p(1)$. If p is a path in X, the **reverse path** $\bar{p}$ is the path defined by*

$$\bar{p}(t) = p(1 - t), \quad t \in [0, 1].$$

Note that the initial point of a path is the terminal point of its reverse path, and conversely. Note also that since $[0, 1]$ is connected and connectedness is a continuous invariant (Theorem 5.2), then the image of every path is a connected set.

Definition: *A space X is **path connected** provided that for each pair a, b of points of X there is a path in X with initial point a and terminal point b. A subspace A of X is **path connected** provided that A is a path connected space with its subspace topology. The terms **path connected set** and **path connected subset** are sometimes used for path connected space and path connected subspace, respectively.*

Example 5.5.1

Every interval on the real line is path connected. For a, b in an interval K, the path

$$p(t) = (1 - t)a + tb, \quad t \in [0, 1],$$

is a path in K with initial point a and terminal point b.

Example 5.5.2

The preceding example generalizes to the subsets of $\mathbb{R}^n$ called "convex sets."

For a point $a = (a_1, \ldots, a_n) \in \mathbb{R}^n$ and $r \in \mathbb{R}$, the *scalar multiple* ra is defined by

$$ra = (ra_1, \ldots, ra_n).$$

The *line segment* from $a = (a_1, \ldots, a_n)$ to $b = (b_1, \ldots, b_n)$ is the set $\{(1 - t)a + tb: 0 \le t \le 1\}$. A subset C of $\mathbb{R}^n$ is *convex* provided that for all a, b in C, the line segment from a to b is contained in C.

Since the line segment from a to b is simply the image of the path

$$p(t) = (1 - t)a + tb, \quad 0 \le t \le 1,$$

then each convex subset of $\mathbb{R}^n$ is path connected. In particular, $\mathbb{R}^n$ is path connected.

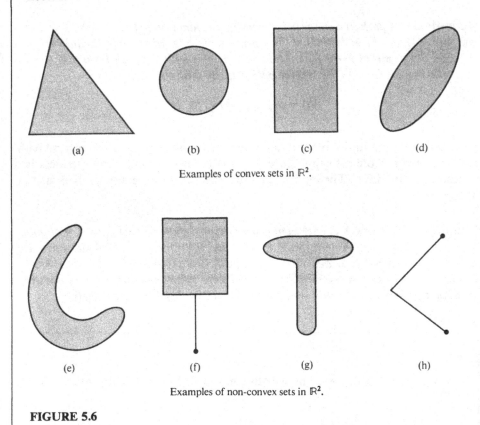

(a)   (b)   (c)   (d)

Examples of convex sets in $\mathbb{R}^2$.

(e)   (f)   (g)   (h)

Examples of non-convex sets in $\mathbb{R}^2$.

FIGURE 5.6

Theorem 5.11:   *Every path connected space is connected.*

Proof: *Suppose that X is a path connected space and let a be a member of X. For each x in X, let p_x be a path in X with initial point a, terminal point x, and image $C_x = p_x([0, 1])$. Since each image C_x is connected and a belongs to each C_x, then, by Theorem 5.5,*

$$X = \bigcup_{x \in X} C_x$$

is connected. □

The connected subsets of $\mathbb{R}$ are the intervals. By Example 5.5.1, every interval is path connected. Combining this fact with Theorem 5.11 shows that connectedness and path connectedness are equivalent properties for subsets of $\mathbb{R}$; a subset of $\mathbb{R}$ is connected if and only if it is path connected. The next example shows a connected subset of $\mathbb{R}^2$ that is not path connected and therefore justifies the statement made earlier that path connectedness is stronger than connectedness.

Example 5.5.3

Recall from Example 5.2.3 that the topologist's sine curve is the subspace $T = A \cup B$ of $\mathbb{R}^2$ for which $A = \{(0, y) \in \mathbb{R}^2: -1 \leq y \leq 1\}$ and $B = \{(x, y) \in \mathbb{R}^2: 0 < x \leq 1, y = \sin(\pi/x)\}$. Refer to Figure 5.1 for the picture and to Example 5.2.3 for the proof that T is connected.

It is intuitively plausible that there should be no path in T joining a point of A to a point of B. The following lemma will aid in the proof.

Lemma: *In the topologists' sine curve T, any connected subset C containing a point a in A and a point b in B has diameter greater than 2.*

Proof: *The lemma is based on the fact that any such set C must contain all of B to the left of $b = (b_1, b_2)$. The diameter of this set exceeds 2 because $\sin(\pi/x)$ oscillates in value between the extremes 1 and −1 as x approaches 0. For the proof, suppose there is a point $d = (d_1, d_2) \in B$ for which $d_1 \leq b_1$ and $d \notin C$. Then*

$$U = \{(x, y) \in \mathbb{R}^2: x < d_1\}, \quad V = \{(x, y) \in \mathbb{R}^2: d_1 < x\}$$

are disjoint open sets in $\mathbb{R}^2$, each containing at least one point of C, whose union contains C. Thus, by Theorem 5.3, C is not connected. This contradiction shows that C must contain all points of B to the left of b and hence must have diameter greater than 2.

Returning now to the argument that T is not path connected, suppose that there is a path p in T with initial point a in A and terminal point b in B. The following argument shows that p cannot be continuous at the value of $t \in [0, 1]$ where the curve "passes from A into B." Let

$$W = \{t \in [0, 1]: p([0, t]) \subset A\},$$

and let w be the least upper bound of W. Then $w \in \bar{W}$, so by continuity of p,

$$p(w) \in p(\bar{W}) \subset \overline{p(W)} \subset \bar{A} = A.$$

Also by continuity of p at w there must be a positive number δ such that if $t \in [0, 1]$ and $|w - t| < \delta$, then the distance in $\mathbb{R}^2$ from $p(w)$ to $p(t)$ is less than $1/2$. Since $w + \delta$ exceeds the least upper bound of W, there is a number $v \in [0, 1]$ such that $w < v < w + \delta$ and $p(v) \in B$. Then the image $p([w, v])$ is a connected subset of T containing a point $p(w) \in A$ and a point $p(v) \in B$. By the choice of δ, the diameter of $p([w, v])$ cannot exceed 1. This contradicts the lemma, however, and completes the proof that T is not path connected.

Example 5.5.4

The subspace X of the plane shown in Figure 5.7 is also connected but not path connected. The reader is left the exercise of showing that there is no path in X from a to b. (The point c indicated by the hollow dot does not belong to X.)

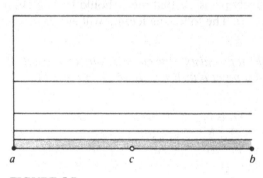

a c b

FIGURE 5.7

If there is a path in a space X with initial point a and terminal point b and a second path with initial point b and terminal point c, then it seems reasonable that one should be able to "put the paths together" to obtain a path from a to c. This will be proved with the aid of the following Gluing Lemma, which shows how two continuous functions can be "glued" together.

The Gluing Lemma: *Let A and B be closed subsets of a space $X = A \cup B$ and f: A → Y and g: B → Y continuous functions into a space Y for which f(x) = g(x) for all $x \in A \cap B$. Then the function h: X → Y defined by*

$$h(x) = \begin{cases} f(x) & if\ x \in A \\ g(x) & if\ x \in B \end{cases}$$

is continuous.

Proof: *For a closed subset C of Y,*

$$h^{-1}(C) = \{x \in X\colon h(x) \in C\} = \{x \in A\colon f(x) \in C\} \cup \{x \in B\colon g(x) \in C\}$$
$$= f^{-1}(C) \cup g^{-1}(C).$$

Since f is continuous, $f^{-1}(C)$ is a closed set in the subspace topology on A:

$$f^{-1}(C) = D \cap A$$

where D is closed in X. Since A is closed in X, then $f^{-1}(C)$ is the intersection of two closed sets and is therefore closed in X. The same reasoning shows that $g^{-1}(C)$ is closed in X. Therefore,

$$h^{-1}(C) = f^{-1}(C) \cup g^{-1}(C)$$

is a closed set in X for each closed set C in Y, so h is continuous.   □

Definition: *Let p_1 and p_2 be paths in a space X for which $p_1(1) = p_2(0)$. The **path product** $p_1 * p_2$ of p_1 and p_2 is the path in X defined by*

$$p_1 * p_2(t) = \begin{cases} p_1(2t) & 0 \le t \le 1/2 \\ p_2(2t - 1) & 1/2 \le t \le 1. \end{cases}$$

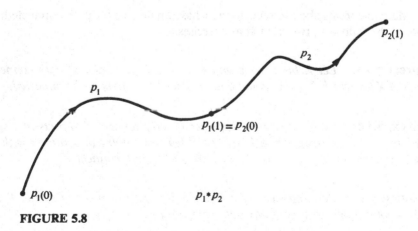

FIGURE 5.8

The continuity of $p_1 * p_2$ in the preceding definition follows easily from the Gluing Lemma with $A = [0, 1/2]$, $B = [1/2, 1]$, $f(t) = p_1(2t)$, and $g(t) = p_2(2t - 1)$. The agreement of f and g on $A \cap B$ is provided by

$$f(1/2) = p_1(2 \cdot 1/2) = p_1(1) = p_2(0) = p_2(2 \cdot 1/2 - 1) = g(1/2).$$

Theorem 5.12: *Every open, connected subset of $\mathbb{R}^n$ is path connected.*

Proof: *For an open, connected subset O of $\mathbb{R}^n$ and $a \in O$, let $U_a = \{x \in O$: there is a path in O joining a and $x\}$.*

Note that U_a is an open set containing a for each $a \in O$: If $x \in U_a$, let $B(x, r)$ be an open ball centered at x and contained in O. Since $B(x, r)$ is convex, Example 5.5.2 guarantees that there is a path in $B(x, r)$ from x to any member y of $B(x, r)$. Since there is a path in O from a to x, the method of forming path products shows that there is a path in O from a to each point of $B(x, r)$. Thus $B(x, r) \subset U_a$, so U_a is an open set.

For any a, b in O, U_a and U_b are either identical or disjoint. To prove this, suppose that x belongs to $U_a \cap U_b$. Then there are paths in O from a to x and from b to x and, consequently, a path in O from a to b. But then any point that can be joined by a path in O to point a can be joined by a path in O to point b, and vice versa, so $U_a = U_b$. Thus either $U_a = U_b$ or $U_a \cap U_b = \varnothing$.

Suppose now that O is not path connected and let a, b be members of O which cannot be joined by a path in O. Then $U_a \neq O$ since $b \notin U_a$, and U_a and U_b are disjoint open sets in $\mathbb{R}^n$. Then

$$V = \cup \{U_x : x \in O \quad and \quad x \notin U_a\}$$

and U_a are disjoint, non-empty open sets whose union is O, contradicting the fact that O is connected. □

Many theorems about connectedness have analogues for path connectedness. Proofs of the following two are left as exercises.

Theorem 5.13: *Let X be a space and $\{A_\alpha: \alpha \in I\}$ a family of path connected subsets of X for which $\cap_{\alpha \in I} A_\alpha$ is not empty. Then $\cup_{\alpha \in I} A_\alpha$ is path connected.*

Theorem 5.14: *Let $\{A_n\}_{n=1}^\infty$ be a sequence of path connected subsets of a space X such that for each integer $n \geq 1$, A_n has at least one point in common with one of the preceding sets $A_1, \ldots, A_{n-1}$. Then $\cup_{n=1}^\infty A_n$ is path connected.*

Definition: *A **path component** of a space X is a path connected subset P of X which is not a proper subset of any path connected subset of X.*

The following properties of path components should be noted.

(1) Each point a in X belongs to exactly one path component. The path component P_a containing a is the union of all the path connected subsets of X which contain a and is therefore the largest path connected subset of X which contains a.

(2) For points a, b in X, the path components P_a and P_b are either identical or disjoint.

(3) Every path connected subset of X is contained in a path component.

(4) X is path connected if and only if it has only one path component.

The preceding properties of path components are quite analogous to corresponding properties of components of a space X. Note, however, that it is not stated above that a path component must be a closed set. It is left as an exercise to show that one of the path components of the topologist's sine curve is not a closed set.

Each topological space is divided into components and into path components. Is each component a subset of a path component, or is it the other way around?

EXERCISE 5.5

1. Identify two path components of the topologist's sine curve (Examples 5.2.3 and 5.5.3). Show that one of the path components is closed and the other is not.

2. Prove that in $\mathbb{R}^n$ each path component of an open set is an open set.

3. Let X be a space and P a path component of X. Prove that P is a subset of some component of X.

4. Let O be an open set in $\mathbb{R}^n$. Prove that the components of O and the path components of O are identical.

5. Prove Theorems 5.13 and 5.14.

6. Prove that path connectedness is a continuous invariant.

7. Prove that a space X is path connected if and only if there is a point a in X such that each point of X can be joined to a by a path in X.

8. Give an example of a space X with a path connected subset A whose closure $\bar{A}$ is not path connected.

9. (a) Give the definitions for the terms *scalar multiple*, *line segment*, and *convex set* in Hilbert space H.

    (b) Prove that every convex subset of H is path connected.

10. Let X be a space, and define a relation $\sim$ on X as follows: $x \sim y$ if and only if there is a path in X from x to y. Show that $\sim$ is an equivalence relation whose equivalence classes are the path components of X.

11. Prove that convexity is not a topological property.

12. Give an example of a path connected subset of $\mathbb{R}^2$ that is not convex.

5.6 LOCALLY CONNECTED AND LOCALLY PATH CONNECTED SPACES

Connectedness is called a *global topological property* because it describes a characteristic of an entire topological space. Path connectedness, second countability, separability, and the Hausdorff property are also global properties. *Local topological properties* deal with the characteristics of a space "near" a particular point. Our first example of such a property was first countability, which describes the open sets containing a particular point. This section introduces local properties corresponding to connectedness and path connectedness.

Definition: *A topological space X is **locally connected** at a point p in X if every open set containing p contains a connected open set which contains p. The space X is **locally connected** provided that it is locally connected at each point.*

The first theorem on local connectedness is essentially a restatement of the definition. The proof is left as an exercise.

Theorem 5.15:

(a)   *A space X is locally connected at a point p in X if and only if there is a local basis at p consisting of connected open sets.*

(b)   *A space X is locally connected if and only if it has a basis of connected open sets.*

Example 5.6.1

(a)   Any interval in $\mathbb{R}$ is both connected and locally connected.

(b) $\mathbb{R}^n$ is connected and locally connected for each positive integer n.

(c)   The subspace $X = [0, 1] \cup [2, 3]$ of $\mathbb{R}$ is locally connected but not connected.

(d)   The *topologist's comb* C is the subspace of $\mathbb{R}^2$ shown in Figure 5.9: C consists of the interval $[0, 1]$ on the real line with vertical segments of length 1 attached at the origin and at each point $1/n$, n a positive integer.

   The topologist's comb C is obviously connected; in fact, it is path connected since paths can run up and down the vertical segments and across the base $[0, 1]$. However, C is not locally connected at any point $(0, t)$, $0 < t \leq 1$, since small open sets containing such points consist of collections of open vertical intervals. A typical one is shown in Figure 5.10.

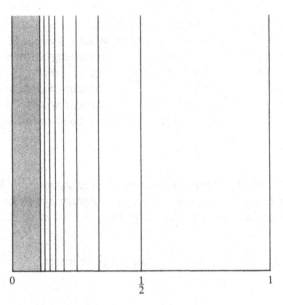

FIGURE 5.9   The topologist's comb.

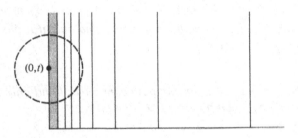

FIGURE 5.10

(e)   The set of rational numbers, as a subspace of $\mathbb{R}$, is neither connected nor locally connected.

As the preceding examples show, one cannot predict local connectedness or connectedness on the basis of the other. This is commonly the case with local and global properties; generally speaking, a local property does not necessarily imply the corresponding global property, and vice versa. Recall, however, that second countability does imply first countability, so this is not an infallible rule.

Theorem 5.16:   *A space X is locally connected if and only if for each open subset O of X, each component of O is an open set.*

Proof: *Suppose that C is a component of an open subset O of a locally connected space X. For x in C, there is a connected open set U_x with $x \in U_x$ and $U_x \subset O$. Since C is the largest connected subset of O containing x, then $U_x \subset C$. Thus C contains an open set about each of its points, so C is open.*

For the converse, suppose that each component of each open subset of X is open, and let a be a point of X and O an open set containing a. Then the component C of O which contains a is an open set containing a and contained in O, so X must be locally connected at a.   □

Definition: *A space X is **locally path connected** at a point p in X if every open set containing p contains a path connected open set containing p. The space X is **locally path connected** provided that it is locally path connected at each of its points.*

Proofs of the next two theorems are left as exercises.

Theorem 5.17:

(a)   *A space X is locally path connected at a point p in X if and only if there is a local basis at p consisting of path connected open sets.*

(b)   *A space X is locally path connected if and only if it has a basis of path connected open sets.*

Theorem 5.18:   *A space X is locally path connected if and only if for each open subset O of X, each path component of O is an open set.*

Both local connectedness and local path connectedness are topological properties. Furthermore, it is easily observed that every locally path connected space is locally connected. One of the exercises for this section is to find an example of a locally connected space that is not locally path connected.

Theorem 5.19:   *If X is a connected, locally path connected space, then X is path connected.*

Proof: *For each a in X, let P_a denote the path component of X to which a belongs. Since X is an open set, Theorem 5.18 shows that each P_a is open. Recall that for path components P_a and P_b of X, either $P_a = P_b$ or P_a and P_b are disjoint.*

For a particular point a in X, suppose $P_a \neq X$. Then P_a and the union of all P_x for which $x \notin P_a$ are disjoint, non-empty open subsets of X whose union is X.

Thus we conclude that X is disconnected, a contradiction. Since X is connected, then $P_a = X$ and X must be path connected as well.   □

EXERCISE 5.6

1. Prove Theorem 5.15.

2. Give an example of a space X which is

    (a) path connected but not locally path connected;

    (b) locally path connected but not path connected.

3. Prove that every locally path connected space is locally connected.

4. (a) Prove that $\mathbb{R}^n$ is locally path connected.

    (b) Prove that every open set in $\mathbb{R}^n$ is locally path connected.

5. Prove Theorems 5.17 and 5.18.

6. Give an example of a space X and a point a in X for which X is locally connected at a but not locally path connected at a.

7. Discuss the properties of local connectedness and local path connectedness for the topologist's sine curve (Examples 5.2.3 and 5.5.3).

8. Let X be the subset of the plane consisting of a circle and a spiral in the circle winding outward with the circle as limit, as shown in Figure 5.11. Prove that X fails to be locally connected at points on the circle.

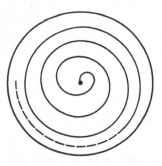

FIGURE 5.11

9. The *broom space B* is the subset of $\mathbb{R}^2$ consisting of segments from the origin of unit length and slope $1/n$, $n = 1, 2, \ldots$, together with the limiting segment $[0, 1]$ on the x-axis. Prove that B is locally path connected at each point of $B\backslash(0, 1]$ and is not locally connected at any point of $(0, 1]$.

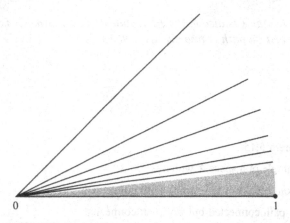

FIGURE 5.12   Broom space.

10. Let X be a locally path connected space. Prove that each path component of X is both open and closed.

11. **Definition:** *Let X be a space and x a point of X. Then X is **connected im kleinen** at x provided that for each open set U containing x there is an open set V containing x with the property that for each point y in V, there is a connected subset of U containing both x and y.*

  (a) Prove that a space which is locally connected at a given point is connected *im kleinen* at that point.

  (b) Show that the space X shown below, which is sometimes called a "sequence of shrinking broom spaces," is connected *im kleinen* at the point x but not locally connected at x.

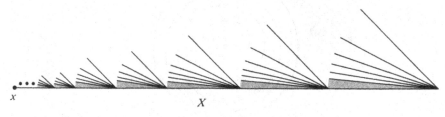

FIGURE 5.13   Sequence of shrinking broom spaces.

  (c) Let X be a space that is connected *im kleinen* at each point x in X. Prove that X is locally connected. (*Hint:* Show that each component of each open subset of X is open.)

12. (a) Show that local connectedness is not a continuous invariant by describing a continuous function $f: [0, 1) \to Y$ from $[0, 1)$ onto the space Y shown in Figure 5.14.

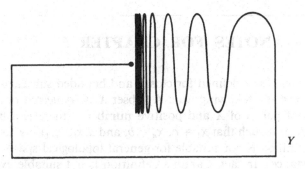

FIGURE 5.14

(b) Let X be a locally connected space, Y a space, and $f: X \to Y$ a continuous function from X onto Y. Prove:

   (i) If f is an open function, then Y is locally connected.

   (ii) If f is a closed function, then Y is locally connected.

SUGGESTIONS FOR FURTHER READING

Most of the general topology textbooks listed in the Bibliography contain information on connectedness and related properties. The treatments given in Dugundji's *Topology* and Willard's *General Topology* are particularly recommended.

HISTORICAL NOTES FOR CHAPTER 5

Connectedness was defined for closed and bounded subsets of $\mathbb{R}^n$ by Georg Cantor in 1883 in the following way: A subset X is *connected* provided that for each pair a, b of points of X and positive number ϵ, there is a finite sequence $\{x_i\}_{i=1}^n$ of points of X such that $x_1 = a$, $x_n = b$, and $d(x_i, x_{i+1}) < \epsilon$ for $i = 1, 2, \ldots,$ $n - 1$. This definition is not suitable for general topological spaces because of its reliance on distances. In fact, Cantor's definition is not suitable even for general metric spaces.

The modern definition of connectedness is due to Camille Jordan (1838–1922), who defined the property in 1892 for closed and bounded subsets of $\mathbb{R}^n$ as follows: A set X is *connected* provided that it has no proper subset A for which $\bar{A}$ and $\overline{X \backslash A}$ are disjoint. This definition is property (6) of the Corollary to Theorem 5.1. Jordan showed the equivalence of his definition to Cantor's and proved Theorems 5.2 and 5.5 for the situation covered by his definition. Jordan's definition was extended to abstract spaces by N. J. Lennes in 1911.

The systematic study of connectedness began with Hausdorff's *Grundzüge der Mengenlehre* in 1914. The notion of separated sets is due to Hausdorff, who defined a set to be connected provided that it is not the union of two separated sets. Hausdorff's book contains Theorem 5.4 and its corollary, as well as Theorem 5.5. The definition of component is due to Hausdorff, as is the idea of a totally disconnected space, which was called "hereditarily disconnected" at that time.

Local connectedness was defined by Hans Hahn (1879–1934) in 1914. Similar properties were considered by Pia Nalli in 1911 and Stephan Mazurkiewicz in 1913. Theorem 5.16 was proved independently by Hahn in 1921 and Kuratowski in 1920.

Path connectedness is the oldest of the connectedness properties; the joining of points by continuous curves can be traced back for centuries. In a stricter topological sense, path connectedness was used by Weierstrass prior to 1890 for subsets of $\mathbb{R}^n$.

6

Compactness

Compactness is the topological property that generalizes to topological spaces a property of closed, bounded intervals [a, b] developed in Chapter 2. Recall that the Heine-Borel Theorem (Theorem 2.12) showed that if O is a collection of open intervals whose union contains [a, b], then there is a finite subcollection of O whose union contains [a, b]. A topological space which has this property for every covering by open sets is called *compact*. Compactness is a subtle property whose ramifications are not immediately apparent; do not let the term "compact" suggest simply smallness of size. The intervals [0, 1] and (0, 1) have the same size, but [0, 1] is compact and (0, 1) is not.

One of the main results of this chapter is that a subset of $\mathbb{R}^n$ is compact if and only if it is closed and bounded. Historically, compactness was intended to generalize to topological spaces the properties which characterize the closed and bounded subsets of $\mathbb{R}^n$. Several different properties that will be introduced in this chapter were put forward with varying degrees of success until it was recognized that compactness is the desired property.

The reader may already know from calculus that a continuous, real-valued function $f\colon [a, b] \to \mathbb{R}$ whose domain is a closed interval [a, b] attains a maximum and a minimum value. This result is proved in the present chapter as a corollary to the more general theorem that any continuous, real-valued function whose domain is a compact space attains a maximum and a minimum value.

6.1 COMPACT SPACES AND SUBSPACES

Definition: *Let A be a subset of a topological space X. An **open cover** of A is a collection O of open subsets of X whose union contains A. A **subcover** derived from an open cover O is a subcollection O' of O whose union contains A.*

Note from the preceding definition that an open cover of a space X is a family of open subsets of X whose union is X.

Example 6.1.1

Consider the subspace $A = [0, 5]$ of $\mathbb{R}$ and the open cover $O = \{(n - 1, n + 1)\}_{n=-\infty}^{\infty}$ of A. In this case the subcollection O' consisting of (−1, 1), (0, 2), (1, 3), (2, 4), (3, 5), (4, 6) is a subcover and happens to be the smallest subcover

161

for A that can be derived from $\mathcal{O}$. There are many other subcovers for A, however, since it is only required that the union of the members of the subcover contain A.

Definition: *A space X is **compact** provided that every open cover of X has a finite subcover. Equivalently, X is **compact** provided that for every collection $\mathcal{O}$ of open sets whose union equals X, there is a finite subcollection $\{O_i\}_{i=1}^{N}$ of $\mathcal{O}$ whose union equals X. A subspace A of a space X is **compact** provided that A is a compact topological space in its subspace topology.*

Since relatively open sets in the subspace topology for a subset A of a space X are the intersections of A with open sets in X, the definition of compactness for subspaces can be restated as follows:

Alternate Definition: *A subspace A of a space X is **compact** if and only if every open cover of A by open sets in X has a finite subcover.*

Example 6.1.2

(a) Any space consisting of a finite number of points is compact.

(b) Each closed and bounded interval $[a, b]$ is compact. This follows from the Heine-Borel Theorem (Theorem 2.12) since each open set is a union of open intervals. Let $\mathcal{O}$ be an open cover of $[a, b]$ by open subsets of $\mathbb{R}$, and for each x in $[a, b]$ let O_x be a member of $\mathcal{O}$ which contains x. Since O_x is open, there is an open interval I_x containing x and contained in O_x. By Theorem 2.12, the collection $\{I_x: x \in [a, b]\}$ has a finite subcover $\{I_{x_j}\}_{j=1}^{N}$. Since

$$I_x \subset O_x, \quad x \in [a, b],$$

it follows that the corresponding collection $\{O_{x_j}\}_{j=1}^{N}$ is a finite subcover of $[a, b]$ derived from $\mathcal{O}$.

(c) We shall see in Theorem 6.11 that the compact subsets of $\mathbb{R}^n$ are precisely the sets which are both closed and bounded.

(d) The real line $\mathbb{R}$ with the finite complement topology $\mathcal{T}'$ is compact. To see this, let $\mathcal{O}$ be an open cover of $\mathbb{R}$. Then any non-empty member O_1 of $\mathcal{O}$ contains all but some finite number N of points of $\mathbb{R}$. By choosing from $\mathcal{O}$ an open set containing each of these remaining points, one obtains a subcover derived from $\mathcal{O}$ having at most $N + 1$ members.

Example 6.1.3

(a) An infinite set X with the discrete topology is not compact. The open cover $\mathcal{O} = \{\{x\}: x \in X\}$ is an open cover with no finite subcover.

(b) The open interval $(0, 1)$ is not compact since $\mathcal{O} = \{(1/n, 1)\}_{n=2}^{\infty}$ is an open cover with no finite subcover. Non-compactness for any open interval (a, b) is proved similarly.

(c) $\mathbb{R}^n$ is not compact for any positive integer n since $\mathcal{O} = \{B((\theta, n)\}_{n=1}^{\infty}$ is an open cover having no finite subcover.

Definition: *A family $\mathcal{A}$ of subsets of a space X has the **finite intersection property** provided that every finite subcollection of $\mathcal{A}$ has non-empty intersection.*

Example 6.1.4

The collection $\{(1/n, 1)\}_{n=2}^{\infty}$ is a family of open subsets of $\mathbb{R}$ with the finite intersection property. Note that the intersection of any finite collection of these open intervals is the smallest interval (the interval of largest index) involved in the intersection.

The duality between unions of open sets O_α and intersections of the corresponding closed sets $C_\alpha = X \backslash O_\alpha$ in a space X,

$$X \backslash \cap\, C_\alpha = \cup\, O_\alpha,$$

produces the following characterization of compactness in terms of closed sets.

Theorem 6.1: *A space X is compact if and only if every family of closed sets in X with the finite intersection property has non-empty intersection.*

Proof: *Suppose first that X is compact and let $\mathcal{A} = \{C_\alpha: \alpha \in I\}$, for an index set I, be a family of closed sets in X with the finite intersection property. It must be shown that the intersection of all the members of $\mathcal{A}$ is non-empty. Proceeding by contradiction, suppose that this intersection $\cap_{\alpha \in I}\, C_\alpha$ is empty. Consider the corresponding family $\mathcal{O} = \{O_\alpha = X \backslash C_\alpha: \alpha \in I\}$ of open sets in X. Then*

$$\bigcup_{\alpha \in I} O_\alpha = \bigcup_{\alpha \in I} (X \backslash C_\alpha) = X \backslash \bigcap_{\alpha \in I} C_\alpha = X \backslash \varnothing = X,$$

so $\mathcal{O}$ is an open cover of X. Thus, by compactness of X, $\mathcal{O}$ has a finite subcover $\mathcal{O}' = \{O_{\alpha_i}\}_{i=1}^{n}$ for X. Then

$$X = \bigcup_{i=i}^{n} O_{\alpha_i} = \bigcup_{i=i}^{n} (X \backslash C_{\alpha_i}) = X \backslash \bigcap_{i=i}^{n} C_{\alpha_i},$$

so $\cap_{i=i}^{n} C_{\alpha_i}$ must be empty. But this contradicts the fact that $\mathcal{A}$ has the finite intersection property. Thus, if $\mathcal{A}$ has the finite intersection property, then the intersection of all the members of $\mathcal{A}$ must be non-empty.

The argument for the reverse implication is similar and is left as an exercise.

□

Theorem 6.2:   Cantor's Theorem of Deduction   Let $\{E_n\}_{n=1}^{\infty}$ be a nested sequence of non-empty, closed and bounded subsets of $\mathbb{R}$. Then $\cap_{n=1}^{\infty} E_n$ is not empty.

Proof: The family $\{E_n\}_{n=1}^{\infty}$, being nested and composed of non-empty sets, has the finite intersection property. Since E_1 is bounded, it must be a subset of some closed interval $[a, b]$. Since each E_n is closed in $\mathbb{R}$, then each is closed in the subspace topology for $[a, b]$. Thus $\{E_n\}_{n=1}^{\infty}$ is a collection of closed subsets of the compact space $[a, b]$, and $\{E_n\}_{n=1}^{\infty}$ has the finite intersection property. Theorem 6.1 guarantees that $\cap_{n=1}^{\infty} E_n$ is not empty.   □

Note that the preceding theorem has Cantor's Nested Intervals Theorem (Theorem 2.11) as a special case.

Example 6.1.5

The requirement that the subsets E_n of Cantor's Theorem of Deduction be bounded cannot be removed. Consider, for example, the collection $\{A_n\}_{n=1}^{\infty}$ of infinite intervals

$$A_n = [n, \infty), \quad n = 1, 2, 3, \ldots.$$

Then $\{A_n\}_{n=1}^{\infty}$ is a nested sequence of closed subsets of $\mathbb{R}$ whose intersection is empty. The requirement that the sets E_n be closed cannot be omitted either. The collection $\{(0, 1/n)\}_{n=1}^{\infty}$ is a nested sequence of non-empty bounded sets whose intersection is empty.

Theorem 6.3:   *Each closed subset of a compact space is compact.*

Proof: *Let A be a closed subset of a compact space X and $\mathcal{O}$ an open cover of A by open sets in X.*

Since A is a closed set, then $X \backslash A$ is open and

$$\mathcal{O}^* = \mathcal{O} \cup \{X \backslash A\}$$

is an open cover of X. By compactness of X, $\mathcal{O}^*$ has a finite subcover which contains only finitely many members $O_1, \ldots, O_n$ of $\mathcal{O}$ and may contain $X\backslash A$.
The fact that

$$X = (X\backslash A) \cup \bigcup_{i=i}^{n} O_i$$

implies that

$$A \subset \bigcup_{i=i}^{n} O_i$$

since $X\backslash A$ contains no points of A. Thus $\{O_i\}_{i=i}^{n}$ is a finite subcover for A derived from $\mathcal{O}$, and A is compact.   □

Theorem 6.4:   *Each compact subset of Hausdorff space is closed.*

Proof: *Let A be a compact subset of Hausdorff space X. It will be proved that $X\backslash A$ is open and hence that A is closed. Let $x \in X\backslash A$. Then for each $y \in A$, there exist disjoint open sets U_y and V_y with $y \in U_y$ and $x \in V_y$. The collection $\{U_y: y \in A\}$ of such open sets U_y is an open cover of A which, by compactness, has a finite subcover $\{U_{y_i}\}_{i=i}^{n}$. Then, since U_{y_i} and the corresponding open set V_{y_i} are disjoint, the sets*

$$U = \bigcup_{i=i}^{n} U_{y_i}, \quad V = \bigcap_{i=i}^{n} V_{y_i}$$

are disjoint open sets containing A and x, respectively. In particular, V is an open set containing x which is disjoint from A. Thus $X\backslash A$ is open, so A is closed.   □

Corollary: *Let X be a compact Hausdorff space. A subset A of X is compact if and only if it is closed.*

Theorem 6.5:   *If A and B are disjoint compact subsets of a Hausdorff space X, then there exist disjoint open sets U and V in X such that $A \subset U$ and $B \subset V$.*

Proof: *Consider the relation between a point x in B and the compact set A. By the proof of Theorem 6.4, there exist disjoint open sets U and V in X containing A and x, respectively. Since we shall require such a pair of open sets for each x in B, we add a subscript x to the notation for indexing. For each x in B, there are disjoint open sets U_x and V_x in X such that*

$$A \subset U_x, \quad x \in V_x.$$

The collection $\{V_x: x \in B\}$ is an open cover of B and, by compactness, has a finite subcover $\{V_{x_i}\}_{i=1}^n$ for B. Then since U_{x_i} and V_{x_i} are disjoint, the sets

$$U = \bigcap_{i=1}^{m} U_{x_i}, \quad V = \bigcup_{i=1}^{m} V_{x_i}$$

are disjoint open sets containing A and B, respectively.   □

Corollary: *If A and B are disjoint closed subsets of a compact Hausdorff space X, then there exist disjoint open sets U and V in X with $A \subset U$ and $B \subset V$.*

Proof: *Since A and B are closed subsets of the compact space X, then A and B are compact by Theorem 6.3. Then A and B are disjoint compact subsets of a Hausdorff space, so the desired conclusion follows from Theorem 6.5.*   □

Theorem 6.5 is sometimes rephrased by saying that in a Hausdorff space, every pair A, B of compact sets can be *separated* by disjoint open sets. This extends to pairs of disjoint compact subsets the analogous Hausdorff property for pairs of points. Note that this use of the term "separated" refers to inclusion in disjoint open sets and is not to be confused with the idea of a separated or disconnected space. The Corollary to Theorem 6.5 says that in a compact Hausdorff space every pair of disjoint closed sets can be separated by disjoint open sets. We shall have a more complete account of such separation theorems in Chapter 8.

EXERCISE 6.1

1. In Example 6.1.1, explain the sense in which the open subcover $\mathcal{O}'$ is the smallest sub-collection of $\mathcal{O}$ which contains A.

2. Complete the proof of Theorem 6.1.

3. Give examples of each of the following:

    (a) A closed subspace that is not compact.

    (b) A compact subspace that is not closed.

    (c) An open, compact subspace.

    (d) Two compact subsets whose intersection is not compact.

4. Prove:

    (a) The union of a finite number of compact subsets of a space X is compact.

    (b) If X is Hausdorff, then the intersection of any family of compact subspaces is compact.

5. Prove that a space X is compact if and only if X has a basis $\mathcal{B}$ for which every open cover of X by members of $\mathcal{B}$ has a finite subcover.

6. Show that the real line with the countable complement topology is not compact.

6.2 COMPACTNESS AND CONTINUITY

Compactness is of importance in topology largely because of its relationships with continuity. Some of these relationships are examined in this section.

Theorem 6.6: *Let X be a compact space, Y a space and $f: X \to Y$ a continuous function from X onto Y. Then Y is compact.*

Proof: *Let $\mathcal{O}$ be an open cover of Y. Then, for each O in $\mathcal{O}$, $f^{-1}(O)$ is open in X so the collection $\mathcal{O}^* = \{f^{-1}(O): O \in \mathcal{O}\}$ of inverse images of members of $\mathcal{O}$ is an open cover of X. Since X is compact, $\mathcal{O}^*$ has a finite subcover $\{f^{-1}(O_i)\}_{i=1}^n$ for X corresponding to a finite subcollection $\{O_i\}_{i=1}^n$ of $\mathcal{O}$. Since*

$$X = \bigcup_{i=1}^{n} f^{-1}(O_i)$$

and f maps X onto Y, then

$$Y = f(X) = f\left(\bigcup_{i=1}^{n} f^{-1}(O_i)\right) = \bigcup_{i=1}^{n} f(f^{-1}(O_i) \subset \bigcup_{i=1}^{n} O_i,$$

Thus $\{O_i\}_{i=1}^n$ is a finite subcover for Y derived from $\mathcal{O}$, so Y is compact. □

Corollary: *Let X be a compact space, Y a space and $f: X \to Y$ a continuous function from X into Y. Then the image $f(X)$ is a compact subspace of Y.*

Corollary: *Compactness is a topological invariant.*

In the language of invariants, Theorem 6.6 simply says that compactness is a continuous invariant. This theorem is often paraphrased by saying that the continuous image of a compact space is compact.

Theorem 6.7: *Let X be a compact space, Y a Hausdorff space, and $f: X \to Y$ a continuous function. Then f is a closed function.*

Proof: *It must be proved that if C is a closed set in X then the image $f(C)$ is closed in Y. This is merely a combination of previous results. By Theorem 6.3, the closed subset C of X is compact. By the first corollary to Theorem 6.6, $f(C)$ is a compact subspace of Y. Since Y is Hausdorff, Theorem 6.4 applies to show that the compact subspace $f(C)$ is a closed set.* □

Theorem 6.8: *Let X be a compact space, Y a Hausdorff space, and $f\colon X \to Y$ a continuous one-to-one function from X onto Y. Then f is a homeomorphism.*

Proof: *Since f is hypothesized to be a continuous bijection, it remains only to be shown that f^{-1} is continuous. This follows from Theorem 6.7. For any closed subset C of X,*

$$(f^{-1})^{-1}(C) = f(C)$$

is closed in Y, so f^{-1} is continuous. □

Theorem 6.8 can be interpreted as saying that the continuity of f^{-1} is produced "free" by the fact that f is a continuous bijection from a compact space onto a Hausdorff space.

Example 6.2.1

Theorem 6.8 would be of great value in elementary calculus if its proof were accessible at that level. Many laborious proofs of continuity could be avoided.

(a) Suppose, for example, that we know that the squaring function

$$f(x) = x^2, \quad x \in \mathbb{R},$$

is continuous and want to prove the continuity of the square root function

$$g(x) = \sqrt{x}, \quad x \geq 0.$$

Consider a non-negative real number a. Let b be a real number greater than a and consider

$$f\colon [0, \sqrt{b}] \to [0, b].$$

Now f is a continuous, one-to-one function from $[0, \sqrt{b}]$ onto $[0, b]$. By Theorem 6.8, its inverse is continuous as a function from $[0, b]$

to $[0, \sqrt{b}]$. In particular, the square root function g is continuous at a and must be continuous on its entire domain $[0, \infty)$.

(b) Since the sine function is a continuous bijection from $[-\pi/2, \pi/2]$ onto $[-1, 1]$, its inverse (the arcsine function) is continuous. A similar argument applies to the arccosine function.

The trick of restricting to closed intervals, as in part (a), can be used to prove continuity for the other inverse trigonometric functions.

Example 6.2.2

The method of restricting to compact sets to argue the continuity of an inverse function, as in Example 6.2.1, is not applicable in every case. This example shows a continuous bijection from a space X onto a Hausdorff space Y whose inverse is *not* continuous.

Let $X = [0, 1)$ and let $Y = S^1$, the unit circle in $\mathbb{R}^2$. Then the function f: $X \to Y$ defined by

$$f(x) = (\cos 2\pi x, \sin 2\pi x), \quad x \in [0, 1),$$

is a continuous, one-to-one function from X onto Y whose inverse function is not continuous at the point $(1, 0)$ in S^1. Note that f maps 0 to $(1, 0)$ and wraps the interval $[0, 1)$ around S^1 in the counterclockwise direction. In Figure 6.1,

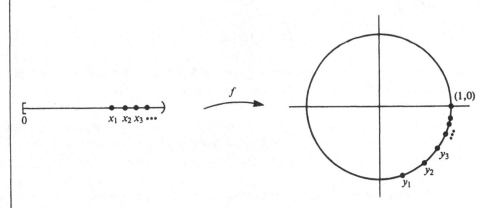

FIGURE 6.1

the sequence $\{y_n\}_{n=1}^{\infty}$ in Y converges to $(1, 0)$, but the corresponding sequence $\{x_n\}_{n=1}^{\infty}$, $x_n = f^{-1}(y_n)$, does not converge to $f^{-1}(1, 0) = 0$ in X.

The reader should know from calculus that a continuous function $f: [a, b] \to \mathbb{R}$ assumes maximum and minimum values on the bounded, closed interval $[a, b]$. In other words, there are points c and d in $[a, b]$ such that

$$f(c) \le f(x) \le f(d)$$

for all x in $[a, b]$. This property is a consequence of the next theorem.

Theorem 6.9:   *Let X be a compact space and $f: X \to \mathbb{R}$ a continuous real-valued function on X. Then there are members c and d of X such that, for all $x \in X$,*

$$f(c) \le f(x) \le f(d).$$

Proof: *Since X is compact and f is continuous, the first corollary to Theorem 6.6 insures that $f(X)$ is a compact subset of $\mathbb{R}$. Since $\mathbb{R}$ is Hausdorff, Theorem 6.4 shows that $f(X)$ is closed.*

Now $\{(-n, n)\}_{n=1}^{\infty}$ is an open cover of $f(X)$ and, since $f(X)$ is compact, this open cover has a finite subcover $\{(-n_i, n_i)\}_{i=1}^{N}$ for $f(X)$. If K is the largest value of n_i for $1 \le i \le N$, then $f(X)$ must be a subset of $(-K, K)$. From this we conclude that $f(X)$ is bounded.

Since $f(X)$ is bounded and closed, it contains its least upper bound U and greatest lower bound L. Thus there are members c and d of X such that

$$f(c) = L, \quad f(d) = U.$$

Then, for any $x \in X$,

$$f(c) \le f(x) \le f(d). \qquad \square$$

Corollary: *A continuous function $f: [a, b] \to \mathbb{R}$ whose domain is a closed and bounded interval assumes a maximum value and a minimum value.*

Definition: *Let (X, d) and (Y, d') be metric spaces and $f: X \to Y$ a function. Then f is **uniformly continuous** if for each positive number ϵ there is a positive number δ such that if x_1, x_2 are points of X for which $d(x_1, x_2) < \delta$, then $d'(f(x_1), f(x_2)) < \epsilon$.*

Note that every uniformly continuous function is continuous.

Theorem 6.10:   *Let (X, d) be a compact metric space, (Y, d') a metric space, and $f: X \to Y$ a continuous function. Then f is uniformly continuous.*

Proof: *Let ϵ be a positive number. Since f is continuous, there is for each x in X a positive number δ_x such that if $x' \in X$ and $d(x, x') < \delta_x$, then $d'(f(x), f(x')) < \epsilon/2$.*

The collection of open balls $\{B(x, \frac{1}{2}\delta_x) : x \in X\}$ is an open cover of X. Since X is compact, this open cover has a finite subcover $\{B(x_i, \frac{1}{2}\delta_{x_i})\}_{i=1}^N$. Let δ be the minimum value of $\frac{1}{2}\delta_{x_i}$, $i = 1, 2, \ldots, N$.

Consider any pair x, x' of points of X with $d(x, x') < \delta$. Then x belongs to $B(x_j, \frac{1}{2}\delta_{x_j})$ for some j with $1 \le j \le N$. Since

$$d(x, x') < \delta \le \frac{1}{2}\delta_{x_j},$$

then x' must belong to $B(x_j, \delta_{x_j})$. Thus

$$d(x_j, x) < \delta_{x_j}, \quad d(x_j, x') < \delta_{x_j},$$

so

$$d'(f(x_j), f(x)) < \epsilon/2, \quad d'(f(x_j), f(x')) < \epsilon/2.$$

Then

$$d'(f(x), f(x')) \le d'(f(x), f(x_j)) + d'(f(x_j), f(x')) < \epsilon/2 + \epsilon/2 = \epsilon.$$

Thus δ is a positive number such that $d'(f(x), f(x')) < \epsilon$ for all x, x' in X with $d(x, x') < \delta$, and f is uniformly continuous. $\qquad\square$

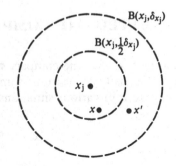

FIGURE 6.2

EXERCISE 6.2

1. Let arctan x denote the inverse of the function tan $x: (-\pi/2, \pi/2) \to \mathbb{R}$. Assume that tan x is continuous and prove that arctan x is continuous.

2. **Definition:** *A function $f: \mathbb{R} \to \mathbb{R}$ is a **strictly increasing function** provided that for all x, y in $\mathbb{R}$ with $x < y$, $f(x) < f(y)$.*
   Prove:

   (a) Every strictly increasing function is one-to-one.

   (b) Let $f: \mathbb{R} \to \mathbb{R}$ be a continuous, strictly increasing function from $\mathbb{R}$ onto $\mathbb{R}$. Then f^{-1} is also continuous and strictly increasing.

3. Let $f: [a, b] \to \mathbb{R}$ be a continuous function from a closed and bounded interval $[a, b]$ into $\mathbb{R}$. Show that $f([a, b])$ is a closed and bounded interval.

4. (a) Prove that every compact subset of a metric space X is closed and bounded.

   (b) Give an example of a metric space having a closed and bounded subset that is not compact. (*Hint:* A bounded metric space which is not compact will suffice. Consider Hilbert space for more sophisticated examples.)

5. Let $\mathcal{B}[a, b]$ denote the collection of all bounded functions from a closed interval $[a, b]$ into $\mathbb{R}$; i.e., $f \in \mathcal{B}[a, b]$ if and only if $f([a, b])$ is a bounded subset of $\mathbb{R}$. Assign $\mathcal{B}[a, b]$ the supremum metric ρ: for f, g in $\mathcal{B}[a, b]$,

   $$\rho(f, g) = \text{lub } \{|f(x) - g(x)|: x \in [a, b]\}.$$

   Prove the following facts about $(\mathcal{B}[a, b], \rho)$

   (a) $(\mathcal{B}[a, b], \rho)$ is a metric space which contains the space $(\mathcal{C}[a, b], \rho')$ of Example 3.1.5 as a subspace.

   (b) A sequence $\{f_n\}_{n=1}^{\infty}$ in $\mathcal{B}[a, b]$ converges to a member $f \in \mathcal{B}[a, b]$ with respect to the metric ρ if and only if $\{f_n\}_{n=1}^{\infty}$ converges to f uniformly. (For this reason, ρ is often called the *uniform metric* for $\mathcal{B}[a, b]$.)

   (c) $\mathcal{C}[a, b]$ is a closed subspace of $\mathcal{B}[a, b]$, but $\mathcal{C}[a, b]$ is not compact.

   (d) $\mathcal{C}[a, b]$ is nowhere dense in $\mathcal{B}[a, b]$.

6.3 PROPERTIES RELATED TO COMPACTNESS

It will be shown in this section that the compact subsets of $\mathbb{R}^n$ are precisely the closed and bounded sets. We shall also examine some other properties related to compactness and equivalent to it in various situations.

Lemma: *The unit n-cube $I^n = \{x = (x_1, \ldots, x_n) \in \mathbb{R}^n: 0 \le x_i \le 1$ for $i = 1, \ldots, n\}$ is a compact subspace of $\mathbb{R}^n$.*

Proof: *The case $n = 1$ is proved by the Heine-Borel Theorem (Theorem 2.12). The following argument, which is parallel to the proof given for the Heine-Borel Theorem, proves the lemma for $n = 2$. The analogous argument for $n > 2$ is left to the reader.*

For a square $[a, b] \times [c, d]$ in $\mathbb{R}^2$, we shall refer to

$$\left[a, \frac{a+b}{2}\right] \times \left[c, \frac{c+d}{2}\right], \quad \left[a, \frac{a+b}{2}\right] \times \left[\frac{c+d}{2}, d\right]$$

$$\left[\frac{a+b}{2}, b\right] \times \left[c, \frac{c+d}{2}\right], \quad \left[\frac{a+b}{2}, b\right] \times \left[\frac{c+d}{2}, d\right]$$

as the four "quarters" of the square.

(a, d) (b, d)

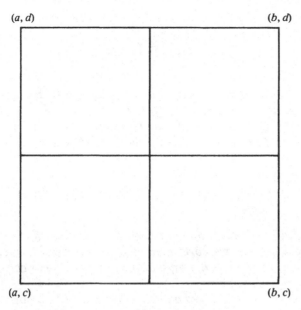

(a, c) (b, c)

FIGURE 6.3 **The four quarters of a square.**

Let $\mathcal{O}$ be an open cover of I^2 and, proceeding by contradiction, suppose that $\mathcal{O}$ has no finite subcover for I^2. Then there is at least one quarter

$$Q_1 = [a_1, b_1] \times [c_1, d_1]$$

of I^2 for which $\mathcal{O}$ has no infinite subcover. Note that the coordinate intervals $[a_1, b_1]$ and $[c_1, d_1]$ have length 1/2. Since Q_1 is not contained in the union of any finite subfamily of $\mathcal{O}$, then at least one of its quarters

$$Q_2 = [a_2, b_2] \times [c_2, d_2]$$

must have the same property. Note that $[a_2, b_2]$ and $[c_2, d_2]$ have length 1/4. Proceeding inductively, we define a nested sequence $\{Q_n\}_{n=1}^{\infty}$ of squares in $\mathbb{R}^2$,

$$Q_n = [a_n, b_n] \times [c_n, d_n] \, ,$$

whose coordinate intervals $\{[a_n, b_n]\}_{n=1}^{\infty}$ and $\{[c_n, d_n]\}_{n=1}^{\infty}$ are nested and of length

$$|a_n - b_n| = |c_n - d_n| = 1/2^n, \quad n \geq 1.$$

By Cantor's Nested Intervals Theorem (Theorem 2.11) there are real numbers

$$p \in \bigcap_{n=1}^{\infty} [a_n, b_n] \, , \quad q \in \bigcap_{n=1}^{\infty} [c_n, d_n] \, .$$

Then

$$(p, q) \in \bigcap_{n=1}^{\infty} Q_n,$$

so (p, q) must belong to some member O of the open cover 𝒪. But since the diameters of the squares Q_n approach zero, then $Q_n \subset O$ for some integer n. This contradicts the assumption that Q_n is not contained in the union of any finite number of members of 𝒪. Thus 𝒪 must have a finite subcover for I^2, so I^2 is compact. □

Theorem 6.11:   *A subset A of $\mathbb{R}^n$ is compact if and only if it is closed and bounded.*

Proof: *Suppose first that A is compact. Since $\mathbb{R}^n$ is Hausdorff, Theorem 6.4 insures that A is closed. Since the open cover $\{B(\theta, n)\}_{n=1}^{\infty}$ of open balls centered at the origin must have a finite subcover, then A must also be bounded. Thus each compact subset of $\mathbb{R}^n$ is closed and bounded.*

   Suppose now that A is closed and bounded. Let M be a positive number such that

$$\|x\| \leq M, \quad x \in A.$$

Let J^n be the cube in $\mathbb{R}^n$ each of whose coordinate subspaces is the interval $[-M, M]$:

$$J^n = \prod_{i=1}^{n} [a_i, b_i], \quad [a_i, b_i] = [-M, M], \quad i = 1, \ldots, n.$$

Then $A \subset J^n$. Since J^n is homeomorphic to the unit n-cube I^n, and since the lemma insures that I^n is compact, then J^n is compact. Thus A is a closed subset of a compact space and is itself compact by Theorem 6.3. □

Definition: *A topological space X is **countably compact** provided that every countable open cover of X has a finite subcover.*

The reader is left the following exercises:

(a)   Every compact space is countably compact.
(b)   Countable compactness is a topological property.
(c)   (A challenging problem) There are countably compact spaces that are not compact.

The next definition introduces a condition under which compactness and countable compactness are equivalent.

Definition: *A space X has the **Lindelöf property** or is a **Lindelöf space** if every open cover of X has a countable subcover.*

Theorem 6.12:   *If X is a Lindelöf space, then X is compact if and only if it is countably compact.*

Proof: *Since compactness always implies countable compactness, it is necessary to prove only the converse. Let $\mathcal{O}$ be an open cover of a Lindelöf space X. Then $\mathcal{O}$ has a countable subcover $\mathcal{O}'$ for X. Since X is countably compact, $\mathcal{O}'$ has a finite subcover $\mathcal{O}''$ for X. Then $\mathcal{O}''$ is a finite subcover for X derived from $\mathcal{O}$, so X is compact.*   □

Theorem 6.12 may appear to be merely a play on words. It appears that the Lindelöf property was designed expressly to make Theorem 6.12 obvious. The next theorem shows, however, that the Lindelöf property holds in an important class of spaces.

Theorem 6.13:   The Lindelöf Theorem   *Every second countable space is Lindelöf.*

Proof: *Let $\mathcal{B}$ be a countable basis for a second countable space X and $\mathcal{O}$ an open cover of X. For each x in X, let O_x be a member of $\mathcal{O}$ containing x and B_x a member of $\mathcal{B}$ such that*

$$x \in B_x \subset O_x.$$

Since $\mathcal{B}$ is a countable basis, the set $\{B_x : x \in X\}$ defined in this way is a countable open cover of X. For each B_x, let O'_x be a member of $\mathcal{O}$ such that

$$B_x \subset O'_x.$$

Then the collection $\mathcal{O}'$ of open sets O'_x so defined is a countable subcover for X derived from $\mathcal{O}$.   □

Since $\mathbb{R}^n$ is second countable (Theorem 4.8), Theorem 6.13 insures that $\mathbb{R}^n$ is Lindelöf, and Theorem 6.12 shows that the concepts of compactness and countable compactness coincide for subsets of $\mathbb{R}^n$: A subset of $\mathbb{R}^n$ is compact if and only if it is countably compact.

Definition: *A space X has the **Bolzano-Weierstrass property** provided that every infinite subset of X has a limit point.*

Note that a subspace A of a space X has the Bolzano-Weierstrass property means that every infinite subset of A has a limit point *in A*. The Bolzano-Weierstrass property is evidently a topological invariant.

Theorem 6.14:   *Every compact space has the Bolzano-Weierstrass property.*

Proof: *Suppose to the contrary that X is a compact space with an infinite subset B having no limit point. Then B is closed, and for each x in B there is an open set O_x in X containing x which contains no other member of B. Then $\mathcal{O} = \{O_x:$ $x \in B\}$ is an open cover of B, $\mathcal{O}$ is infinite because B is infinite, and $\mathcal{O}$ has no proper subcover for B because the members of $\mathcal{O}$ each contain only one member of B. Thus B is not compact, contradicting Theorem 6.3. This contradiction establishes the theorem.*   □

Example 6.3.1

(a)   According to both Theorem 2.13 and 6.14, each closed and bounded interval $[a, b]$ has the Bolzano-Weierstrass property.

(b)   An open interval (a, b) fails to have the Bolzano-Weierstrass property. Note than an infinite sequence converging to a or to b has no limit point in (a, b).

(c)   The real line does not have the Bolzano-Weierstrass property. The set of integers, for example, has no limit point.

(d)   Let S denote the unit sphere in Hilbert space:

$$S = \{x = (x_1, x_2, \ldots) \in H: \|x\| = 1\}.$$

Then S is bounded and closed but it does not have the Bolzano-Weierstrass Property. The set of points

$$P_1 = (1, 0, 0, \ldots), \quad P_2 = (0, 1, 0, 0, \ldots),$$

$$P_3 = (0, 0, 1, 0, 0, \ldots), \ldots,$$

where P_n has nth coordinate 1 and all other coordinates 0, has no limit point. (The distance from P_i to P_j is $\sqrt{2}$ if $i \neq j$.)

Theorem 6.15:   *A metric space is compact if and only if it has the Bolzano-Weierstrass property.*

Proof: *In view of Theorem 6.14, it is only necessary to prove that each metric space (X, d) with the Bolzano-Weierstrass property is compact. This proof will be* *given in a sequence of lemmas. We assume in each lemma that (X, d) satisfies the Bolzano-Weierstrass property.*

Lemma 1:   *Let $\mathcal{O}$ be an open cover of X. Then there is a positive number ϵ such that, for each x in X, the open ball $B(x, \epsilon)$ is a subset of some member of $\mathcal{O}$.*

Proof of Lemma 1:   *Assuming that the conclusion of the lemma is false, there must be for each positive integer n a point x_n in X for which $B(x_n, 1/n)$ is not a subset of any member of the open cover $\mathcal{O}$.*

   For each point y in X, y belongs to some member U_y of $\mathcal{O}$. Since U_y is open, there is a positive number ϵ_y such that $B(y, \epsilon_y) \subset U_y$. Thus x_n must be distinct from y whenever $1/n < \epsilon_y$. This indicates that a point of X cannot equal x_n for an infinite number of values of n and hence that $\{x_n\}_{n=1}^{\infty}$ is an infinite subset of X.

   By the Bolzano-Weierstrass property $\{x_n\}_{n=1}^{\infty}$ has a limit point a in X. Then $a \in O$ for some member O of $\mathcal{O}$, and there is a positive number δ such that $B(a, \delta) \subset O$. By the corollary to Theorem 3.9, $B(a, \delta/2)$ must contain infinitely many members of $\{x_n\}_{n=1}^{\infty}$. Thus $B(a, \delta/2)$ contains some x_n for which $1/n < \delta/2$. Then for z in $B(x_n, 1/n)$,

$$d(a, z) \leq d(a, x_n) + d(x_n, z) < \delta/2 + \delta/2 = \delta$$

so

$$B(x_n, 1/n) \subset B(a, \delta) \subset O,$$

contradicting the fact that $B(x_n, 1/n)$ is not a subset of any member of $\mathcal{O}$. This completes the proof of Lemma 1.

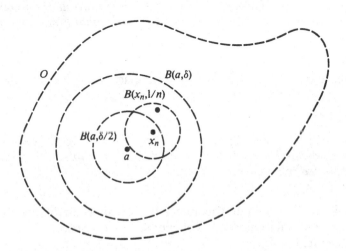

FIGURE 6.4

Lemma 2:   *For each positive number ϵ there is a finite subset A_ϵ of X such that each member of X is within distance ϵ of some member of A_ϵ.*

Proof of Lemma 2:   *Proceeding by contradiction, as usual, suppose that the conclusion of the lemma fails to hold for a particular positive number ϵ. Let $a_1 \in X$. By hypothesis, there is a point a_2 in X such that*

$$d(a_1, a_2) \geq \epsilon.$$

Suppose that a set $\{a_i\}_{i=1}^n$ has been defined for which

$$d(a_i, a_j) \geq \epsilon, \quad i \neq j.$$

Then there must be a point a_{n+1} in X for which

$$d(a_{n+1}, a_i) \geq \epsilon, \quad 1 \leq i \leq n.$$

Thus there is an infinite sequence $A = \{a_i\}_{i=1}^\infty$ of points of X such that the distance between any two members of A is at least ϵ. This set has no limit point, contradicting the Bolzano-Weierstrass property for X. This completes the proof of Lemma 2.

Turning now to the proof that X is compact, let $\mathcal{O}$ be an open cover of X. By Lemma 1 there is a positive number ϵ such that, for each x in X, the open ball $B(x, \epsilon)$ is contained in some member of $\mathcal{O}$. By Lemma 2, there is a finite subset $A_\epsilon = \{x_i\}_{i=1}^n$ of X for which $\{B(x_i, \epsilon)\}_{i=1}^n$ is an open cover of X. For each $i = 1, \ldots, n$, let O_i be a member of $\mathcal{O}$ which contains $B(x_i, \epsilon)$. Then $\{O_i\}_{i=1}^n$ is a finite subcover of X derived from $\mathcal{O}$, so X is compact. $\square$

Definition: *Let X be a metric space and ϵ a positive number. An **ϵ-net** for X is a finite subset A_ϵ of X such that each member of X is within distance ϵ of some member of A_ϵ. The metric space X is **totally bounded** provided that it has an ϵ-net for each positive number ϵ.*

Lemma 2 above states that every metric space satisfying the Bolzano-Weierstrass property, and hence every compact metric space, is totally bounded.

Definition: *Let X be a metric space and $\mathcal{O}$ an open cover of X. A **Lebesgue number** for $\mathcal{O}$ is a positive number ϵ with the property that every subset of X of diameter less than ϵ is contained in some member of $\mathcal{O}$.*

Lemma 1 in the proof of Theorem 6.15 essentially establishes the existence of Lebesgue numbers for open covers of compact metric spaces. This is formalized in the next theorem.

Theorem 6.16:   *If X is a compact metric space, then every open cover of X has a Lebesgue number.*

Proof: *Let $\mathcal{O}$ be an open cover of the compact metric space X. By Lemma 1 in the proof of Theorem 6.15, there is a positive number ϵ such that each open ball of radius ϵ is contained in some member of $\mathcal{O}$. Since each set of diameter less than ϵ is a subset of an open ball of radius ϵ, then ϵ is a Lebesgue number for $\mathcal{O}$.*   □

The information about compact subsets of $\mathbb{R}^n$ provided by Theorems 6.11, 6.12, and 6.15 is summarized as follows:

Theorem 6.17:   *For a subset A of $\mathbb{R}^n$, the following statements are equivalent:*

(1)   A is compact.

(2)   A has the Bolzano-Weierstrass property.

(3)   A is countably compact.

(4)   A is closed and bounded.

Example 6.3.2

Every compact subset of a metric space is closed and bounded, but a subset may be closed and bounded without being compact. According to Example 6.3.1, the unit sphere S in Hilbert space is closed and bounded but does not satisfy the Bolzano-Weierstrass property. Thus S is closed and bounded but not compact.

Problem 9(c) for this section establishes a criterion comparable to being closed and bounded which is equivalent to compactness in general metric spaces: *A metric space is compact if and only if it is complete and totally bounded.*

There are many interesting relationships among the properties of compactness, connectedness, path connectedness, and their corresponding local properties. One example that illustrates a famous characterization theorem is described here without proof. The interested reader is encouraged to pursue this topic in the Suggestions for Further Reading at the end of the chapter.

Definition: *A compact, connected, locally connected metric space is called a **Peano space** or a **Peano continuum**.*

For example, closed and bounded intervals in $\mathbb{R}$, closed squares and closed disks in $\mathbb{R}^2$, closed cubes and closed balls in $\mathbb{R}^3$, and their higher dimensional analogues in $\mathbb{R}^n$ are all Peano spaces. The remarkable fact about Peano spaces is that for any Peano space X, there is a continuous function from the closed unit interval

$I = [0, 1]$ onto X. Such a function was first discovered for the case of the unit square in the late nineteenth century by the Italian mathematician Guiseppe Peano and was called a "space-filling curve."

In fact, the properties of being a Hausdorff space and the image of the closed unit interval under a continuous function characterize the Peano spaces: In order that a topological space X be a Peano space it is necessary and sufficient that X be Hausdorff and that there exist a continuous function from I onto X. This celebrated result is called the Hahn-Mazurkiewicz Theorem. One of its surprising consequences is the existence of space-filling curves from I onto closed cubes and closed balls of any finite dimension.

EXERCISE 6.3

1. (a) Show that countable compactness, the Lindelöf property, and the Bolzano-Weierstrass property are topological invariants but are not hereditary.

   (b) Show that the three properties of part (a) are inherited by closed subspaces.

2. Give an example of a Lindelöf space that is not second countable.

3. (a) Prove that every uncountable subset of the real line has a limit point. (*Hint:* The union of a countable family of finite sets is countable.)

   (b) Prove that every uncountable subset of $\mathbb{R}^n$ has a limit point.

4. (a) Let (X, d) be a metric space. Prove that every subset of X of diameter less than ϵ is contained in an open ball of radius ϵ, $\epsilon > 0$.

   (b) Give an example to show that a set of diameter less than ϵ may not be contained in an open ball of radius $\epsilon/2$.

5. (a) Prove that a Hausdorff space X is countably compact if and only if it has the Bolzano-Weierstrass property.

   (b) Prove that a metric space is compact if and only if it is countably compact.

6. Prove that a subset A of $\mathbb{R}^n$ is compact if and only if every nested sequence $\{A_n\}_{n=1}^{\infty}$ of relatively closed, non-empty subsets of A has non-empty intersection.

7. Prove that if a metric space (X, d) has an ϵ-net for some positive number ϵ, then (X, d) is bounded. Conclude that every totally bounded metric space is bounded.

8. Prove that every compact metric space is totally bounded, separable, and second countable.

9. (a) Show that completeness of metric spaces is not a topological invariant.

   (b) Show that every compact metric space is complete.

   (c) Show that a metric space is compact if and only if it is complete and totally bounded. (*Hint:* The hard part is to prove that a complete and totally bounded metric space X is compact. This can be done as follows by showing that X has the Bolzano-Weierstrass property: Let A be an infinite subset of X. Since X is totally bounded, some infinite subset A_1 of A must be contained in an open ball of radius 1. Show that there is a nested sequence $\{A_n\}_{n=1}^{\infty}$ of infinite subsets of A such that A_n is contained in a ball of radius $1/n$. For each n, let $x_n \in A_n$. Then $\{x_n\}_{n=1}^{\infty}$ is a Cauchy sequence.)

10. Let $\{C_\alpha: \alpha \in I\}$ be a family of closed subsets of a compact metric space X such that $\bigcap_{\alpha \in I} C_\alpha = \varnothing$. Prove that there is a positive number ϵ such that every subset of X of diameter less than ϵ fails to intersect at least one member of $\{C_\alpha: \alpha \in I\}$.

11. Use a Lebesgue number argument to make a new proof for Theorem 6.10.

12. Let X denote the real line with the half-open interval topology of Example 4.3.4. Show that X is a Lindelöf space.

13. Let X be a metric space. Prove that second countability, separability, and the Lindelöf property are equivalent for X.

14. (a) Let X be a second countable space. Prove that X is separable and Lindelöf.

    (b) Give an example of a separable Hausdorff space that is not second countable.

15. Prove:

    (a) Every compact space is countably compact.

    (b) Countable compactness is a topological property.

    (c) (A challenging problem) There are countably compact spaces that are not compact.

6.4 ONE-POINT COMPACTIFICATION

In this section we consider one construction for answering the question "When can a topological space be considered to be a subspace of a compact topological space?" We shall see that the question can always be answered affirmatively by adding one additional point. It will be noted that this construction is of little value, however, unless the given space satisfies a local compactness condition.

Definition: *A space X is **locally compact at a point** x in X provided that there is an open set U containing x for which $\bar{U}$ is compact. A space is **locally compact** provided that it is locally compact at each of its points.*

Local compactness is a topological property. If X is compact, then X itself is an open set with compact closure. Thus every compact space is locally compact. It should be clear that local compactness does not imply compactness.

Local compactness and local connectedness are modifications of global properties to local ones. Note, however, that a locally compact space need only have for each point p at least one open set containing p whose closure is compact, but a locally connected space must have for each point p an entire local basis of connected open sets. It is left as an exercise for the reader to show that this difference is only superficial by proving that each point p in a locally compact space has a local basis of open sets with compact closures.

Example 6.4.1

(a) $\mathbb{R}^n$ is locally compact since the open ball $B(x, r)$ is an open set containing X whose closure $B[x, r]$ is compact.

(b)   Hilbert space H is not locally compact. The following argument shows that H is not locally compact at the origin θ, but it adapts easily to any point.

   Suppose H is locally compact at θ and let U be an open set containing θ whose closure $\bar{U}$ is compact. Let r be a positive number such that $B(\theta, r) \subset U$. Then

$$\overline{B(\theta, r)} = B[\theta, r] \subset \bar{U}$$

and $B[\theta, r]$ is compact since it is a closed subset of the compact set $\bar{U}$. However, the set $A = \{p_i\}_{i=1}^{\infty}$ of points p_i having ith coordinate r and all other coordinates 0 is an infinite subset of $B[\theta, r]$ with no limit point ($d(p_i, p_j) = \sqrt{2}\, r$ for $i \neq j$). In view of the fact that compactness is equivalent to the Bolzano-Weierstrass property in metric spaces (Theorem 6.15), we must conclude that $B[\theta, r]$ is not compact. Thus $\bar{U}$ is not compact and H is not locally compact at the origin.

Definition: *Let X be a topological space and ∞, called the **point at infinity**, an object not in X. Let $X_\infty = X \cup \{\infty\}$ and define a topology $\mathcal{T}_\infty$ on X_∞ by specifying the following open sets: (a) the open sets of X, considered as subsets of X_∞; (b) the subsets of X_∞ whose complements are closed, compact subsets of X; and (c) the set X_∞. The space $(X_\infty, \mathcal{T}_\infty)$ is called the **one-point compactification** of X.*

The proof that $\mathcal{T}_\infty$ is indeed a topology for the set X_∞ of the preceding definition is left as an exercise.

Theorem 6.18:   *Let $(X, \mathcal{T})$ be a space and $(X_\infty, \mathcal{T}_\infty)$ its one-point compactification. Then*

(a) $(X_\infty, \mathcal{T}_\infty)$ is compact;
(b) $(X, \mathcal{T})$ is a subspace of $(X_\infty, \mathcal{T}_\infty)$;

(c) X_∞ is Hausdorff if and only if X is Hausdorff and locally compact;

(d) X is a dense subset of X_∞ if and only if X is not compact.

Proof:

(a)   Any open cover $\mathcal{O}$ of X_∞ must have a member U which contains ∞. Since the complement $X_\infty \backslash U$ is required to be compact, it has a finite subcover $\{O_i\}_{i=1}^n$ derived from $\mathcal{O}$. Then the subcollection consisting of U and O_1, ..., O_n is a finite subcover for X_∞ derived from $\mathcal{O}$. Thus X_∞ is compact.

(b)   The fact that the relative topology for X as a subspace of $(X_\infty, \mathcal{T}_\infty)$ is $\mathcal{T}$, the original topology for X, is implicit in the definition of $\mathcal{T}_\infty$; the details are left to the reader. Note that an open set U containing ∞ must be one of two types: (1) $U = X_\infty$, in which case $U \cap X = X$ is open in $\mathcal{T}$; (2) U is a subset of X_∞ for which $X_\infty \backslash U$ is a closed, compact subset of X. In the latter case, $U \cap X$ is an open set in $\mathcal{T}$ since its complement $X_\infty \backslash U$ is closed.

(c)   Suppose first that X_∞ is Hausdorff. Then X is Hausdorff since this property is hereditary. To see that X is locally compact, let $a \in X$. There exist disjoint open sets U and V in X_∞ such that

$$\infty \in U, \quad a \in V.$$

Thus

$$V \subset X_\infty \backslash U$$

and the latter set is closed and compact in X. Hence

$$\bar{V} \subset X_\infty \backslash U$$

so V is compact since it is a closed subset of a compact set. Thus X is locally compact at a.

Suppose that X is Hausdorff and locally compact. To show that X_∞ is Hausdorff, we need only show that ∞ and an arbitrary point $a \in X$ have disjoint open neighborhoods. Since X is locally compact, there is an open set O in X containing a such that $\bar{O}$ is compact. Then O and $X_\infty \backslash \bar{O}$ are disjoint open sets in X containing a and ∞, respectively.

The proof of (d) is left as an exercise, with the following hint. If X is compact, then $\{\infty\} = X_\infty \backslash X$ is open in X. Furthermore, if X is not dense in X_∞, then $\{\infty\}$ must be an open set.   □

If X fails to be a locally compact Hausdorff space, then X_∞ is not Hausdorff. Since non-Hausdorff spaces are of limited interest, many texts define the one-point compactification only for locally compact Hausdorff spaces X.

Example 6.4.2

Consider the open unit interval $X = (0, 1)$ and adjoin an additional point ∞ to form the set X_∞. Open sets which do not contain ∞ are the usual open sets in $(0, 1)$. Open sets containing ∞ are X_∞ itself and sets U whose complements are closed, compact subsets of $(0, 1)$. An example of such an open set U is shown in Figure 6.5.

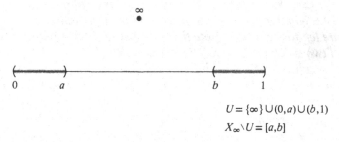

$$U = \{\infty\} \cup (0,a) \cup (b,1)$$
$$X_\infty \backslash U = [a,b]$$

FIGURE 6.5

Thus (X_∞, T_∞) is a compact space. It is an easy exercise to show that the function

$$f: X_\infty \rightarrow S^1$$

from X_∞ to the unit circle S^1 in $\mathbb{R}^2$ defined by

$$f(t) = \begin{cases} (\cos 2\pi t, \sin 2\pi t) & \text{if } 0 < t < 1 \\ (1, 0) & \text{if } t = \infty \end{cases}$$

is a one-to-one continuous function from X_∞ onto S^1. By Theorem 6.8, f is a homeomorphism and the one-point compactification X_∞ of the open interval $(0, 1)$ is topologically equivalent to the unit circle.

Example 6.4.3

(a) The one-point compactification $\mathbb{R}_\infty$ of the real line $\mathbb{R}$ is (homeomorphic to) a circle. Actually, this follows from Example 6.4.2 since $\mathbb{R}$ is homeomorphic to $(0, 1)$, but the following description gives an interesting method of visualizing $\mathbb{R}_\infty$. Note that the compactness of $\mathbb{R}_\infty$ is guaranteed by Theorem 6.18.

Consider $\mathbb{R}$ with a circle C tangent at the origin, as in the following figure.

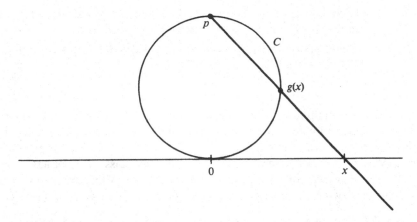

FIGURE 6.6

Define a function $g: \mathbb{R} \to C$ as follows: For $x \in \mathbb{R}$, $g(x)$ is the point of intersection of the line segment from x to the "north pole" p of C with the circle. Note that g is one-to-one and that $g(\mathbb{R}) = C \setminus \{p\}$. We extend g to a function from $\mathbb{R}_\infty$ to C be defining $g(\infty) = p$. Then g is a continuous bijection from $\mathbb{R}_\infty$ onto C. By Theorem 6.8, f is a homeomorphism, so $\mathbb{R}_\infty$ is homeomorphic to C.

(b) The one-point compactification $\mathbb{R}^2_\infty$ of $\mathbb{R}^2$ is a two-dimensional sphere $S^2 = \{x = (x_1, x_2, x_3) \in \mathbb{R}^3 : \|x\| = 1\}$. The details of this example are left to the reader with the suggestive picture in Figure 6.7.

The method of defining the functions g of parts (a) and (b) is

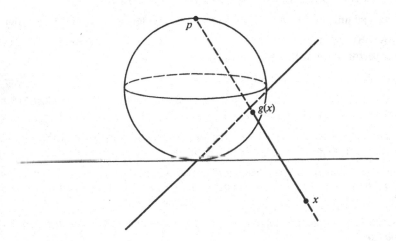

FIGURE 6.7

called *stereographic projection*. The point x is said to be the stereographic projection of the corresponding point $g(x)$.

(c)   The one-point compactification $\mathbb{R}^n_\infty$ of $\mathbb{R}^n$ is the n-dimensional sphere $S^n = \{x = (x_1, \ldots, x_{n+1}) \in \mathbb{R}^{n+1}: \|x\| = 1\}$.

Forming the one-point compactification X_∞ of a space X is the simplest method of embedding a locally compact Hausdorff space in a compact Hausdorff space. We shall return to this problem in Chapter 8 with the Stone-Čech compactification which applies to a more general class of spaces, the class of completely regular spaces. The Stone-Čech compactification $\beta(X)$ of a space X has the following properties, the second of which makes it more useful than the one-point compactification:

(1) $\beta(X)$ is a compact Hausdorff space in which X is a dense subspace.

(2)   Every bounded continuous function $f: X \to \mathbb{R}$ from X to the real line $\mathbb{R}$ can be uniquely extended to a continuous function $f^*: \beta(X) \to \mathbb{R}$.

Several other aspects of the problem of extending continuous functions will also be addressed in Chapter 8.

EXERCISE 6.4

1.   Prove that local compactness is a topological property.

2.   Let X be a space and x a point of X at which X is locally compact. Prove that there is a local basis $\mathcal{B}$ at x such that $\bar{B}$ is compact for each $B \in \mathcal{B}$.

3.   Show that Hilbert space is not locally compact at any point.

4.   Is the real line with the finite complement topology locally compact? Prove your answer.

5.   Prove that the family of sets $\mathcal{T}_\infty$, which was claimed in the definition of one-point compactification to be a topology, really is a topology.

6.   Complete parts (b) and (c) of Example 6.4.3.

7.   If X is a Hausdorff space, show that the requirement that $X_\infty \setminus U$ be closed in X can be omitted in the definition of the topology for X_∞.

8.   Prove that a space X is locally compact if and only if for each x in X there is a subspace A of X such that $\bar{A}$ is compact and $x \in \text{int } A$.

9.   Let X be a locally compact Hausdorff space, A a closed subset of X, and p a point not in A. Prove that there are disjoint open sets U and V in X such that $p \in U$ and $A \subset V$. (*Hint:* Consider the one-point compactification of X.)

10.   **Definition:**   *A point p in a space X is an **isolated point** provided that $\{p\}$ is an open set.*

Prove that a space X is compact if and only if ∞ is an isolated point of X_∞.

11. Prove that a space X is a dense subset of its one-point compactification if and only if X is not compact.

12. (a) Let X be a Hausdorff space. Show that X is locally compact if and only if each point of X has a compact neighborhood; i.e., X is locally compact if and only if each point x of X belongs to the interior of a compact set K_x.

     (b) Give an example of a space X for which each point has a compact neighborhood, but X is not locally compact. (*Hint:* By (a), X cannot be a Hausdorff space.)

6.5 THE CANTOR SET

Of all the subsets of the real line, the Cantor set is probably the most fertile source of examples and counterexamples in topology. In this section we define this remarkable set and examine some of its properties.

Definition: *The **Cantor set** is the subset of $I = [0, 1]$ defined as follows: Let*

$$F_1 = [0, 1/3] \cup [2/3, 1]$$

be the subset of $[0, 1]$ formed by removing the open middle third $(1/3, 2/3)$. Let

$$F_2 = [0, 1/9] \cup [2/9, 1/3] \cup [2/3, 7/9] \cup [8/9, 1]$$

be the subset of F_1 formed by removing the open middle thirds $(1/9, 2/9)$ and $(7/9, 8/9)$ of the two components of F_1. Continuing in this manner, let F_{n+1} be the subset of F_n obtained by removing the open middle third of each of the components of F_n. Then the set

$$C = \bigcap_{n=1}^{\infty} F_n$$

*is the **Cantor set**.*

Note that the Cantor set is a closed subset of $\mathbb{R}$ since it is the intersection of closed sets.

For an alternative definition of the Cantor set, recall that a *ternary expansion* of a real number x, $0 \le x \le 1$, is an expression $x = 0.x_1 x_2 x_3 \ldots$ representing x as a sum of powers of 3,

$$x = \sum_{n=1}^{\infty} x_n/3^n,$$

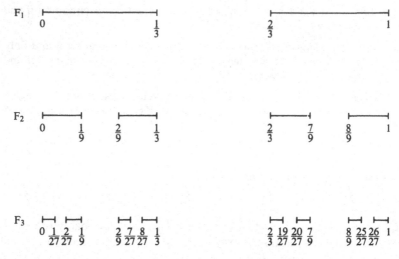

FIGURE 6.8   The first three stages in the construction of the Cantor set.

where x_n is restricted to the values 0, 1, and 2. Thus 1/3 has the two ternary expansions

$$1/3 = 1/3 + 0/3^2 + 0/3^3 + 0/3^4 + \cdots,$$

and

$$1/3 = 0/3 + 2/3^2 + 2/3^3 + 2/3^4 + \cdots,$$

and 1/9 can be similarly expressed as

$$1/9 = 0/3 + 1/3^2 + 0/3^3 + 0/3^4 + \cdots$$

and

$$1/9 = 0/3 + 0/3^2 + 2/3^3 + 2/3^4 + \cdots.$$

Observe that the numbers which absolutely require 1 in the first place of their ternary expansions are the numbers strictly between 1/3 and 2/3. As shown above, 1/3 has a ternary expansion without 1 in the first place. Thus, F_1 excludes all members of [0, 1] which require 1 in the first place of the ternary expansion. Similarly, F_2 excludes those members of [0, 1] which require 1 in the second place. In general, F_n excludes those members of [0, 1] which require 1 in the nth place of their ternary expansions. Thus the Cantor set is the set of real numbers x which have a ternary expansion $x = 0.x_1x_2x_3\ldots$ where x_n is restricted to the values 0 and 2.

Example 6.5.1

The Cantor set contains the end points 0, 1, 1/3, 2/3, 1/9, 2/9, 7/9, 8/9, $\cdots$ of the open intervals deleted to define the sets $\{F_n\}_{n=1}^{\infty}$. However, these are not the only members of the Cantor set. In particular, the number 1/4 belongs to C. To see this consider the convergent series

$$x = \sum_{n=1}^{\infty} 2/3^{2n} = 2/3^2 + 2/3^4 + 2/3^6 + \cdots$$

By factoring out $1/3^2$ from each term, it follows that x satisfies the equation

$$x = 1/9 \, (2 + x).$$

Solving gives $x = 1/4$. Thus 1/4 is a member of C since it has a ternary expansion which does not require 1 in any term.

Actually, the Cantor set is uncountable. The proof of this fact, which follows from the ternary expansion representation, is left as an exercise.

Example 6.5.2

In the construction of the Cantor set from the intersection of the sets F_1, F_2, $F_3, \ldots,$ the open intervals removed were of length

$$L = 1/3 + 2/9 + 4/27 + 8/81 + \cdots + 2^{n-1}/3^n + \cdots.$$

The sum of this geometric series, which represents the total length of the intervals removed, can be calculated by several different methods. First, there is the formula

$$1 + r + r^2 + \cdots + r^n + \cdots = \frac{1}{1-r}, \quad |r| < 1.$$

This gives

$$L = 1/3(1 + 2/3 + 4/9 + \cdots + 2^n/3^n + \cdots) = 1/3 \left(\frac{1}{1 - 2/3}\right) = 1.$$

For a second method, note that

$$L = 2/3(1/2 + 1/3 + 2/9 + \cdots + 2^{n-1}/3^n + \cdots) = 2/3 \, (1/2 + L)$$

so that

$$L = 1/3 + (2/3)L,$$

which again gives $L = 1$.

> By either method, we conclude that the sum of the lengths of the intervals removed in the construction of C is 1, the total length of [0, 1]! Thus the Cantor set is revealed to have the following rather bizarre property: C is a closed, uncountable subset of [0, 1] which contains no proper interval. In other words, the Cantor set is an uncountable and nowhere dense subset of $\mathbb{R}$.

Note   Students of analysis may observe that the Cantor set C, as defined here, has measure 0. The measure of a set is not a topological property, however. There are other representations of the Cantor set in which the measure is positive.

Definition: *A closed subset A of a topological space X is **perfect** provided that every point of A is a limit point of A.*

Theorem 6.19:   *The Cantor set is a compact, perfect, totally disconnected metric space.*

Proof: *Since C is closed and bounded, Theorem 6.17 assures compactness. The Cantor set is a metric space since it is a subspace of the real line. The fact that C is totally disconnected is an easy consequence of the fact that it contains no proper intervals; since the connected subsets of $\mathbb{R}$ are precisely the intervals, this means that the components of C consist of single points.*

It remains to be proved that C is perfect. Let x be a member of C and ϵ a positive number. Let N be a positive integer for which $2/3^N < \epsilon$. Since $x = 0.x_1x_2x_3\ldots$ has a ternary expansion where each x_n is 0 or 2, we let $y = 0.y_1y_2y_3\ldots$ be the real number having the indicated ternary expansion with $y_n = x_n$ for $n \neq N$ and y_N differing from x_N as follows: y_N is 0 if x_N is 2, and y_N is 2 if x_N is 0. Then y is a member of C, and

$$|x - y| = 2/3^N < \epsilon.$$

Thus y is a member of C within distance ϵ of x, so x is a limit point of C and C is a perfect set. $\square$

There is an important extension of the preceding theorem that illustrates a case in which the classification problem has been solved: Not only is the Cantor set a compact, perfect, totally disconnected metric space, but any topological space with these four properties is homeomorphic to the Cantor set. Thus any two compact, perfect, totally disconnected metric spaces are homeomorphic, and the Cantor set may be considered the prototype of this genre. In order to move on to other subjects, the proof of this classification theorem is omitted. Further details can be found in the suggested reading list for this chapter.

EXERCISE 6.5

1. Show that 1/36 is a member of the Cantor set.

2. Show that each set F_n defined in the construction of the Cantor set has 2^n components.

3. (a) If x is a member of $[0, 1)$ with a binary expansion

$$x = \sum_{n=1}^{\infty} a_n/2^n, \quad a_n \in \{0, 1\},$$

show that the ternary expansion

$$\sum_{n=1}^{\infty} 2a_n/3^n$$

represents a point of the Cantor set C.

   (b) Use (a) to define a one-to-one function from $[0, 1)$ into C. Take into account the fact that some members of $[0, 1)$ have more than one binary expansion.

   (c) Conclude that C is uncountable.

4. Example 6.5.2 showed that the Cantor set is nowhere dense in $\mathbb{R}$. Conclude, however, that as a metric space in its own right, with $d(a, b) = |a - b|$ for a, b in C, the Cantor set is of the second category.

5. Prove that a subspace A of a space X is perfect if and only if A is closed and has no isolated points.

6. (a) Prove that every perfect subset of $[0, 1]$ is uncountable. (*Hint:* Assume to the contrary that $\{a_n\}_{n=1}^{\infty}$ is a countable, closed, perfect subset of $[0, 1]$. Show that there is a nested sequence $\{V_n\}_{n=1}^{\infty}$ of non-empty open subsets of $[0, 1]$ such that $\bar{V}_n$ does not contain a_n. Consider $\cap_{n=1}^{\infty} \bar{V}_n$.)

   (b) Use part (a) to give a new proof that every non-degenerate interval in $\mathbb{R}$ is uncountable.

7. Prove that every perfect, compact Hausdorff space is uncountable.

8. Let $a = .a_1a_2a_3\ldots$ and $b = .b_1b_2b_3\ldots$ be points of the Cantor set whose indicated ternary expansions into 0's and 2's agree for the first N terms and disagree at the next term: $a_n = b_n$, $1 \le n \le N$, and $a_{N+1} \ne b_{N+1}$. How close can a and b be?

9. Define a function $f: C \to [0, 1]$ from the Cantor set to $[0, 1]$ as follows: For x in C, consider the ternary expansion

$$x = \sum_{n=1}^{\infty} x_n/3^n, \quad x_n \in \{0, 2\}.$$

Define

$$F(x) = \sum_{n=1}^{\infty} x_n/2^{n+1}.$$

(a) Show that f is a non-decreasing function (i.e., prove that if $x \le y$, then $f(x) \le f(y)$).

(b) Show that f maps C onto $[0, 1]$. Conclude from this that C is uncountable.

(c) Show that

$$f(1/3) = f(2/3), \quad f(1/9) = f(2/9), \quad f(7/9) = f(8/9).$$

(d) Show, in general, that $f(a_n) = f(b_n)$ where a_n, b_n are endpoints of one of the middle third intervals deleted from F_n in the construction of C.

(e) Define an extension F: $[0, 1] \rightarrow [0, 1]$ of f as follows: If x belongs to an interval (a_n, b_n) of part (d), then $f(a_n) = f(b_n)$. Define $F(x)$ to have this common value. Thus, we extend f to $[0, 1]$ by defining the extension to be constant on the intervals deleted to form C. The function F is called the *Cantor function*.

Show that the Cantor function is continuous and sketch its graph.

SUGGESTIONS FOR FURTHER READING

Most textbooks on point-set and general topology include compactness and related properties. For a readable account somewhat more advanced than that in this text, *General Topology* by Kelley and *Topology: A First Course* by Munkres are recommended.

For a proof that every compact, perfect, totally disconnected, metric space is homeomorphic to the Cantor set, see *Topology* by Hocking and Young or *General Topology* by Willard. These texts also contain accessible proofs of the Hahn-Mazurkiewicz Theorem.

HISTORICAL NOTES FOR CHAPTER 6

The term *compact* was introduced in 1904 by Maurice Fréchet to describe those spaces in which every sequence has a convergent subsequence. Such spaces are now called *sequentially compact*. Sequential compactness is equivalent to compactness in metric spaces.

The property of compactness has a long and complicated history. Basically, the purpose of defining compactness was to generalize the properties of closed and bounded intervals to general topological spaces. The early attempts to achieve this goal included sequential compactness, countable compactness, the Bolzano-Weierstrass property, and, finally, the modern property of compactness, which was introduced by P. S. Alexandroff and Paul Urysohn in 1923.

The first theorem on compactness was the Heine-Borel Theorem (Theorem 2.12), which states that a closed and bounded interval $[a, b]$ is compact. Actually, Emile Borel proved in his doctoral thesis in 1894 that every countable open cover of $[a, b]$ has a finite subcover; in other words, that $[a, b]$ is countably compact. The extension to arbitrary open covers was made possible by the work of Ernst Lindelöf (1870–1946), who showed that every open cover of $[a, b]$ has a countable subcover. Eduard Heine (1821–1881), whose name appears in the Heine-Borel Theorem, was not involved in its discovery. Heine's primary mathematical contribution was to prove in 1872 that every continuous real-valued function defined on $[a, b]$ is uniformly continuous. A. M. Schoenflies (1858–1923), in reading Heine's proof, noted a relation to Borel's theorem and gave Borel's result its present name, the Heine-Borel Theorem. It is doubtful that Heine would have claimed any credit for the famous theorem named in his honor.

The Heine-Borel Theorem was easily extended from closed and bounded intervals to closed and bounded subsets of $\mathbb{R}$. W. H. Young (1863–1942) extended the theorem to $\mathbb{R}^2$ in 1902 by proving that every open cover of a closed and bounded subset of $\mathbb{R}^2$ has a finite subcover. Henri Lebesgue (1875–1941) published the same result in 1904 and claimed to have known the extension to $\mathbb{R}^n$ as early as 1898.

Consideration of compactness via the finite intersection property is due to Frigyes Riesz in 1908. Felix Hausdorff showed in *Grundzüge der Mengenlehre* in 1914 that sequential compactness, countable compactness, the Bolzano-Weierstrass property, and compactness are all equivalent in metric spaces. The equivalence of sequential compactness and compactness in the metric case was shown earlier by Fréchet.

The crucial step in the characterization of compact subsets of $\mathbb{R}^n$ (Theorem 6.11) can be traced to Weierstrass, who proved that closed and bounded subsets of $\mathbb{R}^2$ have the Bolzano-Weierstrass property.

Lindelöf spaces were first considered by Ernst Lindelöf, who proved the Lindelöf Theorem (Theorem 6.13) and that every subspace of $\mathbb{R}^n$ has the Lindelöf property in 1903. The term *Lindelöf space* was coined by K. Kuratowski and W. Sierpiński, who initiated the formal study of these spaces in 1921.

The prototype of Theorem 6.10 on uniform continuity is, of course, Heine's Theorem of 1872 which showed that every real-valued continuous function with domain a closed and bounded interval is uniformly continuous. Camille Jordan proved in 1893 the precursors of Theorems 6.6 and 6.8 for compact subsets of $\mathbb{R}^n$.

As mentioned earlier, the definition of compactness in use today was proposed in 1923 by the Russian mathematicians P. S. Alexandroff and Paul Urysohn. They called the property *bicompactness* and developed its properties in a series of papers published in 1923, 1924, and 1929. Their results included Theorems 6.3, 6.4, and 6.5. Alexandroff proved Theorems 6.6, 6.7, and 6.8 in 1927. An equivalent definition of compactness was given in 1921 by Leopold Vietoris, whose results included Theorems 6.1, 6.3, and 6.4.

Local compactness was introduced independently by Heinrich Tietze and Alexandroff. The one-point compactification and Theorem 6.18 are due to Alexandroff. The Stone-Čech compactification was developed in 1937 independently by Eduard Čech and M. H. Stone.

Of the other ideas of Chapter 6, Lebesgue numbers were first used by Henri Lebesgue, and total boundedness was defined by Hausdorff in *Grundezüge der Mengenlehre*. The Cantor set was studied by Cantor in 1883, and the extension of Cantor's Nested Intervals Theorem (Theorem 2.11) to Cantor's Theorem of Deduction (Theorem 6.2) was also made by Cantor. As was noted in the text, space-filling curves were first considered by Guiseppe Peano, and Peano spaces are named in his honor. The Hahn-Mazurkiewicz theorem is due to Hans Hahn and Stephan Mazurkiewicz.

7
Product and Quotient Spaces

This chapter will make precise two ideas that were mentioned earlier. The first, the product of topological spaces, was considered for metric spaces in Section 3.6. We shall define the product of topological spaces in such a way that the product of metric spaces can be considered a special case and then examine continuity and topological invariants in product spaces. Products of finite collections of spaces will be considered before arbitrary products.

The second principal idea of this chapter is quotient or identification spaces. In Chapter 1 the Möbius strip was defined by identifying or "gluing together" the ends of a rectangular strip after giving the strip a half twist. This method of identifying points to form a new space is the basis of the quotient construction.

7.1 FINITE PRODUCTS

Example 7.1.1

The Euclidean plane $\mathbb{R}^2$, considered as a set, is the Cartesian product $\mathbb{R} \times \mathbb{R}$ of the real line with itself. This example shows a standard method of defining a topology for $\mathbb{R} \times \mathbb{R}$ from the topology of $\mathbb{R}$, without recourse to metrics, which produces the usual topology of $\mathbb{R}^2$. This construction is generalized in the remainder of the section to arbitrary finite products of topological spaces.

Consider the collection $\mathcal{B}$ of all subsets of the form $O_1 \times O_2$, where O_1 and O_2 are open in $\mathbb{R}$. If O_1 and O_2 are open intervals, then $O_1 \times O_2$ is an open rectangle, as shown in Figure 7.1.

In general, O_1 and O_2 are unions of open intervals, so $O_1 \times O_2$ is a union of open rectangles. The set $\mathcal{B}$ is not a topology because the union of two members of $\mathcal{B}$ may fail to be a member of $\mathcal{B}$. In Figure 7.2, for example, the set $(O_1 \times O_2) \cup (O_3 \times O_4)$ cannot be expressed as the product of two open sets.

Although $\mathcal{B}$ is not a topology, it is a basis for a topology, as will now be demonstrated by showing that $\mathcal{B}$ satisfies the requirements of Theorem 4.8. First, it should clear that the union of the members of $\mathcal{B}$ is $\mathbb{R}^2$. For the second condition, let

$$B_1 = O_1 \times O_2, \quad B_2 = U_1 \times U_2$$

be members of $\mathcal{B}$. Since

$$B_1 \cap B_2 = (O_1 \times O_2) \cap (U_1 \times U_2) = (O_1 \cap U_1) \times (O_2 \cap U_2),$$

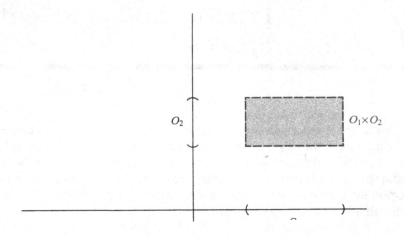

FIGURE 7.1

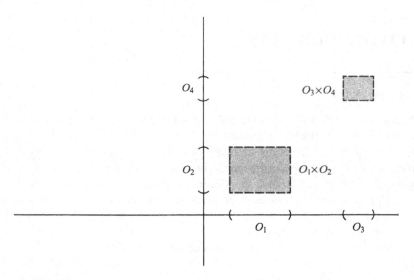

FIGURE 7.2

then $B_1 \cap B_2$ is actually a member of $\mathcal{B}$. This property is stronger than what is required by Theorem 4.8. At any rate, $\mathcal{B}$ is a basis for a topology for $\mathbb{R}^2$. The topology determined by $\mathcal{B}$ is the *product topology* for $\mathbb{R}^2$.

The product topology for $\mathbb{R}^2$ consists of all sets which can be expressed as unions of open rectangles. The usual topology for $\mathbb{R}^2$ consists of all sets which can be expressed as unions of open balls, in this case open disks. Since any open rectangle can be expressed as a union of open balls and any open ball can be expressed as a union of open rectangles, then the product topology for $\mathbb{R}^2$ is precisely the usual topology.

Definition: *Let $(X_1, \mathcal{T}_1), (X_2, \mathcal{T}_2), \ldots, (X_n, \mathcal{T}_n)$ be a finite collection of non-empty topological spaces, and let X denote the Cartesian product*

$$X = \prod_{i=1}^{n} X_i = X_1 \times X_2 \times \cdots \times X_n.$$

Let $\mathcal{B}$ be the family of all subsets of X of the form

$$O = \prod_{i=1}^{n} O_i = O_1 \times O_2 \times \cdots \times O_n$$

*where each set O_i is an open set in the topology $\mathcal{T}_i$ for X_i. Then $\mathcal{B}$ is a basis for a topology for X. This topology is the **product topology,** and the set X with the product topology is a **product space.** The spaces $X_1, X_2, \ldots, X_n$ are called the **coordinate spaces** or **factor spaces** of X.*

Since each point x in X is of the form

$$x = (x_1, x_2, \ldots, x_n), \quad x_i \in X_i, \quad 1 \le i \le n,$$

*there is a function $p_i: X \to X_i$ defined by $p_i(x_1, x_2, \ldots, x_n) = x_i$. The function $p_i: X \to X_i, 1 \le i \le n$, is called the **projection map on the ith coordinate space** or the **ith projection map.***

Lemma: *The projection maps $p_i: X \to X_i$ from a product space $X = X_1 \times X_2 \times \cdots \times X_n$ to the coordinates spaces are continuous.*

Proof: *For an example to indicate the method of proof, consider the first coordinate map $p_1: X \to X_1$ and let O_1 be an open set in X_1. Then*

$$p_1^{-1}(O_1) = \{x = (x_1, x_2, \ldots, x_n) \in X: x_1 \in O_1\} = O_1 \times X_2 \times \cdots \times X_n$$

is the product of O_1 with the coordinate spaces $X_2, \ldots, X_n$. Thus $p_1^{-1}(O_1)$ is open in X, so p_1 is continuous.

For an arbitrary coordinate map $p_i: X \to X_i$ and an open set O_i in X_i, $p_i^{-1}(O_i)$ consists of all points $x = (x_1, x_2, \ldots, x_n)$ of X for which the ith coordinate x_i is a member of O_i. In other words, the values of the coordinates of x are unrestricted except for x_i, and x_i is required to be in O_i. This means that

$$p_i^{-1}(O_i) = A_1 \times A_2 \times \cdots \times A_n$$

where $A_i = O_i$ and $A_j = X_j$ when $i \neq j$. Thus $p_i^{-1}(O_i)$ is a product of open sets and is therefore open in X. Thus each coordinate map is continuous. $\qquad\square$

In the notation of the preceding lemma, note that for a basic open set $\prod_{i=1}^n O_i$ in X,

$$\prod_{i=1}^n O_i = \bigcap_{i=1}^n p_i^{-1}(O_i).$$

This means that each basic open set in the defining basis for the product topology is a finite intersection of sets of the form $p_i^{-1}(O_i)$, where O_i is open in X_i. In other words, the collection

$$\mathscr{S} = \{p_i^{-1}(O_i) : O_i \text{ is open in } X_i, \quad 1 \leq i \leq n\}$$

is a subbasis for the product topology. This fact is useful in characterizing continuity for functions for which the range is a product space.

Theorem 7.1: *Let $f: Y \rightarrow X$ be a function from a space Y into a product space $X = \prod_{i=1}^n X_i$. Then f is continuous if and only if the composition $p_i f$ of f with each projection map is continuous.*

Proof: *If f is continuous, the fact that each composition $p_i f$ is continuous follows from the continuity of p_i and the fact that the composition of continuous functions is continuous.*

Suppose now that $p_i f$ is continuous for each coordinate map p_i. By Theorem 4.11, it is sufficient to show that there is a subbasis $\mathscr{S}$ for X such that $f^{-1}(S)$ is open in Y for each subbasic open set S. Consider the subbasis $\mathscr{S}$ for X defined in the remarks preceding the theorem:

$$\mathscr{S} = \{p_i^{-1}(O_i) : O_i \text{ is open in } X_i, \quad 1 \leq i \leq n\}.$$

For any subbasic open set $p_i^{-1}(O_i)$,

$$f^{-1}(p_i^{-1}(O_i)) = (p_i f)^{-1}(O_i)$$

is open in Y because $p_i f$ is continuous. Thus f is continuous. $\qquad\square$

The reader may have seen Theorem 7.1 in a slightly different form in calculus. Suppose that $f: Y \rightarrow X$ is a function from a space Y into a product space $X = \prod_{i=1}^n X_i$. For y in Y, $f(y)$ is an n-tuple, so

$$f(y) = (f_1(y), f_2(y), \ldots, f_n(y))$$

where each f_i is a function from Y into the ith coordinate space. Then f_i is simply the composition of f with the ith coordinate map, $f_i = p_i f$. Theorem 7.1 asserts that f is continuous if and only if each of the coordinate functions f_i is continuous. This result is used in calculus to show the continuity of functions like

$$f(x) = (x^2, \sin x + \cos x)$$

from $\mathbb{R}$ to $\mathbb{R}^2$.

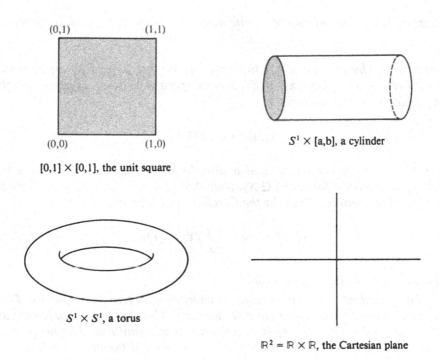

FIGURE 7.3 **Examples of product spaces.**

If each coordinate space has a certain topological property, it is natural to ask if the product space has the property also. The following sequence of theorems shows that the Hausdorff property, connectedness, separability, first countability, second countability, and compactness carry over from coordinate spaces to finite products. The reader should be aware, however, that there are topological properties for which this is not the case. We shall see later, for example, that the product of two Lindelöf spaces may fail to be Lindelöf.

Theorem 7.2: *The product of a finite number of Hausdorff spaces is a Hausdorff space.*

Proof: *Let $\{X_i\}_{i=1}^n$ be a finite collection of Hausdorff spaces and consider distinct points $a = (a_1, \ldots, a_n)$ and $b = (b_1, \ldots, b_n)$ of the product space $X = \prod_{i=1}^n X_i$. Then there is a coordinate space X_i for which $a_i \neq b_i$. Since X_i is Hausdorff, there are disjoint open sets U_i and V_i in X_i containing a_i and b_i, respectively. Then*

$$U = p_i^{-1}(U_i), \quad V = p_i^{-1}(V_i)$$

are disjoint open sets in X containing a and b, respectively.   □

Theorem 7.3:   *The product of a finite number of connected spaces is connected.*

Proof: *Consider the case $n = 2$ first and suppose that X_1 and X_2 are connected spaces. Let $x_1 \in X_1$. Then $\{x_1\} \times X_2$, as a subspace of $X_1 \times X_2$, is homeomorphic to X_2 under the correspondence*

$$(x_1, t) \leftrightarrow t, \quad t \in X_2.$$

Then $\{x_1\} \times X_2$ is connected, and a directly analogous argument shows that $X_1 \times \{t\}$ is connected for each $t \in X_2$. Note that $\{x_1\} \times X_2$ and $X_1 \times \{t\}$ have the point (x_1, t) in common. Then, by the Corollary to Theorem 5.5, the set

$$(\{x_1\} \times X_2) \cup \bigcup_{t \in X_2} (X_1 \times \{t\}),$$

which equals $X_1 \times X_2$, is connected.
    The preceding argument is easier to understand in terms of a picture. Think of X_1 as a horizontal axis and X_2 as a vertical axis. Then $\{x_1\} \times X_2$ is a (connected) vertical line and each $X_1 \times \{t\}$ is a (connected) horizontal line. The union of the one vertical line and all the horizontal lines is connected because each horizontal line crosses the vertical line.
    Proceeding now by induction, suppose that $\prod_{i=1}^{n-1} X_i$ is connected and consider $\prod_{i=1}^n X_i$, where X_i, $i = 1, \ldots, n$, is a connected space. By the argument for $n = 2$, $\prod_{i=1}^n X_i$ is connected since it is the product of the two connected spaces $\prod_{i=1}^{n-1} X_i$ and X_n. Hence the product of any finite collection of connected spaces is connected.   □

If you are content with the inductive proof of Theorem 7.3, then skip this paragraph. Actually there is a small difficulty with the last part of the proof of Theorem 7.3. According to the definition of product spaces, $\prod_{i=1}^n X_i$ and $(\prod_{i=1}^{n-1} X_i) \times X_n$ are technically not the same. Points of the former are n-tuples $(x_1, \ldots, x_n)$ while points of the latter are ordered pairs $((x_1, \ldots, x_{n-1}), x_n)$ whose first coordinates are $(n - 1)$-tuples. However. the obvious correspondence between

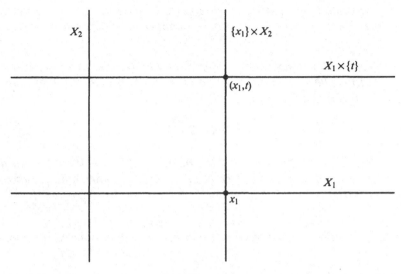

FIGURE 7.4

the points of $\prod_{i=1}^{n} X_i$ and $(\prod_{i=1}^{n} X_i) \times X_n$ is a homeomorphism, and it is customary to consider these spaces as equal. That custom is followed in this text.

Proofs of the next three theorems are left as exercises.

Theorem 7.4:   *The product of a finite number of separable spaces is separable.*

Theorem 7.5:

(a)   *The product of a finite number of first countable spaces is first countable.*

(b)   *The product of a finite number of second countable spaces is second countable.*

Theorem 7.6:   *If (X_1, d_1), (X_2, d_2), . . . , (X_n, d_n) are metric spaces, then the product topology for $\prod_{i=1}^{n} X_i$ is the topology generated by the product metric.*
(Hint for proof: *Look at Examples 3.5.4 and 7.1.1.*)

In view of Theorem 7.6, we may consider $\mathbb{R}^n$ to be the product space $\mathbb{R} \times \mathbb{R} \times \cdots \times \mathbb{R}$ in which the real line $\mathbb{R}$ is used as coordinate space n times.

The final theorem of this section shows that the product of a finite number of compact spaces is compact, a result that seems quite natural. The proof is a bit more difficult for compactness than for the other topological properties we have considered. The following lemma will be needed.

Lemma: *In order that a space X be compact it is sufficient that there exist a basis $\mathcal{B}$ for X such that every open cover of X by members of $\mathcal{B}$ has a finite subcover.*

Proof: *Suppose that X has such a basis $\mathcal{B}$ and let $\mathcal{O}$ be an open cover of X. For each x in X, let O_x be a member of $\mathcal{O}$ containing x. By the definition of basis, there is for each x in X a member B_x of $\mathcal{B}$ such that*

$$x \in B_x \subset O_x.$$

Then $\{B_x : x \in X\}$ is an open cover of X by members of $\mathcal{B}$ and therefore has a finite subcover $\{B_{x_i}\}_{i=1}^n$. The corresponding collection $\{O_{x_i}\}_{i=1}^n$ is then a finite subcover for X derived from $\mathcal{O}$.   □

Theorem 7.7:   *The product of a finite number of compact spaces is compact.*

Proof: *By an inductive argument like that for Theorem 7.3, it is sufficient to prove that if X_1 and X_2 are compact spaces, then $X_1 \times X_2$ is compact.*

Let $\mathcal{B} = \{U \times V : U \text{ is open in } X_1 \text{ and } V \text{ is open in } X_2\}$ be the defining basis for the product topology of $X_1 \times X_2$, and let $\mathcal{O}$ be an open cover of $X_1 \times X_2$ composed of members of $\mathcal{B}$. By the lemma, the compactness of $X_1 \times X_2$ will be proved if it can be shown that there is a finite subcover for $X_1 \times X_2$ derived from $\mathcal{O}$.

For x in X_1, the subset $\{x\} \times X_2$ of $X_1 \times X_2$ is compact and is therefore contained in the union of a finite number of members, say $U_1 \times V_1, U_2 \times V_2, \ldots, U_m \times V_m$ of $\mathcal{O}$, each of which meets $\{x\} \times X_2$. Then

$$U_x = \bigcap_{i=1}^m U_i$$

is an open subset of X_1 containing x. Note that

$$U_x \times X_2 = U_x \times \left(\bigcup_{i=1}^m V_i\right) = \bigcup_{i=1}^m (U_x \times V_i) \subset \bigcup_{i=1}^m (U_i \times V_i),$$

the last inclusion following from the fact that U_x is contained in U_i for $i = 1, \ldots, m$. Thus, for each x in X_1, there is an open set U_x containing x for which the set $U_x \times X_2$ is contained in the union of a finite number of members of $\mathcal{O}$.

Since the family of open sets $\{U_x : x \in X_1\}$ is an open cover of the compact space X_1, then there is a finite subcover $\{U_{x_i}\}_{i=1}^p$ for X_1 derived from $\{U_x : x \in X\}$. We now have the following situation:

$$X_1 \times X_2 = \left(\bigcup_{i=1}^p U_{x_i}\right) \times X_2 = \bigcup_{i=1}^p (U_{x_i} \times X_2)$$

is the union of a finite number of sets $U_{x_1} \times X_2$, $U_{x_2} \times X_2$, ..., $U_{x_p} \times X_2$ *each of which is contained in the union of a finite number of members of* $\mathcal{O}$. *Since the union of a finite collection of finite sets is finite, then* $X_1 \times X_2$ *is contained in the union of a finite number of members of* $\mathcal{O}$. *Thus* $X_1 \times X_2$ *is compact.*   □

The lemma preceding Theorem 7.7 actually gives a necessary and sufficient condition for compactness. The necessity follows from the fact that every open cover of X, and in particular every cover by basic open sets, must have a finite subcover. There is a stronger and more useful condition for compactness involving subbasic open sets: A space X is compact if and only if there is a subbasis $\mathcal{S}$ for X such that every open cover of X by members of $\mathcal{S}$ has a finite subcover. This result, the celebrated Alexander Subbasis Theorem, will be used in the next section to prove compactness for infinite products of compact spaces.

EXERCISE 7.1

1. Let O_i and U_i be subsets of a set X_i for $i = 1, 2$. Prove that

   $$(O_1 \times O_2) \cap (U_1 \times U_2) = (O_1 \cap U_1) \times (O_2 \cap U_2).$$

   Generalize this result to the case of n sets $X_1, X_2, \ldots, X_n$.

2. Show that the set $\mathcal{B}$ in the definition of product topology is actually a basis, as claimed in the definition.

3. Let X_1, X_2, and X_3 be spaces.

   (a) Prove that $(X_1 \times X_2) \times X_3$ is homeomorphic to $X_1 \times (X_2 \times X_3)$.

   (b) Prove that $X_1 \times X_2$ is homeomorphic to $X_2 \times X_1$.

4. Prove Theorems 7.4, 7.5, and 7.6.

5. Use the Alexander Subbasis Theorem to give a different proof of Theorem 7.7.

6. Knowing that a subset A of $\mathbb{R}$ is compact if and only if it is closed and bounded, use product space ideas to prove the same characterization of the compact subsets of $\mathbb{R}^n$.

7. Let X_1 and X_2 be spaces with subsets $A \subset X_1$ and $B \subset X_2$. In the product space $X_1 \times X_2$, prove that

   (a) $\overline{A \times B} = \bar{A} \times \bar{B}$.

   (b) int $(A \times B) = $ int $A \times$ int B.

8. Let X denote the real line with the half-open interval topology of Example 4.3.4 and let $Y = X \times X$.

   (a) Prove that $\mathcal{B} = \{[a, b) \times [c, d): a, b, c, d \in \mathbb{R}, a < b, c < d\}$ is a basis for Y.

   (b) Show that Y is separable.

(c)   Show that the line $y = -x + 1$ is a non-separable subspace of Y.

(d)   Show that X is Lindelöf but Y is not. (*Hint:* The Lindelöf property is inherited by closed subspaces. Show that the line of part (c) is closed and not Lindelöf.)

9.   Prove that the product of a finite family of locally compact spaces is locally compact.

10.   This exercise is intended to show that a result like Theorem 7.1 does not hold for functions $g: \prod X_i \rightarrow Y$ when the *domain* is a product space instead of the range.

Definition: *Let X_1 and X_2 be spaces and g: $X_1 \times X_2 \rightarrow Z$ a function from $X_1 \times X_2$ to a space Z. Then g is* **continuous in the first variable** *provided that for each y in X_2, the function g($\cdot$, y): $X_1 \rightarrow Z$, whose value at x is g(x, y), is continuous. Continuity in the second variable is defined in the analogous way.*

Show that the function $f: \mathbb{R} \times \mathbb{R} \rightarrow \mathbb{R}$ defined by

$$f(x, y) = \begin{cases} xy/(x^2 + y^2) & \text{if } (x, y) \neq (0, 0) \\ 0 & \text{if } (x, y) = (0, 0) \end{cases}$$

is continuous in the first variable and continuous in the second variable but not continuous at (0, 0).

11.   **Definition:**   *In a product space $X \times X$, the set $\{(x, x): x \in X\}$ is called the* **diagonal.**

Prove that a space X is Hausdorff if and only if the diagonal of $X \times X$ is a closed set.

7.2 ARBITRARY PRODUCTS

This section generalizes the idea of product space to the product of an arbitrary family of topological spaces. It will be evident that the finite product considered in Section 7.1 is a special case.

Suppose that $\mathcal{A}$ is an index set and that X_α is a space for each α in $\mathcal{A}$. The first question to be answered in trying to define the product of the spaces X_α, $\alpha \in \mathcal{A}$, is the following: What is meant by the Cartesian product $\prod_{\alpha \in \mathcal{A}} X_\alpha$ when $\mathcal{A}$ is an infinite set? Suppose first that $\mathcal{A}$ is countably infinite. Then $\{X_\alpha: \alpha \in \mathcal{A}\}$ can be considered an infinite sequence $\{X_i\}_{i=1}^{\infty}$. In this case $\prod_{i=1}^{\infty} X_i$ is the collection of all infinite sequences

$$x = (x_1, x_2, \ldots, x_n, \ldots)$$

for which the ith coordinate x_i belongs to the ith coordinate space X_i for all values of i. Actually, the sequence x is a function. Its domain is the set of positive integers (the index set) and its value at the integer i is the ith coordinate x_i. This function idea is used to define the Cartesian product of an arbitrary family of sets.

Definition: *Let $\mathcal{A}$ be an index set and $\{X_\alpha : \alpha \in \mathcal{A}\}$ a family of sets. The **Cartesian product** $X = \prod_{\alpha \in \mathcal{A}} X_\alpha$ is the collection of all functions x with domain $\mathcal{A}$ having the property that the value x_α of x at α belongs to the set X_α:*

$$X = \prod_{\alpha \in \mathcal{A}} X_\alpha = \left\{ x \colon \mathcal{A} \to \bigcup_{\alpha \in \mathcal{A}} X_\alpha \colon x_\alpha \in X_\alpha \text{ for each } \alpha \in \mathcal{A} \right\}$$

For $\alpha \in \mathcal{A}$, the function

$$p_\alpha \colon X \to X_\alpha$$

defined by

$$p_\alpha(x) = x_\alpha, \quad x \in X,$$

*is called the **projection map** of X on the **αth coordinate set** X_α.*

In the preceding definition, the symbol x_α is used to denote $x(\alpha)$, the value of x at α, in analogy with the notation

$$x = (x_1, x_2, \ldots, x_n, \ldots)$$

commonly used for sequences. The definition of Cartesian product set agrees with the definition already given for a finite product $\prod_{i=1}^{n} X_i$ since a point

$$x = (x_1, \ldots, x_n), \quad x_i \in X_i, \quad 1 \le i \le n,$$

may be interpreted as a function with domain $\{1, 2, \ldots, n\}$ whose value x_i at i is a member of the set X_i.

Definition: *Let $\mathcal{A}$ be an index set and $\{X_\alpha : \alpha \in \mathcal{A}\}$ a collection of non-empty topological spaces. The **product topology** for $X = \prod_{\alpha \in \mathcal{A}} X_\alpha$ is the topology generated by the subbasis $\mathcal{S}$ of all sets of the form $p_\alpha^{-1}(O_\alpha)$, $\alpha \in \mathcal{A}$, where O_α is open in X_α. The product set X with the product topology is called a **product space** with X_α as αth coordinate space.*

According to the preceding definition, a basis for the product topology for X consists of all finite intersections

$$\bigcap_{i=1}^{n} p_{\alpha_i}^{-1}(O_{\alpha_i}), \quad \alpha_i \in \mathcal{A}, \quad 1 \le i \le n,$$

where each O_{α_i} is an open subset of X_{α_i}. Such a basic open set may be expressed in product form as

$$\bigcap_{i=1}^{n} p_{\alpha_i}^{-1}(O_{\alpha_i}) = \prod_{\alpha \in \mathcal{A}} A_\alpha$$

where $A_{\alpha_i} = O_{\alpha_i}$ for $i = 1, \ldots, n$ and $A_\alpha = X_\alpha$ otherwise. Hence we may think of a basis for the product topology as consisting of all products $\prod_{\alpha \in \mathcal{A}} A_\alpha$ where each A_α is open in X_α and $A_\alpha = X_\alpha$ for all but finitely many α in $\mathcal{A}$.

Since each set of the form $p_\alpha^{-1}(O_\alpha)$, O_α open in X_α, is a subbasic open set, each projection map p_α is continuous. The next theorem is a direct analogue of Theorem 7.1; its proof is left as an exercise.

Theorem 7.8:   *Let* $f: Y \rightarrow X$ *be a function from a space* Y *into a product space* $X = \prod_{\alpha \in \mathcal{A}} X_\alpha$. *Then* f *is continuous if and only if the composition* $p_\alpha f$ *of* f *with each projection map is continuous.*

Theorem 7.9:   *The product of any family of Hausdorff spaces is a Hausdorff space.*

Proof: *Let* $\{X_\alpha : \alpha \in \mathcal{A}\}$ *be a family of Hausdorff spaces and let* x *and* y *be distinct points of the product space* $X = \prod_{\alpha \in \mathcal{A}} X_\alpha$. *Then there is some member* $\alpha \in \mathcal{A}$ *such that* $x_\alpha \neq y_\alpha$. *Since* X_α *is Hausdorff, there are disjoint open sets* U_α *and* V_α *in* X_α *containing* x_α *and* y_α, *respectively. Then*

$$U = p_\alpha^{-1}(U_\alpha), \quad V = p_\alpha^{-1}(V_\alpha)$$

are disjoint open sets in X *containing* x *and* y, *respectively.*   □

The properties of separability, first countability, and second countability involve countable sets, so it should not be expected that Theorems 7.4 and 7.5 generalize to uncountable products. These theorems do generalize to countably infinite products; the proofs are left as exercises. A bit more can be said for separability. If each space in the set $\{X_\alpha : \alpha \in \mathcal{A}\}$ is separable and the cardinal number of $\mathcal{A}$ is less than or equal to the cardinal number of $[0, 1]$, then $\prod_{\alpha \in \mathcal{A}} X_\alpha$ is separable. Proofs of this result can be found in the references listed at the end of the chapter.

The remainder of this section is devoted to showing that every product of connected spaces is connected and every product of compact spaces is compact, thus extending the corresponding results already proved for finite products.

Theorem 7.10:   *The product of an arbitrary collection of connected spaces is connected.*

Proof: *Let $\{X_\alpha: \alpha \in \mathcal{A}\}$ be a family of connected spaces indexed by $\mathcal{A}$ and let $y = \{y_\alpha: \alpha \in \mathcal{A}\}$ be a particular point in the product space $X = \prod_{\alpha \in \mathcal{A}} X_\alpha$. It will be shown that X is connected by showing that the component C of X which contains y is the entire space.*

First, by an argument analogous to the proof that the product of two connected spaces is connected, it follows that the set Y_1 of all points differing from y in at most one coordinate is connected. It follows by induction that the set Y_n of all points differing from y in at most n coordinates is connected. Since each set Y_n contains y, then the union

$$Y = \bigcup_{n=1}^{\infty} Y_n$$

is connected. It is an easy consequence of the definition of the product topology that Y is dense in X, so Theorem 5.4 applies to show that X is connected.   □

In order to prove that the product of compact spaces is compact, the following extension of the lemma preceding Theorem 7.7 to subbases will be needed. The proof of this result, the Alexander Subbasis Theorem, involves set theoretic considerations which would take us rather far afield. For this reason, the proof is relegated to outline form in the exercises at the end of this section. The interested reader is encouraged to pursue this topic further in the suggested reading at the end of the chapter.

Lemma: The Alexander Subbasis Theorem   *In order that a space X be compact, it is necessary and sufficient that there exist a subbasis $\mathcal{S}$ for X such that every open cover of X by members of $\mathcal{S}$ has a finite subcover.*

Theorem 7.11:   The Tychonoff Theorem   *The product of an arbitrary family of compact spaces is compact.*

Proof: *Let $\{X_\alpha: \alpha \in \mathcal{A}\}$ be a collection of compact spaces and let $X = \prod_{\alpha \in \mathcal{A}} X_\alpha$ be the product space. Let $\mathcal{S}$ be the defining subbasis for the product topology. Recall that $\mathcal{S}$ consists of all subsets of X of the form $p_\alpha^{-1}(O_\alpha)$, $\alpha \in \mathcal{A}$, where O_α is open in the coordinate space X_α. By the Alexander Subbasis Theorem, it is sufficient to show that every open cover of X by members of $\mathcal{S}$ has a finite subcover.*

Proceeding by contradiction, suppose that $\mathcal{U}$ is an open cover of X by members of $\mathcal{S}$ which has no finite subcover. For α in $\mathcal{A}$, let $\mathcal{U}_\alpha$ be the family of open sets O_α in X_α for which $p_\alpha^{-1}(O_\alpha)$ belongs to $\mathcal{U}$. Note that if a finite subfamily $\{O_{\alpha_i}\}_{i=1}^n$ of $\mathcal{U}_\alpha$ covered X_α, then the corresponding finite subfamily $\{p_{\alpha_i}^{-1}(O_{\alpha_i})\}_{i=1}^n$ would cover X. Thus we conclude that no finite subfamily of $\mathcal{U}_\alpha$ covers X_α. From the compactness of X_α, it follows that $\mathcal{U}_\alpha$ must fail to be a cover of X_α. Thus for each α in $\mathcal{A}$, there *is a point x_α in X_α which is not in any member of $\mathcal{U}_\alpha$. It follows that the point $x =$ $\{x_\alpha: \alpha \in \mathcal{A}\}$ defined by these coordinates fails to be in any member of $\mathcal{U}$. This contradicts the fact that $\mathcal{U}$ is an open cover and completes the proof of the theorem.*   □

Example 7.2.1   The Hilbert Cube and Infinite Dimensional Euclidean Space

The *Hilbert cube* I^∞ is the product of a countably infinite family of closed unit intervals:

$$I^\infty = \prod_{n=1}^{\infty} A_n, \quad A_n = [0, 1], \quad n = 1, 2, \ldots$$

Infinite dimensional Euclidean space $\mathbb{R}^\infty$ is the product of a countably infinite family of lines:

$$\mathbb{R}^\infty = \prod_{n=1}^{\infty} B_n, \quad B_n = \mathbb{R}, \quad n = 1, 2, \ldots.$$

Since $[0, 1] \subset \mathbb{R}$, it is clear that I^∞ is a subspace of $\mathbb{R}^\infty$. Since $\mathbb{R}$ is homeomorphic to $(0, 1)$, then $\mathbb{R}^\infty$ is homeomorphic to a subspace of I^∞. However, I^∞ and $\mathbb{R}^\infty$ are not homeomorphic since, for example, I^∞ is compact and $\mathbb{R}^\infty$ is not. This phenomenon of non-homeomorphic spaces embedded in each other is not too unusual; it also occurs for an open interval and a closed interval. Note that both I^∞ and $\mathbb{R}^\infty$ are connected by Theorem 7.10.

The purpose of this example is to show that the infinite products I^∞ and $\mathbb{R}^\infty$ are metric spaces by showing that they can be embedded in Hilbert space H. Define $f: I^\infty \to H$ by

$$f(x_1, x_2, x_3, \ldots) = (x_1, x_2/2, x_3/3, \ldots).$$

Since $\sum_{n=1}^{\infty} 1/n^2$ is a convergent series, it follows that $\sum_{n=1}^{\infty} x_n^2/n^2$, $0 \le x_n \le 1$ for all n, is also convergent and hence that f does map into H. The image of I^∞ under f is the subset

$$f(I^\infty) = \prod_{n=1}^{\infty} [0, 1/n]$$

of H. It is left as an exercise to show that f is one-to-one and continuous. Since I^∞ is compact and H is Hausdorff, Theorem 6.8 insures that f is an embedding of I^∞ in H. Since $\mathbb{R}^\infty$ can be embedded in I^∞ and the composition of two embeddings is an embedding, then $\mathbb{R}^\infty$ can be embedded in H also.

Since I^∞ and $\mathbb{R}^\infty$ are homeomorphic to subspaces of the metric space H, we conclude that their topologies are determined by metrics; that is to say, I^∞ and $\mathbb{R}^\infty$ are metrizable spaces. We shall return to Hilbert space embeddings in the next chapter to prove a general metrization theorem. The spaces $\mathbb{R}^\infty$ and H are actually homeomorphic, but the proof of this is well beyond the scope of an introductory course.

Example 7.2.2   The Cantor Set as an Infinite Product

This example shows that the Cantor set C is homeomorphic to a countably infinite product of discrete, two-point spaces. Consider the product space $A = \prod_{i=1}^{\infty} A_i$, where each set A_i is the discrete space $\{0, 1\}$. The space A is sometimes denoted 2^ω since each coordinate space has two members and ω often denotes a countably infinite set.

Recall that the Cantor set consists of all real numbers x in $[0, 1]$ which have a ternary expansion $x = .x_1 x_2 x_3 \ldots$ where x_i has only the values 0 and 2. The space A consists of all infinite sequences $y = (y_1, y_2, y_3, \ldots)$ where y_i has only the values 0 and 1. Thus we define $f: C \to A$ by

$$F(.x_1 x_2 x_3 \ldots) = (x_1/2, x_2/2, x_3/2 \ldots), \quad x = .x_1 x_2 x_3 \ldots \in C.$$

The preceding discussion establishes the fact that f is a one-to-one correspondence between C and A.

To prove that f is continuous, we shall show that the composition $p_i f$: $C \to \{0, 1\}$ of f with the ith coordinate map is continuous and use Theorem 7.8. For a positive integer i and point x in C, $p_i f(x)$ is one-half of the ith coordinate of the ternary expansion of x into 0's and 2's. Since members of C sufficiently close to x must have identical ternary expansion through the ith term, it follows that

$$p_i f(x) = p_i f(x')$$

when x' is sufficiently close to x. Thus $p_i f$ is continuous. Theorem 7.8 applies to establish the continuity of f. Since $f: C \to A$ is a continuous one-to-one function from a compact space onto a Hausdorff space, Theorem 6.8 shows that f is a homeomorphism.

EXERCISE 7.2

1. For each of the following properties, prove that if each factor space X_i, $i = 1, 2, \ldots$, has the property, then so does the product space $X = \prod_{i=1}^{\infty} X_i$:

    (a) first countability,

    (b) second countability,

    (c) separability.

2. Prove: If $\mathcal{A} = \mathcal{B} \cup \mathcal{C}$, where $\mathcal{B}$, $\mathcal{C}$ are disjoint, non-empty index sets and $\{X_\alpha : \alpha \in \mathcal{A}\}$ is a family of spaces indexed by $\mathcal{A}$, then $\prod_{\alpha \in \mathcal{A}} X_\alpha$ is homeomorphic to $\prod_{\beta \in \mathcal{B}} X_\beta \times \prod_{\gamma \in \mathcal{C}} X_\gamma$.

3. Prove that an infinite product of discrete spaces may not be discrete.

4. Prove that the function $f \colon I^\infty \to H$ of Example 7.2.1 is an embedding.

5. (a) Let $\{X_\alpha : \alpha \in \mathcal{A}\}$ be a collection of spaces with product space $X = \prod_{\alpha \in \mathcal{A}} X_\alpha$. Show that each projection map $p_\alpha \colon X \to X_\alpha$ is an open mapping.

   (b) Show that the projections $p_1, p_2 \colon \mathbb{R}^2 \to \mathbb{R}$ are not closed mappings.

6. **Definition.** *Let $\{X_\alpha : \alpha \in \mathcal{A}\}$ be a family of non-empty spaces. The collection of all subsets O of the product set $X = \prod_{\alpha \in \mathcal{A}} X_\alpha$ of the form*

   $$O = \prod_{\alpha \in \mathcal{A}} O_\alpha,$$

   *where each O_α is open in X_α, is a basis for a topology, called the **box topology** for X. The product set X with the box topology is called a **box product space**.*

   Prove:

   (a) If $\mathcal{A}$ is a finite set then the box topology for X coincides with the product topology.

   (b) The projection maps for a box product space are continuous, open maps.

   (c) If each space X_α, $\alpha \in \mathcal{A}$, is discrete, then the box product space X is also discrete.

   (d) If each space X_α, $\alpha \in \mathcal{A}$, is compact, the box product space may fail to be compact. (*Hint:* Let $X_\alpha = \{0, 1\}$ for each α in an infinite set $\mathcal{A}$.)

7. Let X and A be non-empty sets. The symbol X^A denotes the collection of all functions $f \colon A \to X$.

   (a) Show that the set X^A is the product set $\prod_{\alpha \in A} X_\alpha$, where $X_\alpha = X$ for each $\alpha \in A$.

   (b) Show that if $A = B \cup C$ for disjoint, non-empty subsets B and C, then the spaces X^A and $X^B \times X^C$, with their product topologies, are homeomorphic.

8. Complete the details of the proof of Theorem 7.10. In particular, show that each of the sets Y_n is connected.

9. Although it is not proved in this text, it is true that every compact, perfect, totally disconnected metric space is homeomorphic to the Cantor set. Use this result to prove that a countably infinite product of discrete two-point spaces (Example 7.2.2) is homeomorphic to the Cantor set.

10. (a) Prove that a finite product of discrete spaces is discrete.

    (b) Prove that if each space in an infinite collection of discrete spaces has more than one point, then their product is not discrete.

11. Show that the Hilbert Cube (Example 7.2.1) is nowhere dense in Hilbert space.

12. This exercise presents an outline of the proof of the Alexander Subbasis Theorem.

    Let $\mathcal{S}$ be a subbasis for a space X for which every open cover of X by members of $\mathcal{S}$ has a finite subcover. Suppose X is not compact and obtain a contradiction by showing the following:

(a)   There is a family $\mathcal{O}$ of open subsets of X for which no finite subcollection of $\mathcal{O}$ covers X and $\mathcal{O}$ is maximal with respect to this property. In other words, $\mathcal{O}$ is not properly contained in any family of open sets no finite subfamily of which covers X. (This will involve some form of the Axiom of Choice.)

(b)   If $\{U_i\}_{i=1}^n$ is a finite collection of open sets whose intersection is contained in a member of $\mathcal{O}$, then at least one of the sets U_i belongs to $\mathcal{O}$.

(c)   Conclude that the family $\mathcal{O}$ fails to cover X.

13.   Let $\{(X_n, d_n)\}_{n=1}^{\infty}$ be an infinite sequence of metric spaces. Prove that the product space $X = \prod_{n=1}^{\infty} X_n$ is a metrizable space. (*Hint:* Show first that each metric space is topologically equivalent to a metric space of diameter at most 1. Thus it can be assumed that each metric space under consideration has diameter at most 1. Show that the function d on $X \times X$ defined by

$$d(x, y) = \left(\sum_{n=1}^{\infty} \left(\frac{d_n(x_n, y_n)}{n} \right)^2 \right)^{1/2}$$

for $x = (x_1, x_2, \ldots)$, $y = (y_1, y_2, \ldots)$ in X is a metric which generates the product topology.)

7.3 COMPARISON OF TOPOLOGIES

Definition: *Let X be a set with two topologies, $\mathcal{T}_1$ and $\mathcal{T}_2$. If $\mathcal{T}_1 \subset \mathcal{T}_2$, then $\mathcal{T}_1$ is **weaker** than $\mathcal{T}_2$ and $\mathcal{T}_2$ is **stronger** than $\mathcal{T}_1$.*

The terms *coarser* and *finer* are sometimes used synonymously with weaker and stronger, respectively. It should be clear that the trivial topology is the weakest topology for any set and the discrete topology is the strongest.

It is not hard to verify that the intersection $\mathcal{T}_1 \cap \mathcal{T}_2$ of two topologies is a topology. This topology is weaker than both $\mathcal{T}_1$ and $\mathcal{T}_2$ but stronger than any other topology weaker than both of them. Similarly, the intersection of any family of topologies is a topology that is weaker than every member of the family but stronger than any other topology having this property.

The situation with unions is somewhat more complicated since the union of two topologies may fail to be a topology. However, the union of any family of topologies is a subbasis for a topology, and this topology is the weakest topology which is stronger than each topology in the given family. Verification of this is left as an exercise.

Definition: *Let X be a set and $\{f_\alpha : X \to X_\alpha, \alpha \in \mathcal{A}\}$ a family of functions from X into topological spaces X_α. The **weak topology** for X generated by the functions f_α is the topology determined by the subbasis of open sets $\{f_\alpha^{-1}(O_\alpha) : \alpha \in \mathcal{A}, O_\alpha$ open in $X_\alpha\}$.*

It should be apparent that the weak topology generated by a family of functions is the weakest topology with respect to which each of the given functions is continuous.

Example 7.3.1

Let (X, d) be a metric space. For a in X, define $f_a: X \to \mathbb{R}^+$ from X to the space $\mathbb{R}^+$ of non-negative real numbers by

$$f_a(x) = d(x, a), \quad x \in X.$$

Then for $r > 0$,

$$f_a^{-1}([0, r)) = B(a, r),$$

the open ball centered at a with radius r. It follows that the weak topology generated by $\{f_a: a \in X\}$ is the metric topology for X.

The weak topology gives the following alternative view of the product topology.

Theorem 7.12:   *Let $\{X_\alpha: \alpha \in \mathcal{A}\}$ be a family of spaces and let $X = \prod_{\alpha \in \mathcal{A}} X_\alpha$. Then the weak topology generated by the projection maps $p_\alpha: X \to X_\alpha$ is the product topology for X.*

Proof: *By definition, the product topology for X has as a subbasis the family of all sets of the form $p_\alpha^{-1}(O_\alpha)$, $\alpha \in \mathcal{A}$, O_α open in X_α. This subbasis is precisely the same as that for the weak topology generated by the projection maps.*   □

Theorem 7.13:   *Let $(X, \mathcal{T})$ be a space and $\mathcal{F} = \{f_\alpha: X \to Y_\alpha\}$ a family of continuous functions with domain X. Then the weak topology generated by $\mathcal{F}$ is weaker than $\mathcal{T}$.*

Proof: *By hypothesis, each member of $\mathcal{F}$ is continuous with respect to the given topology. Then the weakest topology with respect to which each member of $\mathcal{F}$ is continuous must be weaker than $\mathcal{T}$.*   □

We shall return in Chapter 8 to the weak topology determined by a given family of functions. In the meantime, the reader is invited to consider the following question: Under what conditions does the weak topology generated by a given family of functions equal the original topology on a space X?

EXERCISE 7.3

1. (a) Show that the intersection of any family of topologies for a set X is a topology for X.

   (b) Use part (a) to conclude that the intersection of a family of topologies for X is the strongest topology for X that is weaker than each topology in the given family.

2. (a) Show that the union of any family of topologies for a set X is a subbasis for a topology for X.

   (b) Show that the topology of part (a) is the weakest topology that is stronger than each topology in the given family.

3. Let X be a set, $\{f_\alpha: X \to X_\alpha,\ \alpha \in \mathcal{A}\}$ a family of functions with domain X, and $\mathcal{T}$ the weak topology generated by the given family of functions. Prove that if Y is a subset of X, then the subspace topology for Y as a subspace of $(X, \mathcal{T})$ equals the weak topology generated by the family of restrictions $\{f_\alpha|_Y: Y \to X_\alpha,\ \alpha \in \mathcal{A}\}$.

4. Describe the weak topology for $\mathbb{R}$ generated by each of the following families of functions. (Assume that the range space has the usual topology and determine the weak topology for the domain.)

   (a) The family of constant functions $f: \mathbb{R} \to \mathbb{R}$.

   (b) The set of all functions continuous with respect to the usual topology.

   (c) The family consisting only of the identity map on $\mathbb{R}$.

   (d) The set of all bounded real-valued functions that are continuous with respect to the usual topology.

5. Let X be a set with three different topologies $\mathcal{S}, \mathcal{T}, \mathcal{U}$ for which $\mathcal{S}$ is weaker than $\mathcal{T}$, $\mathcal{T}$ is weaker than $\mathcal{U}$, and $(X, \mathcal{T})$ is compact and Hausdorff. Show that $(X, \mathcal{S})$ is compact but not Hausdorff, and $(X, \mathcal{U})$ is Hausdorff but not compact.

6. Consider $\mathbb{R}^n$ with the finite complement topology $\mathcal{T}_1$ and the Zariski topology $\mathcal{T}_2$ (Example 4.5.4). Show that $\mathcal{T}_1$ is weaker than $\mathcal{T}_2$.

7.4 QUOTIENT SPACES

The idea of quotient space is one of the more intuitively plausible notions of topology. It originated from the process of pasting or gluing together parts of geometric figures to form new figures.

Example 7.4.1

Consider the process of pasting together two opposite edges of a square to form a cylinder. For definiteness, let us take the unit square $I^2 = \{(x_1, x_2) \in \mathbb{R}^2: 0 \leq$

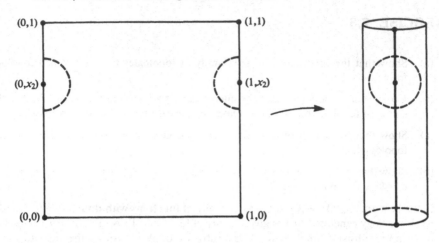

FIGURE 7.5

$x_1 \leq 1, 0 \leq x_2 \leq 1\}$ in the plane and describe the process of pasting together the left and right edges to form a cylinder.

We wish to describe a method by which two points $(0, x_2)$ and $(1, x_2)$ in I^2 may be considered as one point of a new space, the cylinder. The standard method of doing this is to define an equivalence relation on I^2 and to let the new points be the equivalence classes of the relation. Define a relation $\sim$ on I_2 as follows: Each point is related to itself, and each point $(0, x_2)$, $0 \leq x_2 \leq 1$, is related to the point $(1, x_2)$. Thus the equivalence class $[x]$ of a point $x = (x_1, x_2)$ consists of the one point x if x_1 is not 0 or 1. Otherwise, it consists of the two points $(0, x_2)$ and $(1, x_2)$. The set of equivalence classes defined in this way is called a *quotient space* when the following topology is assigned. A collection of equivalence classes is an open set in the quotient space if and only if the points that they contain form an open set in I^2. For a point x in the interior of I^2, a basic open set about x is an open ball, and the same picture applies for a basic open set around $[x]$. For points $(0, x_2)$ and $(1, x_2)$, basic open sets are "open half balls" (open quarter balls at the corners), as shown in Figure 7.5. A basic open set around the corresponding point of the quotient space is the union of the two half balls joined along their straight edges. This heuristic description is formalized in the next definition.

Definition: *Let X be a space and $\sim$ an equivalence relation on X. Let $X/\sim$ denote the set of all equivalence classes $[x] = \{y \in X : x \sim y\}$ determined by $\sim$. A collection $\mathcal{A}$ of equivalence classes is an **open set** in $X/\sim$ if the union of the members of $\mathcal{A}$ is an open set in X. The collection of such open sets in $X/\sim$ is a topology called the **quotient topology** for $X/\sim$, and the set $X/\sim$ with its quotient topology is called the **quotient space of X modulo** $\sim$.*

It is an easy exercise to show that the open sets for $X/\sim$ in the preceding definition actually form a topology.

To define a particular quotient space of a given space X, one need only specify the equivalence relation to be used. From that point on, the construction is standard. Since for every equivalence relation each point must be related to itself, this is often not stated explicitly. The common practice is to specify which unequal points are to be equivalent, with it being understood that each point is always to be considered equivalent to itself.

The simplest kind of quotient space has its own name, as explained in the next definition.

Definition: *Let $\sim$ be an equivalence relation on a space X for which there is one equivalence class $[x_0] = A$ having more than one member and for which every other equivalence class has only one member, $[x] = \{x\}$. Then the quotient space $X/\sim$ is denoted X/A and called the* **quotient space** *or* **identification space obtained by identifying the members of A to a single point.**

Example 7.4.2

(a)   The quotient space of $[0, 1]$ obtained by identifying the two points 0 and 1 is homeomorphic to a circle.

(b)   The quotient space of I^2 obtained by identifying the boundary to a single point is homeomorphic to a sphere in $\mathbb{R}^3$.

(c)   The quotient space of I^2 obtained by identifying the pairs of points $(0, x_2)$ and $(1, 1 - x_2)$, $0 \le x_2 \le 1$, is homeomorphic to the Möbius strip.

(d)   Let $X = I^2$ with equivalence relation $\sim$ defined as follows: (1) $(x_1, 0) \sim (x_1, 1)$ for each $x_1 \in [0, 1]$, and (2) $(0, x_2) \sim (1, x_2)$ for each $x_2 \in [0, 1]$. Then, thinking of the quotient space being defined in two steps as in Figure 7.6, $X/\sim$ is homeomorphic to a torus in $\mathbb{R}^3$.

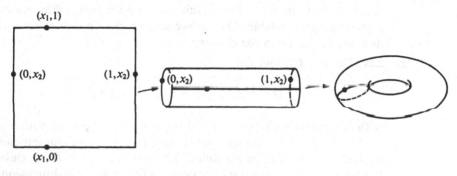

FIGURE 7.6

(e) Let $\sim$ be the equivalence relation on $\mathbb{R}$ defined by $x \sim y$ if and only if $x - y$ is an integer. In other words, $x \sim (n + x)$ for all integers n. In order to get an intuitive picture of $\mathbb{R}/\sim$, consider the equivalence classes in terms of real numbers t with $0 \le t \le 1$:

$$[t] = \{t + n : n \text{ is an integer}\}.$$

Every real number belongs to such a class, so we have reduced the problem from a statement about $\mathbb{R}$ to a statement about $[0, 1]$. Realizing that $[0] = [1]$, the quotient space $\mathbb{R}/\sim$ is homeomorphic to the quotient space of $[0, 1]$ obtained by identifying 0 and 1. Thus, $\mathbb{R}/\sim$ is topologically equivalent to a circle.

(f) Let $X = I^2$ with equivalence relation $\sim$ defined by: (1) $(x_1, 0) \sim (x_1, 1)$ for each $x_1 \in [0, 1]$, and (2) $(0, x_2) \sim (0, 1 - x_2)$ for each x_2 in $[0, 1]$. This example is similar to part (c) except that in the second stage of the identification the circular ends of the cylinder are identified with orientations reversed; the pairs A and A', B and B', and C and C' are to be identified in Figure 7.7.

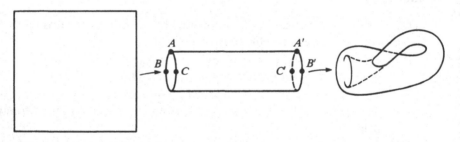

FIGURE 7.7

The resulting space $X/\sim$, called the *Klein bottle,* can be embedded in $\mathbb{R}^4$ but not in $\mathbb{R}^3$. The picture shown is the best 3-dimensional representation available. The surface actually does not intersect itself as it appears to do in the drawing.

(g) Let D denote the unit disk

$$D = \{(x_1, x_2) \in \mathbb{R}^2 : x_1^2 + x_2^2 \le 1\}$$

with equivalence relation $x \sim -x$ for $x = (x_1, x_2)$ on the bounding circle $x_1^2 + x_2^2 = 1$. In other words, each pair $x, -x$ of diametrically opposite points is to be identified. The resulting space $D/\sim$, called the *projective plane,* cannot be reasonably drawn in three dimensions.

The foregoing examples indicate that quotient spaces go considerably beyond what can be accomplished with paper and glue.

Definition: *Let X be a space, $\sim$ an equivalence relation on X, and $X/\sim$ the resulting quotient space. The function $q\colon X \to X/\sim$ which maps each point x in X to its equivalence class [x] in X is called the **quotient map** of $\sim$.*

The definition of the quotient topology insures that the quotient map $q\colon X \to X/\sim$ is continuous. In fact, the quotient topology for $X/\sim$ can be defined as follows: A subset A of $X/\sim$ is open if and only if $q^{-1}(A)$ is an open set in X. The proof of the next theorem is left as an exercise.

Theorem 7.14:   *Let $X/\sim$ be a quotient space of X with quotient map $q\colon X \to X/\sim$. Then a function $h\colon X/\sim \to Y$ from $X/\sim$ into a space Y is continuous if and only if the composite function $hq\colon X \to Y$ is continuous.*

Definition: *Let X be a space, Y a set, and $f\colon X \to Y$ a function from X onto Y. Define a subset O of Y to be **open** provided that its inverse image $f^{-1}(O)$ is open in X. The family of open sets defined in this way is a topology for Y called the **quotient topology determined by f**.*

Remark   The quotient topology determined by $f\colon X \to Y$ is a topology for Y. To see this, note that

$$f^{-1}(\varnothing) = \varnothing, \quad f^{-1}(Y) = X$$

are open in X, so $\varnothing$ and Y are open in Y. If $\{O_\alpha \colon \alpha \in \mathcal{A}\}$ is a family of open sets in Y, then

$$f^{-1}\left(\bigcup_{\alpha \in \mathcal{A}} O_\alpha\right) = \bigcup_{\alpha \in \mathcal{A}} f^{-1}(O_\alpha)$$

is open in X, so $\bigcup_{\alpha \in \mathcal{A}} O_\alpha$ is open in Y. If $\{O_i\}_{i=1}^{n}$ is a finite family of open sets in Y, then

$$f^{-1}\left(\bigcap_{i=1}^{n} O_i\right) = \bigcap_{i=1}^{n} f^{-1}(O_i)$$

is open in X, so $\bigcap_{i=1}^{n} O_i$ is open in Y. Thus the quotient topology is indeed a topology for Y.

A comparison of definitions shows that the quotient topology for a quotient space $X/\sim$ is identical with the quotient topology determined by the quotient map $q: X \to X/\sim$. It is left as an exercise for the reader to prove that the quotient topology is the strongest topology with respect to which the quotient map is continuous.

Theorem 7.15:   *Let X and Y be spaces and $f: X \to Y$ a continuous function from X onto Y. If f is either open or closed, then Y has the quotient topology determined by f.*

Proof: *Let T denote the given topology in Y and T_f the quotient topology determined by f. If $O \in T$, then, since f is continuous, $f^{-1}(O)$ is open in X. But this means that $O \in T_f$. Thus $T \subset T_f$.*
    Suppose in addition that f is an open function, and let $U \in T_f$. Then $f^{-1}(U)$ is open in X and f is an open function with respect to T, so

$$f(f^{-1}(U)) = U$$

is an open set in the topology T. Then $T_f \subset T$, so $T = T_f$. The analogous proof for closed functions is left as an exercise.   □

Definition: *Let $f: X \to Y$ be a function from space X onto space Y. The relation $\sim_f$ defined by $x_1 \sim_f x_2$ if and only if $f(x_1) = f(x_2)$ is an equivalence relation on X called **equivalence modulo f**.*

It should be easily observed that equivalence modulo f is an equivalence relation. Note also that the equivalence class of x under $\sim_f$ is simply the set of points which have the same image as x:

$$[x] = f^{-1}(f(x)), \quad x \in X.$$

Since f is required to be surjective, then the correspondence

$$[x] \leftrightarrow f(x)$$

is a one-to-one correspondence between the quotient space $X/\sim_f$ and Y. The next theorem gives a necessary and sufficient condition for this correspondence to be a homeomorphism.

Theorem 7.16:   *Let X and Y be spaces and $f: X \to Y$ a continuous function from X onto Y. In order that the natural correspondence $h: X/\sim_f \to Y$ defined by*

$$h([x]) = f(x), \quad x \in X,$$

be a homeomorphism, it is necessary and sufficient that Y have the quotient topology determined by f.

Proof: *The diagram below may be of help in following the proof.*

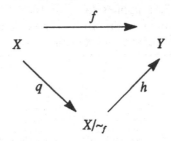

The diagram is called commutative *because hq = f. Suppose first that h is a homeomorphism and consider a subset O of Y. Then O is open in Y if and only if $h^{-1}(O)$ is open in $X/\sim_f$. By the definition of the quotient map q and the quotient topology for $X/\sim_f$, $h^{-1}(O)$ is open in $X/\sim_f$ if and only if $q^{-1}h^{-1}(O)$ is open in X. Since hq = f, then*

$$q^{-1}h^{-1}(O) = f^{-1}(O).$$

Thus O is open in Y if and only if $f^{-1}(O)$ is open in X. Thus Y has the quotient topology determined by f.

For the converse, suppose that Y has the quotient topology determined by f. Then h is continuous by Theorem 7.14 since the composite map hq = f is continuous. To see that h is an open function, let U be open in $X/\sim_f$ and consider h(U):

$$h(U) = \{f(x) \colon [x] \in U\} = \{f(x) \colon x \in q^{-1}(U)\} = f(q^{-1}(U)).$$

Now Y has the quotient topology determined by f, so the test for openness of h(U) is to determine whether or not $f^{-1}(h(U))$ is open in X. Since

$$f^{-1}(h(U)) = f^{-1}(f(q^{-1}(U))) = q^{-1}(U)$$

and $q^{-1}(U)$ is open in X, then h(U) is open in Y. Thus h is an open mapping. Summarizing, h is a continuous, open bijection and is therefore a homeomorphism.   □

Example 7.4.3

Theorem 7.16 can be used to replace the heuristic geometric arguments given earlier in determining quotient spaces.

Consider, for example, the function $f \colon \mathbb{R} \to S^1$ from $\mathbb{R}$ to the unit circle S^1 defined by

$$f(x) = (\cos 2\pi x, \sin 2\pi x), \quad x \in \mathbb{R}.$$

Then $f(x) = f(y)$ if and only if x and y differ by an integer, so the equivalence relation $\sim_f$ is precisely the relation of Example 7.4.2(e). Since f is a continuous, open surjection, Theorem 7.15 insures that S^1 has the quotient topology determined by f. Theorem 7.16 applies to show that $X/\sim_f$, the quotient space of Example 7.4.2(e), is homeomorphic to S^1. The reader should perform a similar analysis for parts (a) through (d) of Example 7.4.2.

EXERCISE 7.4

1. In the definition of quotient space, show that the quotient topology really is a topology for $X/\sim$.

2. Let $f: X \to Y$ be a continuous function from X onto Y and suppose that Y has the quotient topology determined by f. Prove that a function $g: Y \to Z$ from Y to a space Z is continuous if and only if the composite function $gf: X \to Z$ is continuous.

3. Let $X/\sim$ be a quotient space of X and $q: X \to X/\sim$ the quotient map. Prove:

   (a) A subset O of $X/\sim$ is open in $X/\sim$ if and only if $q^{-1}(O)$ is open in X.

   (b) A subset C of $X/\sim$ is closed if and only if $q^{-1}(C)$ is closed in X.

4. Consider the relation $\sim$ on $\mathbb{R}^2$ under which two points are related if and only if they have the same first coordinate. Prove that $\mathbb{R}^2/\sim$ is homeomorphic to $\mathbb{R}$ and interpret geometrically.

5. Describe an equivalence relation for $\mathbb{R}^2$ for which the resulting quotient space is homeomorphic to a circle.

6. (a) Let $\sim$ be the equivalence relation on the unit circle S^1 defined by $x \sim -x$, $x \in S^1$. Show that $S^1/\sim$ is homeomorphic to S^1 and interpret geometrically.

   (b) Let $\approx$ be the equivalence relation on the unit sphere S^2 defined by $x \approx -x$, $x \in S^2$. Show that $S^2/\approx$ is homeomorphic to the projective plane defined in Example 7.4.2(g).

7. By defining suitable functions and using Theorem 7.16, redo parts (a) through (d) of Example 7.4.2.

8. Let X and Y be spaces and $f: X \to Y$ a continuous function from X onto Y. Prove:

   (a) The topology of Y is a subset of the quotient topology determined by f.

   (b) The quotient topology for Y determined by f is the finest topology for Y with respect to which f is continuous.

9. Prove Theorem 7.15 for the case in which f is a closed mapping.

10. Show that the relation $=$, equality for the points of a space X, determines a quotient space $X/=$ homeomorphic to X.

11. Let X be a space and $\sim$ an equivalence relation on X for which $X/\sim$ and each equivalence class $[x]$, $x \in X$, are connected. Prove that X is connected.

12. Prove:

    (a) If X is connected, then every quotient space of X is connected.

    (b) If X is compact, then every quotient space of X is compact.

    (c) If X is a Lindelöf space, then every quotient space of X is Lindelöf.

    (d) If X is locally connected, then every quotient space of X is locally connected.

13. Give an example of a Hausdorff space which has a quotient space that is not Hausdorff.

14. Give examples to show that:

    (a) A quotient map may not be an open function.

    (b) A quotient map may not be a closed function.

7.5 SURFACES AND MANIFOLDS

Many of the most interesting and important topological spaces are "locally like Euclidean spaces" in the sense that each point has neighborhoods that are homeomorphic to open sets in Euclidean spaces. This includes curves, which are locally like $\mathbb{R}$, and surfaces such as the two-sphere and torus, which are locally like $\mathbb{R}^2$. The purpose of this section is to define an important class of such spaces, called *manifolds,* and to develop some of their properties. In the latter part of the section we shall see a more specialized type of manifold that requires "smoothness" as well as the locally Euclidean property.

Definition: *A **topological n-dimensional manifold** or **n-manifold** is a second countable Hausdorff space in which each point has a neighborhood homeomorphic to an open set in Euclidean n-space $\mathbb{R}^n$. A 1-manifold is called a **curve,** and a 2-manifold is called a **surface**.*

For the dimension of an n-manifold to be well defined, it must be noted that an open set in $\mathbb{R}^n$ is not homeomorphic to an open set in $\mathbb{R}^m$ unless $m = n$. This fact, called the Invariance of Domain Theorem, is proved in textbooks on geometric and algebraic topology. References are given in the suggested reading list at the end of the chapter.

Note that the definition of n-manifold requires uniformity of dimension throughout the manifold. Note also that, without loss of generality, the open set in $\mathbb{R}^n$ may be taken to be an n-dimensional open ball. Since such a ball is homeomorphic to $\mathbb{R}^n$, the definition could have specified that each point have a neighborhood homeomorphic to $\mathbb{R}^n$. Different choices are useful in various applications.

Example 7.5.1

(a) The unit circle S^1 is a 1-manifold. The required neighborhoods are defined as follows: The open upper half circle $C_1 = \{(x, y) \in S^1: y > 0\}$ is homeomorphic to the open interval $(-1, 1)$ by the map p_1 which projects to the first coordinate. The inverse of this projection is the function $q_1: (-1, 1) \rightarrow C_1$ defined by

$$q_1(x) = (x, \sqrt{1 - x^2}), \quad x \in (-1, 1).$$

Similar considerations show that the open lower half circle $C_2 = \{(x, y) \in S^1: y < 0\}$ is also homeomorphic to $(-1, 1)$. The two open sets C_1 and C_2 suffice for all points in S^1 except $(1, 0)$ and $(-1, 0)$. For these the open right half circle and open left half circle are suitable neighborhoods.

(b) In general, S^n is an n-manifold. The reader is left the exercise of extending the argument of part (a) to the general case.

Example 7.5.2

(a) The graph of a continuous function $f: \mathbb{R} \rightarrow \mathbb{R}$ is always a 1-manifold. Other familiar curves in the sense of calculus, such as those defined parametrically, are also 1-manifolds if they do not have points of self-intersection. In Figure 7.8, the space of part (iii) is not a 1-manifold since the point of intersection has no neighborhood homeomorphic to an open interval.

The reader should be aware that mathematicians use the term "curve" in several different ways. For example, some authors refer to

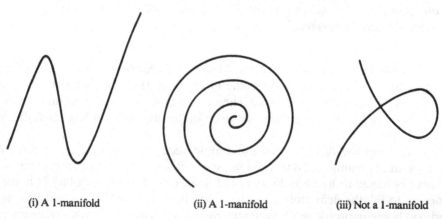

(i) A 1-manifold   (ii) A 1-manifold   (iii) Not a 1-manifold

FIGURE 7.8

the image of a path as a curve. This usage allows for "space-filling curves," which were discussed briefly in Chapter 6, and should not be confused here with the usage of the term to denote a 1-manifold.

(b) The torus, Klein bottle, and projective plane are additional examples of surfaces or 2-manifolds. An "open annulus," composed of an annulus with the inner and outer bounding circles removed, is also a 2-manifold.

(c) An *n-dimensional torus* T^n is the product of n copies of the circle S^1. *Projective n-space* P^n is the quotient space of S^n obtained by identifying pairs of antipodal points. Both T^n and P^n are n-manifolds. Of course, the most immediate example of an n-manifold is $\mathbb{R}^n$ itself.

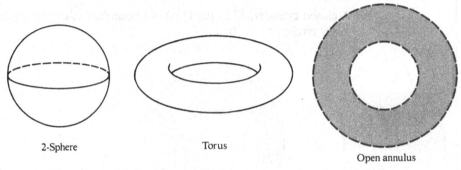

2-Sphere   Torus

Open annulus

FIGURE 7.9   Examples of 2-manifolds.

Theorem 7.17:   *The product of an n-manifold and an m-manifold is an $(n + m)$-manifold.*

Proof: *Let X and Y be an n-manifold and m-manifold, respectively. Note first that X × Y is a second countable Hausdorff space since it is the product of two spaces having these properties. To establish the locally Euclidean condition, consider a point (x, y) in X × Y and let U and V be neighborhoods of x and y in X and Y which are homeomorphic to $\mathbb{R}^n$ and $\mathbb{R}^m$, respectively. Then U × V is a neighborhood of (x, y) which is homeomorphic to $\mathbb{R}^n \times \mathbb{R}^m$. Since this space is homeomorphic to $\mathbb{R}^{n+m}$, then X × Y is a manifold of dimension n + m.*   □

Definition: *An **n-dimensional topological manifold with boundary** is a second countable Hausdorff space X with two types of points: (a) **interior points,** each of which has a neighborhood homeomorphic to $\mathbb{R}^n$, and (b) **boundary points** b, each of which has a neighborhood homeomorphic to the upper half space $U^n = \{x = (x_1, x_2, \ldots, x_n) \in \mathbb{R}^n : x_n \geq 0\}$ by a homeomorphism which maps b to a point of U^n*

*whose nth coordinate is 0. The set of interior points of X is called the **interior**, Int X, and the set of boundary points is called the **boundary**, ∂X, of X.*

Example 7.5.3

(a)   A closed interval $[a, b]$ is a 1-manifold with boundary. The interior is (a, b) and the boundary is the set $\{a, b\}$ of endpoints.

(b)   As examples of 2-manifolds with boundary, note the following:

(i)   A rectangle, whose boundary is its four bounding line segments.

(ii)   An annulus, whose boundary consists of the inner and outer circles.

(iii)   A closed cylinder, $S^1 \times [0, 1]$ whose boundary is the upper and lower circles, $S^1 \times \{0, 1\}$.

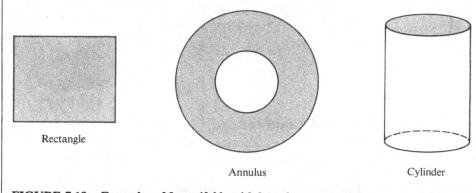

Rectangle

Annulus

Cylinder

FIGURE 7.10   Examples of 2-manifolds with boundary.

Note that a manifold, which is also called a manifold without boundary, may be considered the special case of a manifold with boundary in which the boundary happens to be empty. The non-empty boundary of an n-manifold with boundary is an $(n-1)$-manifold without boundary. A rigorous proof of this intuitively plausible fact requires the Invariance of Domain Theorem, which states that an open set in $\mathbb{R}^n$ is not homeomorphic to an open set in $\mathbb{R}^m$ unless $m = n$.

Definition: *A **closed manifold** is a compact, connected manifold with empty boundary.*

The interior and boundary of a manifold are intrinsic properties of that manifold and not of the space in which the manifold is embedded. Thus they must not be confused with the terms "interior" and "boundary" for subsets of a topological space. For example, S^2 is a 2-manifold whose interior is S^2 and whose boundary is empty. However, as a subspace of $\mathbb{R}^3$, the set S^2 has empty interior and each of its points is a boundary point.

Theorem 7.18:   *Let X be an n-manifold without boundary and Y an m-manifold with boundary. Then $X \times Y$ is an (n + m)-manifold with boundary and*

$$\partial(X \times Y) = X \times \partial Y.$$

Proof: *Note first that $X \times Y$ is second countable and Hausdorff since both factor spaces are. Since $Y = Int\ Y \cup \partial Y$, we have*

$$X \times Y = (X \times Int\ Y) \cup (X \times \partial Y).$$

The preceding theorem shows that each point of $X \times Int\ Y$ has a neighborhood homeomorphic to $\mathbb{R}^{n+m}$. It remains to be proved that each point of $X \times \partial Y$ has a neighborhood homeomorphic to U^{n+m}. This follows easily, however, since if (x, y) belongs to $X \times \partial Y$, then x has a neighborhood homeomorphic to $\mathbb{R}^n$ and y has a neighborhood homeomorphic to U^m. Then (x, y) has a neighborhood homeomorphic to $\mathbb{R}^n \times U^m$, which is homeomorphic to U^{n+m} by an obvious homeomorphism. $\square$

This section has given only the barest of introductions to the subject of topological manifolds. An advanced theorem, which will not be proved here, shows that it is always possible to consider an n-manifold as a subspace of some Euclidean space. It should be clear, however, that the dimension of the manifold and the dimension of the containing space may be different. A 2-sphere, for example, cannot be embedded in $\mathbb{R}^2$. References for more advanced treatments of topological manifolds are given in the Suggestions for Further Reading at the end of the chapter.

A *topological immersion* of a closed manifold X in $\mathbb{R}^m$ is a continuous function $f: X \rightarrow \mathbb{R}^m$ such that each point x in X has a neighborhood U that is mapped by f homeomorphically onto $f(U)$. The usual model of the Klein bottle, which appears in Figure 7.7, represents an immersion of the Klein bottle in $\mathbb{R}^3$. The Klein bottle is a 2-manifold that cannot be embedded in $\mathbb{R}^3$.

Since manifolds share many geometric characteristics with $\mathbb{R}^n$, the study of manifolds and related objects is often called *geometric topology*. Manifolds can be specialized further by requiring a smoothness condition similar to that required for differentiable functions. The resulting manifolds, which are defined below, are called *smooth manifolds* and their study is called *differential topology*. Introductory textbooks for both geometric and differential topology appear in the suggested reading list at the end of the chapter.

Definition: *A function $f: U \rightarrow \mathbb{R}^m$ from an open set U in $\mathbb{R}^n$ into $\mathbb{R}^m$ is **smooth** provided that f has continuous partial derivatives of all orders. A function $f: A \rightarrow$*

$\mathbb{R}^m$ *from an arbitrary subset A of* $\mathbb{R}^n$ *to* $\mathbb{R}^m$ *is **smooth** provided that for each x in A there is an open set U containing x and a smooth function F: U → $\mathbb{R}^m$ such that F agrees with f on U ∩ A.*

In the preceding definition, it is necessary that smooth functions be local restrictions of similar functions defined on open sets so that partial differentiation makes sense.

Definition: *Let A and B be subsets of Euclidean spaces* $\mathbb{R}^n$ *and* $\mathbb{R}^m$, *respectively. A **diffeomorphism** from A to B is a one-to-one function f: A → B from A onto B for which both f and the inverse function* f^{-1} *are smooth. If there is a diffeomorphism from A onto B, then A and B are called **diffeomorphic spaces.***

Definition: *A subset X of a Euclidean space is an **n-dimensional smooth manifold** or a **smooth n-manifold** if each point of X has a neighborhood which is diffeomorphic to* $\mathbb{R}^n$.

Example 7.5.4

The functions defined for S^1 in Example 7.5.1 are smooth and show that S^1 is a smooth 1-manifold. It follows similarly that S^n is a smooth manifold. The torus and projective plane are smooth surfaces, and the graph of an infinitely differentiable function f: $\mathbb{R} → \mathbb{R}$ is a smooth curve. The graph of the absolute value function $y = |x|$, which has a sharp point at the origin, is not smooth.

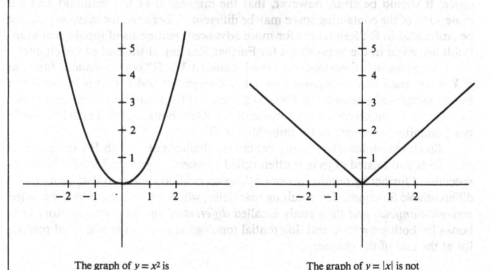

The graph of $y = x^2$ is
a smooth curve.

The graph of $y = |x|$ is not
a smooth curve.

FIGURE 7.11

The proof of the following theorem, which can be patterned after the proof of Theorem 7.17, is left as an exercise.

Theorem 7.19: *The product of a smooth n-manifold and a smooth m-manifold is a smooth (n + m)-manifold.*

Definition: *An **n-dimensional smooth manifold with boundary** is a subset X of a Euclidean space for which the points of X are of two types: (a) **interior points,** each of which has a neighborhood diffeomorphic to $\mathbb{R}^n$, and (b) **boundary points** b, each of which has a neighborhood diffeomorphic to the upper half-space U^n under a diffeomorphism which maps b to a point of U^n whose nth coordinate is 0.*

The analogue of Theorem 7.18 carries over to smooth manifolds, and its proof is left as an exercise.

Theorem 7.20: *Let X be a smooth n-manifold without boundary and Y a smooth m-manifold with boundary. Then $X \times Y$ is a smooth (n + m)-manifold with boundary and*

$$\partial(X \times Y) = X \times \partial Y.$$

Example 7.5.5

The product of two smooth manifolds with boundary may fail to be smooth. The difficulty arises from the fact that the product may fail to be smooth at points of $\partial X \times \partial Y$. For example, $I = [0, 1]$ is a smooth 1-manifold, with $\partial I = \{0, 1\}$. However, $I \times I$ is not smooth at any of the corner points $(0, 0)$, $(0, 1)$, $(1, 0)$, or $(1, 1)$.

EXERCISE 7.5

1. Consider the 2-sphere S^2, torus T, projective plane P, and closed unit interval I. Show that each of the following is a topological manifold. Find the dimension, interior, and boundary in each case and determine whether or not the manifold is smooth.

   (a) $S^2 \times T$
   (b) $S^2 \times I$
   (c) $P \times T$
   (d) $I \times I \times I$
   (e) $S^2 \times I \times I$
   (f) $P \times S^2 \times I$

2. Show that the subspace of $\mathbb{R}^2$ composed of the two coordinate axes is not a manifold.

3. Show that every manifold is locally compact.

4. Explain why each of the following is not a manifold:

   (a) The topologist's sine curve

   (b) The topologist's comb (Example 5.4.2)

   (c) The Cantor set

   (d) Hilbert space

5. Let X be an n-manifold with non-empty interior Int X and non-empty boundary ∂X. Show that Int X and ∂X are manifolds without boundary having dimensions n and $n - 1$, respectively. (Assume the Invariance of Domain Theorem: *An open set in $\mathbb{R}^n$ and an open set in $\mathbb{R}^m$ cannot be homeomorphic unless $m = n$.*)

6. Show that the graph of an infinitely differentiable function $f: \mathbb{R} \to \mathbb{R}$ is a smooth 1-manifold.

7. Show that the relation of being diffeomorphic is an equivalence relation for subspaces of Euclidean spaces.

8. Prove Theorems 7.19 and 7.20.

9. Let X be an n-manifold with boundary and Y an m-manifold with boundary. Is $X \times Y$ a manifold with boundary? Explain.

SUGGESTIONS FOR FURTHER READING

The texts *General Topology* by Kelley and *Topology* by Hocking and Young are recommended for additional reading on product spaces, especially for detailed exposition on the Tychonoff Theorem (Theorem 7.11). For additional work on quotient spaces, see Schurle's *Topics in Topology*. Moise's *Geometric Topology in Dimensions Two and Three* is an excellent introduction to geometric topology. For differential topology, *Differential Topology* by Guillemin and Pollack and *Topology from a Differential Viewpoint* by Milnor are highly recommended.

Proofs of the Invariance of Domain Theorem can be found in *Basic Concepts of Algebraic Topology* by Croom and *Topology: A First Course* by Munkres.

HISTORICAL NOTES FOR CHAPTER 7

Maurice Fréchet introduced finite products of abstract spaces in 1910, but a special case had been considered two years earlier by Ernst Steinitz (1871–1928). The extension to countably infinite products was made by a variety of researchers during the 1920s, and the general definition of product space was formulated by A. N. Tychonoff in 1930.

Tychonoff proved in 1935 that the product of any family of compact spaces is compact. This result, now called the Tychonoff Theorem, established compactness as the proper generalization of the properties of closed and bounded subsets of $\mathbb{R}^n$ to general topological spaces. Prior to the work of Tychonoff, compactness, countable compactness, sequential compactness, the Bolzano-Weierstrass property, and other properties had been proposed as the proper generalization. The Stone-Čech compactification, which was mentioned at the end of Chapter 6, was also inspired by the work of Tychonoff on product spaces.

Theorem 7.3, showing that connectedness is preserved by products, was proved by Hans Hahn in 1932. Infinite dimensional Euclidean space $\mathbb{R}^\infty$ was introduced by Fréchet and is sometimes called *Fréchet space*. The remarkable fact that Fréchet space is homeomorphic to Hilbert space was proved by R. D. Anderson in 1966. The Alexander Subbasis Theorem was proved in 1939 by J. W. Alexander (1888–1971).

Since the quotient space construction developed from the idea of pasting one part of a figure to another part, it is probably impossible to single out the originator. The basic quotient construction was used by A. F. Möbius in 1858 and by Felix Klein (1849–1925) in 1882 in defining the Möbius strip and Klein bottle, respectively. Explicit use of the quotient space construction beyond the identification idea appeared, for a special case, in the work of R. L. Moore in 1925 and P. S. Alexandroff in 1927. The general quotient space and quotient map were introduced by R. W. Baer and F. Levi in 1932.

The systematic study of surfaces and manifolds dates back to the work of Bernard Riemann, A. F. Möbius, Enrico Betti, and others in the mid–nineteenth century.

8 Separation Properties and Metrization

The term "separation property" refers to a characteristic of topological spaces describing those pairs of points or those pairs of sets which can be enclosed in different open sets. The Hausdorff property is an example since it states that any two distinct points are contained in disjoint open sets.

There are some separation properties that are weaker than the Hausdorff property and others that are stronger. The property of being a metric space is stronger, for example, since Example 4.5.3 guarantees that every metric space is Hausdorff. In the latter part of the chapter we shall consider combinations of topological properties which insure that the topology of a given space is generated by a metric.

8.1 T_0, T_1, AND T_2-SPACES

The separation properties to be studied in this chapter are denoted sequentially T_0, T_1, T_2, T_3, T_4 in order of increasing strength. The Hausdorff property is property T_2. Its definition is repeated here to emphasize the sequential progression of the separation properties or *separation axioms,* as they are sometimes called.

Definition: *A space X is a T_0-space if for each pair a, b of distinct points of X, there is an open set containing one of the points but not the other.*

Definition: *A space X is a T_1-space if for each pair a, b of distinct points of X, there are open sets U and V in X such that a belongs to U but b does not, and b belongs to V but a does not.*

Definition: *A space X is a T_2-space or **Hausdorff space** if for each pair a, b of distinct points of X, there are disjoint open sets U and V such that a belongs to U and b belongs to V.*

It should be clear that every T_2-space is T_1, and that every T_1-space is T_0. In other words, T_0, T_1, T_2 is the arrangement of the properties in order of increasing strength.

Theorem 8.1: *A space X is a T_1-space if and only if each finite subset of X is closed.*

Proof: *Suppose X is T_1. It is sufficient to prove that each singleton set $\{a\}$ is closed since any finite set is the union of a finite number of such sets. If b is a point of X different from a, there is an open set V containing b but not a. Thus b is not a limit point of $\{a\}$, so $\{a\}$ is a closed set.*

For the converse, suppose each finite subset of X is closed and consider distinct points a, b in X. Then

$$U = X\backslash\{b\}, \quad V = X\backslash\{a\}$$

are open sets, U contains a but not b, and V contains b but not a. Thus X is T_1. $\square$

The following examples show that a T_0-space may fail to be T_1 and that a T_1-space may fail to be T_2.

Example 8.1.1

    (a)   A T_0-space which is not T_1.

            Let $X = \{a, b\}$ be a two-point set with open sets $\emptyset$, $\{a\}$, and X. Then given two distinct points of X, one of them (namely a) is contained in an open set which does not contain the other. However, every open set containing b also contains a, so X is not T_1.

    (b)   A T_1-space which is not T_2.

            Let X denote the set of real numbers with the finite complement topology. Then for distinct points a, b in X,

$$U = X\backslash\{b\}, \quad V = X\backslash\{a\}$$

are open sets such that U contains a but not b and V contains b but not a. Thus X is T_1.

Since non-empty open sets in X must have finite complements, there cannot exist disjoint open sets for any pair of distinct points of X.

The proof of the following theorem is left to the reader. The Hausdorff case has already been proved in Theorem 7.9.

Theorem 8.2   *The product of T_i-spaces is a T_i-space for $i = 0, 1, 2$.*

EXERCISE 8.1

1. Prove that the T_0 and T_1 properties are hereditary and topological properties.

2. Prove Theorem 8.2.

3. Let X be a space, Y a T_2-space and $f, g: X \to Y$ continuous functions. Prove:

   (a) $\{x \in X: f(x) = g(x)\}$ is a closed subset of X.

   (b) If f and g agree on a dense subset of X, then $f = g$.

4. Let $f: X \to Y$ be a continuous function and assume that Y is T_2. Prove that $\{(x_1, x_2) \in X \times X: f(x_1) = f(x_2)\}$ is a closed subset of $X \times X$.

5. **Definition:** *A $T_{1\frac{1}{4}}$-space is a space in which each sequence has at most one limit. A $T_{1\frac{1}{3}}$-space is a space in which each compact set is closed.*

   Prove that:

   (a) Each T_2-space is $T_{1\frac{1}{4}}$.

   (b) Each $T_{1\frac{1}{3}}$-space is $T_{1\frac{1}{4}}$.

   (c) Each $T_{1\frac{1}{4}}$-space is T_1.

6. **Definition:** *A **simple order relation** for a set X is a relation $<$ on X satisfying:*

   (1) If x, y are distinct members of X, then either $x < y$ or $y < x$.

   (2) If $x < y$, then $y < x$ is false.

   (3) If $x < y$ and $y < z$, then $x < z$.

   *A set X which has a simple order relation is called a **simply ordered set**.*

   Definition: *Let X be a simply ordered set with respect to the simple order relation $<$, and let $\mathcal{B}$ be the family of subsets of X consisting of X and all subsets of the following types:*

   (1) $\{y \in X: x < y\}$, $x \in X$;

   (2) $\{y \in X: y < x\}$, $x \in X$;

   (3) $\{y \in X: z < y < x\}x$, $z \in X$.

   *Then $\mathcal{B}$ is a basis for a topology for X called the **order topology** generated by $<$.*

   (a) Show that the family $\mathcal{B}$ of the preceding definition is actually a basis.

   (b) Show that a simply ordered set with its order topology is a Hausdorff space.

   (c) Define the terms *upper bound* and *least upper bound* for subsets of a simply ordered set.

7. This problem involves the order topology defined in the preceding problem.

   Definition: *Let X be a simply ordered set with its order topology. The space X has a **gap** if there are points x, y in X with $x < y$ for which there is no z in X with $x < z < y$. The space X is called **order complete** provided that each non-empty subset of X that has an upper bound has a least upper bound.*

   Prove: If $\mathcal{T}$ is the order topology for a simply ordered set X, then $(X, \mathcal{T})$ is connected if and only if it is order complete and has no gaps.

8. This problem involves the order topology defined in Problem 6 above. Let X be a compact, connected T_2-space with exactly two non-cut points a, b. Show that the topology of X is an order topology.

8.2 REGULAR SPACES

The properties T_0, T_1, T_2 describe the separation of pairs of points by open sets. The next properties in the sequence describe the separation of a point from a closed set and the separation of a pair of disjoint closed sets. Matters are simplified by requiring that each point be a closed set; in other words, it is required that each space under consideration be a T_1-space.

Definition: *A T_3-space or regular space is a T_1-space X such that for each closed subset C of X and each point a not in C, there exist disjoint open sets U and V in X such that $a \in U$ and $C \subset V$.*

If X is T_3 and a, b are distinct points of X, then $C = \{b\}$ is a closed set which does not contain a. Thus there are disjoint open sets U and V with $a \in U$ and $b \in V$. Thus each T_3-space is T_2.

Theorem 8.3: *A T_1-space X is regular if and only if for each point a in X and each open set U containing a, there is an open set W containing a whose closure is contained in U.*

Proof: *Suppose first that X is regular and let a be a point of X and U an open set containing a. Then $X \backslash U$ is a closed set which does not contain a, so there are disjoint open sets W and V such that*

$$a \in W, \quad X \backslash U \subset V.$$

Since

$$W \subset X \backslash V.$$

and $X \backslash V$ is closed, then

$$\bar{W} \subset X \backslash V.$$

Thus

$$\bar{W} \subset X \backslash V \subset X \backslash (X \backslash U) = U,$$

so W is the required open set.

Suppose now that the latter condition of the theorem holds and let a be a point and C a closed set not containing a. Then $X \backslash C$ is an open set containing a, so there is an open set W such that

$$a \in W, \quad \bar{W} \subset X \backslash C.$$

Then W and $X\backslash\bar{W}$ are disjoint open sets containing a and C, respectively, so X is regular. □

Theorem 8.4: *A T_1-space X is regular if and only if for each point a in X and closed set C not containing a, there exist open sets U and V in X such that $a \in U$, $C \subset V$, and $\bar{U}$ and $\bar{V}$ are disjoint.*

Proof: *The condition clearly implies regularity for a T_1-space X since it requires more than disjointness of the open sets U and V; it requires that their closures be disjoint as well. Thus it need only be proved that each regular space satisfies the condition of the theorem.*

Suppose then that a is a point and C a closed set which does not contain a. By Theorem 8.3, there is an open set W such that

$$a \in W, \quad \bar{W} \subset X\backslash C.$$

Applying the same theorem again, there is an open set U containing a with $\bar{U} \subset W$. Let $V = X\backslash\bar{W}$.
Then

$$\bar{U} \subset W \subset \bar{W} \subset X\backslash C,$$

so

$$C \subset X\backslash\bar{W} = V.$$

Since

$$\bar{U} \cap \bar{V} = \bar{U} \cap \overline{(X\backslash\bar{W})} \subset W \cap \overline{(X\backslash\bar{W})} = \varnothing,$$

then U and V are the required open sets. □

Example 8.2.1   A Hausdorff Space which is Not Regular

Consider the closed upper half-plane $U = \{x = (x_1 x_2) \in \mathbb{R}^2; \ x_2 \geq 0\}$ with the topology defined as follows: For $x = (x_1, x_2)$ in U with $x_2 > 0$, a local basis at x consists of open balls $B(x, r)$, $r < x_2$, in the usual metric d. For a point $z = (z_1, 0)$ in U and positive number r, let

$$D(z, r) = \{z\} \cup \{y = (y_1, y_2) \in U: d(y, z) < r \quad \text{and} \quad y_2 > 0\}.$$

Thus $D(z, r)$ consists of the point z on the horizontal axis and the half of the ball of radius r centered at z which is strictly above the axis. The collection of sets $D(z, r)$, $r > 0$, is a local basis at z. The two types of basic open sets are shown in Figure 8.1.

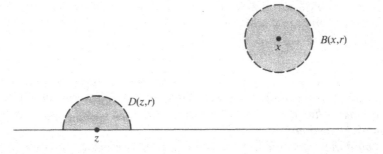

FIGURE 8.1

The collection of all sets $B(x, r)$, $D(z, r)$, $r > 0$, described above is a basis for the topology for U.

The space U is easily seen to be Hausdorff. It is not regular since for a point a on the horizontal axis $\mathbb{R}$, $C = \mathbb{R} \backslash \{a\}$ is a closed set for which there do not exist disjoint open sets containing a and C.

Theorem 8.5:   *The product of any family of regular spaces is regular.*

Proof: *Let $\{X_\alpha : \alpha \in \mathcal{A}\}$ be a family of regular spaces and consider the product space $X = \prod_{\alpha \in \mathcal{A}} X_\alpha$. Let a be a point of X and U an open set containing a. By Theorem 8.3, it is sufficient to show that there is an open set V in X containing a whose closure is contained in U. Let*

$$\bigcap_{i=1}^{n} p_{\alpha_i}^{-1}(U_{\alpha_i})$$

be a basic open set in X which contains a and is contained in U, where each U_{α_i} is an open set in X containing $p_\alpha(a)$. Since each space X_{α_i} is regular, there is for each $i = 1, \ldots, n$ an open set V_{α_i} in X_{α_i} for which

$$p_{\alpha_i}(a) \in V_{\alpha_i}, \quad \bar{V}_{\alpha_i} \subset U_{\alpha_i}.$$

Then

$$V = \bigcap_{i=1}^{n} p_{\alpha_i}^{-1}(V_{\alpha_i})$$

is an open set in X, V contains a, and

$$\bar{V} = \bigcap_{i=1}^{n} p_{\alpha_i}^{-1}(\overline{V_{\alpha_i}}) \subset \bigcap_{i=1}^{n} p_{\alpha_i}^{-1}(U_{\alpha_i}) \subset U.$$

Thus X is regular.   □

EXERCISE 8.2

1. Show that regularity is an hereditary and topological property.

2. Let $A = \{1/n: n = 1, 2, 3, \ldots\}$ and let $\mathcal{T}$ denote the usual topology for ℝ.

    (a) Show that $\mathcal{B} = \mathcal{T} \cup \{\mathbb{R}\backslash A\}$ is a subbasis for a topology $\mathcal{T}'$ for ℝ.

    (b) Show that $(\mathbb{R}, \mathcal{T}')$ is Hausdorff.

    (c) Show that $(\mathbb{R}, \mathcal{T}')$ is not regular. (*Hint:* There do not exist disjoint open sets containing 0 and A.)

3. Prove that every compact Hausdorff space is regular.

4. **Definition:** *A space X is a $T_{2\frac{1}{2}}$-space or* **Urysohn space** *provided that for each pair a, b of distinct points of X, there exist open sets U and V with disjoint closures such that $a \in U$ and $b \in V$.*

    (a) Prove that each regular space is a Urysohn space.

    (b) Prove that each Urysohn space is Hausdorff.

    (c) Give an example of a Hausdorff space that is not Urysohn and an example of a Urysohn space that is not regular.

5. In Example 8.2.1, show that the subspace topology for ℝ as a subset of U is the discrete topology. Use this fact to justify the statement that $\mathbb{R}\backslash\{a\}$ is a closed subset of U.

8.3 NORMAL SPACES

Definition: *A T_1-space X is a* **T_4-space** *or* **normal space** *provided that for each pair A, B of disjoint closed sets in X there exist disjoint open sets U and V such that A is contained in U and B is contained in V.*

Since each singleton set in a normal space is closed, it follows easily that every normal space is regular. The following characterizations of normality can be proved by methods parallel to those used for Theorems 8.3 and 8.4.

Theorem 8.6:   *A T_1-space X is normal if and only if for each closed subset A of X and open set U containing A, there is an open set W containing A whose closure is contained in U.*

Theorem 8.7:   *A T_1-space X is normal if and only if for each pair A, B of disjoint closed sets in X there exist open sets U and V such that*

$$A \subset U, \quad B \subset V, \quad \bar{U} \cap \bar{V} = \varnothing.$$

In our new terminology, the corollary to Theorem 6.5 can be rephrased as follows:

Theorem 8.8:   *Every compact Hausdorff space is normal.*

The next theorem is considerably stronger.

Theorem 8.9:   *Every regular Lindelöf space is normal.*

Proof: *Let A and B be disjoint closed subsets of a regular Lindelöf space X. By Theorem 8.4, there is for each a in A an open set O_a whose closure does not intersect B. Let $\mathcal{A}$ denote the resulting open cover of A. By the same argument, there is also an open cover $\mathcal{B}$ of B by open sets whose closures do not intersect A. Then $\mathcal{A} \cup \mathcal{B} \cup \{X \backslash (A \cup B)\}$ is an open cover of X. Since X is Lindelöf, this open cover has a countable subcover. Thus there are countable sequences $\{U_n\}_{n=1}^{\infty}$ and $\{V_n\}_{n=1}^{\infty}$ of open sets such that*

$$A \subset \bigcup_{n=1}^{\infty} U_n, \quad B \subset \bigcup_{n=1}^{\infty} V_n, \quad \bar{U}_n \cap B = \varnothing, \quad \bar{V}_n \cap A = \varnothing, \quad n = 1, 2, \dots.$$

For each positive integer n, let

$$U'_n = U_n \backslash \bigcup_{i=1}^{n} \bar{V}_i, \quad V'_n = V_n \backslash \bigcup_{i=1}^{n} \bar{U}_i.$$

Note that U'_n and V'_m are disjoint for all integers m and n. (For $m \le n$, $\bar{V}_m$ was subtracted in the construction of U'_n, so $U'_n \cap V'_m = \varnothing$.) Since $\bar{U}_n$ is disjoint from B and $\bar{V}_n$ is disjoint from A for all n, then

$$A \subset \bigcup_{n=1}^{\infty} U'_n, \quad B \subset \bigcup_{n=1}^{\infty} V'_n.$$

Since each finite union of closed sets is closed, then U'_n and V'_n are open sets for each integer n. Thus

$$U = \bigcup_{n=1}^{\infty} U'_n, \quad V = \bigcup_{n=1}^{\infty} V'_n$$

are disjoint open sets containing A and B, respectively, so X is normal.   □

Corollary: *Every second countable regular space is normal.*

The next theorem will be useful in several examples. The proof uses facts about cardinal numbers developed in Problems 9 and 10 of Exercise 2.2. The following terminology will also be helpful. A set A which is equipotent to [0, 1] is said to have the *cardinal number of the continuum*, and we write card $A = c$. Since [0, 1] and $\mathbb{R}$ have the same cardinal number, then card $\mathbb{R} = c$.

Theorem 8.10:   *If X is a separable normal space and E a subset of X with card $E \geq c$, then E has a limit point in X.*

Proof: *Suppose to the contrary that X is a normal space with countable dense subset D and subset E such that card $E \geq c$ and E has no limit point. Then for each subset Y of E, Y and $E \backslash Y$ are disjoint closed sets. Since X is normal, there exist disjoint open sets U_Y and V_Y containing Y and $E \backslash Y$, respectively. Consider the function $h: \mathcal{P}(E) \rightarrow \mathcal{P}(D)$ from the power set of E to the power set of D defined as follows:*

$$h(Y) = U_Y \cap D, \quad Y \in \mathcal{P}(E).$$

The denseness of D will allow us to conclude that h is one-to-one. Suppose Y_1, Y_2 are distinct members of $\mathcal{P}(E)$. Then there is some point y in one of the two sets but not in the other. For definiteness, suppose $y \in Y_1$ and $y \notin Y_2$. Then $y \in U_{Y_1}$ and $y \in V_{Y_2}$, so $U_{Y_1} \cap V_{Y_2}$ is a non-empty open set in X. Since D is dense, then

$$U_{y_1} \cap V_{y_2} \cap D \neq \emptyset.$$

But any point in this set is in $h(Y_1) = U_{Y_1} \cap D$ but not in $h(Y_2)$, since $h(Y_2)$ is disjoint from V_{y_2}. Thus if $Y_1 \neq Y_2$, then $h(Y_1) \neq h(Y_2)$ and h is one-to-one. Then

$$card\ \mathcal{P}(E) \leq card\ \mathcal{P}(D).$$

But since D is countable and card $E \geq c$,

$$card\ \mathcal{P}(D) \leq card\ \mathbb{R} < card\ \mathcal{P}(\mathbb{R}) \leq card\ \mathcal{P}(E),$$

and the relation for the power sets should be

$$\text{card } \mathcal{P}(D) < \text{card } \mathcal{P}(E).$$

This contradiction shows that E must have a limit point.   □

Example 8.3.1   A Regular Space that is Not Normal

Let X denote the real line with the half-open interval topology of Example 4.3.4. As we shall see, the half-open interval space X is both regular and normal. It is the product space $S = X \times X$, called the *Sorgenfrey plane,* that is regular but not normal.

In order to establish that X is regular, note first that each basic open set $[a, b)$, $a < b$, is also closed. If $a \in X$ and C is a closed set not containing a, then there is a basic open set $[a, b)$ contained in $X\backslash C$. Then $[a, b)$ and $X\backslash [a, b)$ are disjoint open sets containing a and C, respectively. Thus X is regular.

It is left as an exercise for the reader to show that X is Lindelöf. In view of Theorem 8.9, X must be normal.

Now consider the Sorgenfrey plane $S = X \times X$, which has as a basis all product sets of the form

$$[a, b) \times [c, d).$$

Inspection of Figure 8.2 reveals that this basis imposes the discrete topology on the diagonal line

$$E = \{(x, y) \in X \times X : x + y = 1\}.$$

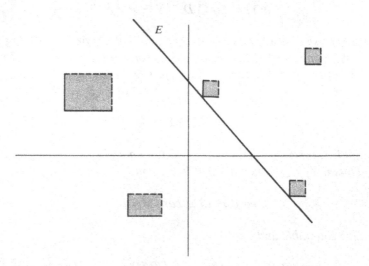

FIGURE 8.2

Since no point outside E is a limit point of E, then E is a subset of S such that card $E = c$ and E has no limit point.

The set of rational numbers is dense in X, so X is separable. Since the product of two separable spaces is separable, then S is separable. Theorem 8.10 now applies to show that S is not normal. If S were normal, the set E would have a limit point. Note, however, that S is regular since the product of regular spaces is regular.

Thus the Sorgenfrey plane is a regular space that is not normal. In addition, the Sorgenfrey plane is first countable and separable but not Lindelöf. (If S were Lindelöf, then it would be normal by Theorem 8.9.) Since X is Lindelöf and normal, this example also shows that the product of two Lindelöf spaces can fail to be Lindelöf and that the product of two normal spaces can fail to be normal.

Theorem 8.11: *Every metric space is normal.*

Proof: *Let A and B be disjoint closed subsets of a metric space (X, d). For each x in A, let δ_x be a positive number such that the open ball $B(x, \delta_x)$ is disjoint from B. For each y in B, let δ_y be a positive number such that $B(y, \delta_y)$ is disjoint from A. Then*

$$U = \bigcup_{x \in A} B(x, \tfrac{1}{2}\delta_x), \quad V = \bigcup_{y \in B} B(y, \tfrac{1}{2}\delta_y)$$

are open sets containing A and B, respectively. To see that U and V are disjoint, suppose $U \cap V \neq \varnothing$. Then there is an $x \subset A$ and a $y \in B$ such that

$$B(x, \tfrac{1}{2}\delta_x) \cap B(y, \tfrac{1}{2}\delta_y) \neq \varnothing.$$

Then

$$d(x, y) \leq \tfrac{1}{2}\delta_x + \tfrac{1}{2}\delta_y < \tfrac{1}{2}d(x, y) + \tfrac{1}{2}d(x, y) = d(x, y),$$

an obvious contradiction. Thus U and V are disjoint, so X is normal. $\square$

Theorem 8.11, under slightly different terminology, was assigned as a problem in Exercise 3.2.

There are separation properties other than T_0, T_1, T_2, T_3, T_4, and the metric property. We shall see a most important one, sometimes called the $T_{3\frac{1}{2}}$ property, in the next section. Several others are introduced in the exercises for this chapter. Incidentally, metric spaces are not called T_5-spaces. The T_5 designation is usually applied to completely normal spaces, which are defined in the exercise for this section. The Suggestions for Further Reading at the end of the chapter contain additional information about the separation properties.

EXERCISE 8.3

1. Show that every closed subspace of a normal space is normal.

2. Prove Theorems 8.6 and 8.7.

3. Prove that every locally compact Hausdorff space is regular. (*Hint:* Consider the one-point compactification.)

4. Prove:

   (a) Every closed subspace of a Lindelöf space is Lindelöf.

   (b) Every uncountable subset of a Lindelöf space has a limit point.

5. **Definition:** *A space X is **completely normal** if for each pair A, B of separated subsets of X there exist disjoint open sets U and V such that A is contained in U and B is contained in V.*

   Prove:

   (a) Every completely normal space is normal.

   (b) Every metric space is completely normal.

   (c) A space X is completely normal if and only if every subspace of X is normal.

6. Consider the closed upper half plane $M = \{(x_1, x_2) \in \mathbb{R}^2 : x_2 \geq 0\}$ with the topology defined as follows: For $x = (x_1, x_2) \in M$ with $x_2 > 0$, a local basis at x consists of all open balls $B(x, r)$, $r < x_2$, in the usual metric d. For a point $z = (z_1, 0)$ in M and $r > 0$, let $A(z, r)$ be the union of $\{z\}$ with the open ball of radius r with center at (z_1, r). Thus $A(z, r)$ is an open ball tangent to the horizontal axis together with the point z of tangency. The collection of sets $A(z, r)$, $r > 0$, is a local basis at z.

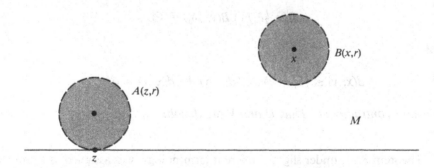

FIGURE 8.3

The set M with the topology generated by all sets of the form $B(x, r)$, $A(z, r)$ described above is called the *Moore plane* or the *Niemytzki plane*.

Prove the following properties of the Moore plane M.

   (a) M is regular and separable.

(b) M is neither Lindelöf nor normal.

(c) M is first countable.

(d) M is not locally compact.

8.4 SEPARATION BY CONTINUOUS FUNCTIONS

Definition: *Let X be a space and f: X → ℝ a continuous real-valued function on X. For subsets A and B of X, f* **separates** *A and B provided that there exist distinct real numbers a and b such that f(A) = a and f(B) = b. A function which separates a singleton set A = {x} from a set B is said to* **separate the point** *x* **from the set** *B; a function which separates the singleton sets A = {x} and B = {y} is said to* **separate the points** *x and y.*

In the terminology of the preceding definition, a function $f: X \to ℝ$ is one-to-one if and only if f separates each pair of distinct points of X.

Example 8.4.1

(a)   The projection map $p_1 \colon ℝ^2 \to ℝ$ defined by

$$p_1(x_1, x_2) = x_1, \quad (x_1, x_2) \in ℝ^2,$$

separates each pair A, B of distinct vertical lines.

(b)   The function $f \colon ℝ \to ℝ$ defined by

$$f(x) = \begin{cases} 0 & \text{if } x \le 0 \\ x & \text{if } 0 < x < 1 \\ 1 & \text{if } x \ge 1 \end{cases}$$

separates the sets $A = (-\infty, 0]$ and $B = [1, \infty)$.

(c)   For each pair $x = (x_1, x_2)$ and $y = (y_1, y_2)$ of distinct points of $ℝ^2$, there is a continuous function $f \colon ℝ^2 \to ℝ$ which separates x and y. In fact, f may be taken to be the projection p_1 on the first coordinate or the corresponding projection p_2 on the second coordinate. Note that this example did not claim that there is one function that separates all distinct pairs of points of $ℝ^2$; the claim was that once the points x, y were specified, then a function could be found to separate them.

Theorem 8.12:   *Let X be a T_1-space.*

(a)   *If for each pair x, y of distinct points of X there is a continuous function $f: X \to \mathbb{R}$ which separates x and y, then X is Hausdorff.*

(b)   *If for each point x in X and closed set C not containing x there is a continuous function $f: X \to \mathbb{R}$ that separates x and C, then X is regular.*

(c)   *If for each pair A, B of disjoint closed sets in X there is a continuous function that separates A and B, then X is normal.*

Proof: *The following argument is for part (c). The completely analogous arguments for (a) and (b) are left to the reader.*

For disjoint closed sets A and B in X, let $f: X \to \mathbb{R}$ be a continuous function that separates A and B. Thus $f(A) = a$ and $f(B) = b$ for some distinct real numbers a and b. Since $\mathbb{R}$ is Hausdorff, there exist disjoint open sets O_a and O_b containing a and b, respectively. Then

$$U = f^{-1}(O_a), \quad V = f^{-1}(O_b)$$

are disjoint open sets in X containing A and B, respectively, so X is normal.   ☐

The reader should note that the only property of $\mathbb{R}$ used in the proof of Theorem 8.12 is the fact that it is Hausdorff, so $\mathbb{R}$ could be replaced in the theorem by an arbitrary T_2-space. In practice, however, most functions used to separate points or sets are real-valued.

The implications of parts (a) and (b) of Theorem 8.12 are not reversible. In other words, there are examples of Hausdorff spaces X with distinct points x and y for which no continuous real-valued function on X separates x and y. There are also examples of regular spaces X with point x and closed set C not containing x for which no continuous real-valued function on X separates x and C. The condition of part (c), however, is equivalent to normality, and the proof of this is the primary object of the present section. This celebrated result, known as Urysohn's Lemma, is one of the most remarkable theorems of topology. A definition and two preparatory lemmas will be needed.

Definition: *A **dyadic number** is a number which can be expressed as a quotient of two integers in which the denominator is a power of 2.*

Thus the dyadic numbers are all the rational numbers which can be expressed in the form r/s where r is an integer and s is $2^0 = 1, 2, 2^2 = 4, 2^3 = 8, \ldots$.

The proof of the first lemma is left as an exercise.

Lemma 1:   *The set of dyadic numbers is dense in $\mathbb{R}$.*

The second lemma is technical in nature, but its utility will be apparent near the end of the proof of Urysohn's Lemma. Those who wish to skip Lemma 2 temporarily and refer to it when it is needed are invited to do so.

Lemma 2: *Let X be a space and D a dense subset of the set $\mathbb{R}^+$ of non-negative real numbers. Suppose that for each member t of D there is an open set U_t in X such that:*

(a) if $t_1 < t_2$, then $\bar{U}_{t_1} \subset U_{t_2}$, and

(b) $\bigcup_{t \in D} U_t = X$.

Then the function $f: X \rightarrow \mathbb{R}$ defined by

$$f(x) = glb\{t \in D : x \in U_t\}, \quad x \in X,$$

is continuous.

Proof: *It should first be noted that f is well-defined. This follows easily from the facts that each point x in X is a member of at least one set U_t and every subset of $\mathbb{R}^+$ has a greatest lower bound.*

By Theorem 4.11, continuity can be proved by showing that there is a subbasis $\mathcal{S}$ for the topology of $\mathbb{R}$ such that $f^{-1}(S)$ is open in X for each S in $\mathcal{S}$. Consider the usual subbasis $\mathcal{S}$ consisting of all subsets of the form $(-\infty, a)$ and (a, ∞), $a \in \mathbb{R}$.

For the first type of subbasic open set,

$$f^{-1}(-\infty, a) = \{x \in X : f(x) < a\} = \bigcup \{U_t : t \in D, \ t < a\}.$$

The last equality follows from the fact that $f(x) < a$ if and only if $x \in U_t$ for some $t < a$. Thus $f^{-1}(-\infty, a)$ is a union of open sets and is therefore open in X.

For a subbasic open set (a, ∞), $a \in \mathbb{R}$, consider the complement

$$X \setminus f^{-1}(a, \infty) = \{x \in X : f(x) \leq a\}.$$

It will be shown that $\{x \in X : f(x) \leq a\}$ is the set $\cap \ \{\bar{U}_t : t \in D, \ t > a\}$. Now if $f(x) \leq a$ and t is a member of D with $a < t$, then there is a member s of D with $s < t$ and $x \in U_s$. Then

$$x \in U_s \subset U_t \subset \bar{U}_t$$

so $x \in \bar{U}_t$ for all members t of D with $t > a$. Thus

$$\{x \in X : f(x) \leq a\} \subset \cap \ \{\bar{U}_t : t \in D \text{ and } t > a\}.$$

Suppose now that x belongs to the intersection of all $\bar{U}_t$ for which $t \in D$ and $t > a$. Let ϵ be a positive number. Since D is dense in $\mathbb{R}$, there is a member s_1 of D with

$a < s_1 < a + \epsilon$. Then $x \in \bar{U}_{s_1}$. Again using the denseness of D, there is a member s_2 of D such that $s_1 < s_2 < a + \epsilon$. Then

$$\bar{U}_{s_1} \subset U_{s_2},$$

so $x \in U_{s_2}$. Hence

$$f(x) \le s_2 < a + \epsilon.$$

Since ϵ was an arbitrary positive number, it follows that $f(x) \le a$. Thus

$$X \setminus f^{-1}(a, \infty) = \{x \in X : f(x) \le a\} = \cap \{\bar{U}_t : t \in D \quad and \quad t > a\}$$

is the intersection of closed sets and is therefore closed. Thus $f^{-1}(a, \infty)$ is open in X, and f is continuous. $\square$

Theorem 8.13:   Urysohn's Lemma   *In order that a T_1-space X be normal it is necessary and sufficient that for each pair A, B of disjoint closed subsets of X there exist a continuous function $f: X \to [0, 1]$ such that $f(A) = 0$ and $f(B) = 1$.*

Proof: *The sufficiency of the condition is proved by Theorem 8.12. For the necessity, consider disjoint closed subsets A and B of X. Let D denote the set of positive dyadic numbers. Since A is contained in the open set $X \setminus B$, Theorem 8.6 guarantees the existence of an open set $U_{\frac{1}{2}}$ such that*

$$A \subset U_{\frac{1}{2}}, \quad \bar{U}_{\frac{1}{2}} \subset X \setminus B.$$

Again applying Theorem 8.6, there exist open sets $U_{\frac{1}{4}}$ and $U_{\frac{3}{4}}$ such that

$$A \subset U_{\frac{1}{4}}, \; \bar{U}_{\frac{1}{4}} \subset U_{\frac{1}{2}}, \; \bar{U}_{\frac{1}{2}} \subset U_{\frac{3}{4}}, \; \bar{U}_{\frac{3}{4}} \subset X \setminus B.$$

By inductive application of the same reasoning, there exists for each dyadic rational number t between 0 and 1 an open set U_t such that

$$A \subset U_t, \quad \bar{U}_t \subset X \setminus B$$

and, for $s < t$, $\bar{U}_s \subset U_t$. We extend this collection to a family of open sets U_t, one for each positive dyadic number t, by defining $U_t = X$ for $t \ge 1$.
    According to Lemma 2, the function $f: X \to \mathbb{R}$ defined by

$$f(x) = glb\{t \in D : x \in U_t\}$$

is continuous. Since $A \subset U_t$ for all $t \in D$, then $f(x) = 0$ for each x in A. Since points of B lie in U_t only for $t \ge 1$, then $f(x) = 1$ for each x in B. $\square$

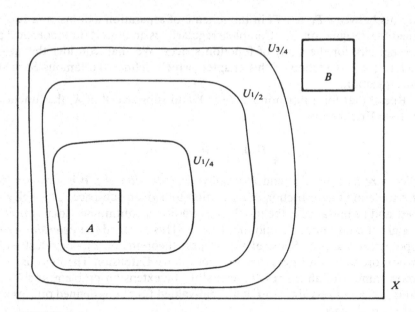

FIGURE 8.4

Corollary: *In order that a T_1-space X be normal it is necessary and sufficient that for each pair A, B of disjoint closed subsets of X there exist a continuous real-valued function f on X which separates A and B.*

Proof: *The sufficiency of the condition is proved by Theorem 8.12, and the necessity is an immediate consequence of Urysohn's Lemma.*   □

A function f of the type described in Urysohn's Lemma is called a *Urysohn function* for the closed sets A and B. Note that a Urysohn function for A and B maps each point of A to 0 and each point of B to 1. This means that

$$A \subset f^{-1}(0), \quad B \subset f^{-1}(1),$$

but it does not necessarily mean that A equals $f^{-1}(0)$ or that B equals $f^{-1}(1)$. Requiring equality produces a stronger property than normality. This property is developed in one of the exercises for this section.

Definition: *A **completely regular space** is a T_1-space X with the property that for each point x in X and each closed subset C with $x \notin C$, there is a continuous real-valued function on X which separates x and C.*

According to Theorem 8.12, every completely regular space is regular. By Urysohn's Lemma, every normal space is completely regular. Thus complete regularity fits between T_3 and T_4 in the scheme of separation axioms, and it is often assigned the designation $T_{3\frac{1}{2}}$. Complete regularity is an important topological property, especially for the study of function spaces. We shall consider this property further in the last section of this chapter, which defines the famous Stone-Čech compactification.

Recall that for a function $F: X \to Y$ and subspace A of X, the function $f = F|_A: A \to Y$ defined by

$$f(x) = F(x), \quad x \in A,$$

is called a *restriction* of F, and F is called an *extension* of f. It is always a simple matter to define the restriction of a function to a given subspace. Of much greater interest and importance is the problem of finding a continuous extension $F: X \to Y$ of a given continuous function $f: A \to Y$. This is called the *extension problem* in topological research. Some of the deepest theorems of topology deal with the extension problem. Our next theorem, the Tietze Extension Theorem, is perhaps the most famous of all the results related to the extension problem; it shows that continuous extensions always exist for real-valued functions defined on closed subsets of normal spaces.

Theorem 8.14:   The Tietze Extension Theorem   *Let X be a normal space, A a closed subset of X and $f: A \to \mathbb{R}$ a continuous function. Then f has a continuous extension $F: X \to \mathbb{R}$.*

Proof: *As the first step of the proof, note that the interval $[0, 1]$ may be replaced in Urysohn's Lemma by any closed interval $[a, b]$ with $a < b$. This is a direct consequence of the fact that $[0, 1]$ is homeomorphic to $[a, b]$ under a homeomorphism which maps 0 to a and 1 to b.*

Suppose then that X is a normal space, A a closed subset of X, and $f: A \to \mathbb{R}$ a continuous function. As a first case, we make the additional assumption that the image $f(A)$ is a subset of $[-1, 1]$. Let

$$A_1 = \{x \in A: f(x) \le -1/3\}, \quad B_1 = \{x \in A: f(x) \ge 1/3\}.$$

Then A_1 and B_1 are closed subsets of A, and since A is closed in X, then A_1 and B_1 are closed subsets of X. By Urysohn's Lemma with $[-1/3, 1/3]$ replacing $[0, 1]$, there is a continuous function

$$f_1: X \to [-1/3, 1/3]$$

such that

$$f_1(A_1) = -1/3, \quad f_1(B_1) = 1/3.$$

Note that for each x in A,

$$|f(x) - f_1(x)| \le 2/3.$$

Now consider the function f' defined on A by

$$f'(x) = f(x) - f_1(x), \quad x \in A.$$

Then f' maps A into the interval [−2/3, 2/3]. Let

$$A_2 = \{x \in A : f'(x) \le -2/9\}, \quad B_2 = \{x \in A : f'(x) \ge 2/9\}.$$

There is a Urysohn function

$$f_2 : X \to [-2/9, 2/9]$$

such that

$$f_2(A_2) = -2/9, \quad f_2(B_2) = 2/9.$$

Note that for each x in A,

$$|f(x) - (f_1(x) + f_2(x))| = |f'(x) - f_2(x)| \le 4/9 = (2/3)^2.$$

Proceeding inductively, there exists a sequence $\{f_n\}_{n=1}^{\infty}$ of continuous functions

$$f_n : X \to [2^{n-1}/3^n, 2^{n-1}/3^n]$$

such that

(1)
$$\left| f(x) - \sum_{n=1}^{N} f_n(x) \right| \le (2/3)^N, \quad x \in A.$$

Since the series $\sum_{n=1}^{\infty} 2^{n-1}/3^n$ is convergent and $|f_n(x)| \le 2^{n-1}/3^n$ for all x in X, then the series $\sum_{n=1}^{\infty} f_n(x)$ converges to a real number F(x) which lies in the interval between

$$\sum_{n=1}^{\infty} -2^{n-1}/3^n = -1 \quad and \quad \sum_{n=1}^{\infty} 2^{n-1}/3^n = 1.$$

Inequality (1) insures that F(x) = f(x) for all x in A.

It remains to be proved that F is continuous. Let x be a member of X and O an open set in $\mathbb{R}$ containing F(x). Let ϵ be a positive number for which the interval $(F(x) - \epsilon, F(x) + \epsilon)$ is contained in O. Let N be a positive integer for which

$$\sum_{n=N+1}^{\infty} (2/3)^n < \epsilon/2.$$

Since the functions $f_1, f_2, \ldots, f_N$ are continuous, there are open sets $V_1, V_2, \ldots, V_N$ in X containing x such that for y in V_n,

$$|f_n(x) - f_n(y)| < \epsilon/2N, \quad 1 \leq n \leq N.$$

Then if $y \in \cap_{n=1}^{N} V_n$,

$$|F(x) - F(y)| = \left| \sum_{n=1}^{\infty} f_n(x) - \sum_{n=1}^{\infty} f_n(y) \right|$$

$$\leq \sum_{n=1}^{N} |f_n(x) - f_n(y)| + \sum_{n=N+1}^{\infty} |f_n(x) - f_n(y)|$$

$$< N(\epsilon/2N) + \sum_{n=N+1}^{\infty} (2/3)^n < \epsilon/2 + \epsilon/2 = \epsilon.$$

Thus $V = \cap_{n=1}^{N} V_n$ is an open set in X containing x such that if $y \in V$,

$$F(y) \in (F(x) - \epsilon, F(x) + \epsilon) \subset O.$$

In other words,

$$F(V) \subset O, \quad V \subset F^{-1}(O),$$

so F is continuous. This completes the proof of the Tietze Extension Theorem under the assumption that $f(A) \subset [-1, 1]$.

To complete the proof, suppose that $f: A \to \mathbb{R}$ is simply a continuous function whose image may not be a subset of $[-1, 1]$. Since $\mathbb{R}$ is homeomorphic to each open interval, there is a homeomorphism $h: \mathbb{R} \to (-1, 1)$. Then $hf: A \to \mathbb{R}$ is a continuous function whose image $hf(A)$ is a subset of $[-1, 1]$. By what has already been proved, hf has a continuous extension $F': X \to [-1, 1]$.

There may be a temptation to claim at this point that $h^{-1}F'$ is the desired extension of f, but this would be wrong. The reason is that F' may take on the values -1 or 1 at which h^{-1} is not defined. This difficulty can be overcome, however, by one more Urysohn function. Let

$$B = \{x \in X: |F'(x)| = 1\}.$$

Then A and B are disjoint closed subsets of X so there is a Urysohn function
$g: X \rightarrow [0, 1]$ *such that*

$$g(B) = 0, \quad g(A) = 1.$$

Then for $x \in X$, *the product* $g(x) \cdot F'(x)$ *belongs to* $(-1, 1)$ *and is therefore in the*
domain of h^{-1}. *Define* $F: X \rightarrow \mathbb{R}$ *by*

$$F(x) = h^{-1}(g(x) \cdot F'(x)), \quad x \in X.$$

Then F is continuous, and for $a \in A$,

$$F(a) = h^{-1}(g(a) \cdot F'(a)) = h^{-1}(1 \cdot hf(a)) = f(a),$$

so F is the desired extension of f.   □

Note in the proof of the Tietze Extension Theorem that if f is bounded, then
the extension F may be chosen to be bounded also.

EXERCISE 8.4

1. Prove parts (a) and (b) of Theorem 8.12.

2. Prove that the set of dyadic numbers is dense in $\mathbb{R}$.

3. Suppose A and B are subsets of a space X and that there is a continuous function
    $f: X \rightarrow \mathbb{R}$ that separates A and B. Show that there is a continuous function $g: X \rightarrow$
    $\mathbb{R}$ for which $g(A) = 0$ and $g(B) = 1$.

4. Prove that a T_1-space X is completely regular if and only if for each point x in X and
    closed set C not containing x there is a continuous function $f: X \rightarrow \mathbb{R}$ such that
    $f(x) = 0$ and $f(C) = 1$.

5. Prove:

    (a)   Complete regularity is hereditary.

    (b)   The product of completely regular spaces is completely regular.

6. Show that the tangent function tan: $(-\pi/2, \pi/2) \rightarrow \mathbb{R}$ has no continuous extension to
    the one-point compactification of $(-\pi/2, \pi/2)$.

7. Let X be a T_1-space satisfying the extension condition of Theorem 8.14: For each
    continuous function $f: A \rightarrow \mathbb{R}$ from closed subset A of X to $\mathbb{R}$ there is a continuous
    extension $F: X \rightarrow \mathbb{R}$.

Prove that X is normal. (*Hint:* For disjoint closed subsets A and B of X, define $f: A \cup B \to \mathbb{R}$ by

$$f(x) = \begin{cases} 0 & \text{if } x \in A \\ 1 & \text{if } x \in B. \end{cases}$$

Extend to a Urysohn function.)

8. **Definition:** *A subset A of a space X is a G_δ in X provided that A is the intersection of a countable collection of open sets. A T_4-space X is **perfectly normal** if each closed set in X is a G_δ.*

   Prove:

   (a) Every metric space is perfectly normal.

   (b) X is perfectly normal if and only if for each pair A, B of disjoint closed sets in X there is a continuous function $f: X \to [0, 1]$ such that

   $$A = f^{-1}(0), \quad B = f^{-1}(1).$$

   (*Hint:* To see that perfect normality implies the stated condition, consider a closed set A. Then $A = \bigcap_{n=1}^{\infty} U_n$ where each set U_n is open, and there is a Urysohn function f_n for A and $X \backslash U_n$, $n \geq 1$. Let

   $$f_A(x) = \sum_{n=1}^{\infty} f_n(x)/2^n.$$

   For disjoint closed sets A and B, let

   $$f(x) = \frac{f_A(x)}{f_A(x) + f_B(x)}, \quad x \in X.)$$

9. Prove that a T_1-space X is completely regular if and only if for each x in X and closed set C not containing x there is a continuous function $f: X \to \mathbb{R}$ such that $f(x) \notin \overline{f(C)}$.

10. Let X be a normal space, A a closed subset of X, and $f: A \to \mathbb{R}^n$ a continuous function. Prove that f has a continuous extension $F: X \to \mathbb{R}^n$.

11. Let (X, d) be a metric space with disjoint, non-empty, closed subsets A and B. For x in X, define

$$f(x) = \frac{d(x, A)}{d(x, A) + d(x, B)}.$$

Show that f is continuous and conclude that every metric space is normal.

12. Let X be a space, $\{a_n\}_{n=1}^{\infty}$ a sequence of positive numbers for which $\sum_{n=1}^{\infty} a_n$ converges to a real number b, and $\{f_n\}_{n=1}^{\infty}$ a sequence of continuous functions $f_n: X \to \mathbb{R}$ for which

$$|f_n(x)| \leq a_n, \quad x \in X.$$

Prove: For each x in X, the series $\sum_{n=1}^{\infty} f_n(x)$ converges to a real number $f(x)$ in $[-b, b]$. The function f so defined is continuous. (*Hint:* Review the continuity part of the proof of the Tietze Extension Theorem.)

13. **Definition:** *Let X be a space, (Y, d) a metric space, and $\{f_n: X \to Y\}_{n=1}^{\infty}$ a sequence of functions from X to Y. Then $\{f_n\}_{n=1}^{\infty}$ **converges uniformly** to a function $f: X \to Y$ provided that for each positive number ϵ there is a positive integer N such that*

$$d(f_n(x), \quad f(x)) < \epsilon$$

for all $n \geq N$ and x in X.
    Prove: If a sequence $\{f_n\}_{n=1}^{\infty}$ of continuous functions from X to Y converges uniformly to a function $f: X \to Y$, then the limit function f is continuous. (*Hint:* See Example 3.7.6.)

14. (a) In the proof of the Tietze Extension Theorem (Theorem 8.14), show that F is continuous by proving it to be the limit of a uniformly convergent sequence of continuous functions.

    (b) In Problem 12 above, prove the continuity of the limit function f by showing that f is the limit of a uniformly convergent sequence of continuous functions.

15. Prove *Dini's Theorem:* Let $\{f_n: X \to \mathbb{R}\}_{n=1}^{\infty}$ be a sequence of continuous functions from a compact space X into $\mathbb{R}$ for which

$$f_n(x) \leq f_{n+1}(x), \quad x \in X, n = 1, 2, 3, \ldots.$$

Suppose that $\{f_n(x)\}_{n=1}^{\infty}$ converges for each x in X to a real number $f(x)$ and that the limit function $f: X \to \mathbb{R}$ is continuous. Then $\{f_n\}_{n=1}^{\infty}$ converges to f uniformly.

16. Let $f: A \to Y$ be a continuous function from a dense subset A of a space X into a Hausdorff space Y. Prove that f has at most one extension to a continuous function $F: X \to Y$.

17. Let X be a T_1-space with the following property: For each closed subset A of X and continuous function $f: A \to \{a, b\}$ from A into a discrete two-point space $\{a, b\}$, f has a continuous extension $F: X \to \{a, b\}$. Prove that X is totally disconnected.

8.5 METRIZATION

Metrizability, the property of being homeomorphic to a metric space, is the strongest of the separation properties considered in this text. Metric spaces are reasonably easy to work with and have been studied extensively. It is therefore important to have criteria which insure that the given topology of a space is generated by a metric. The problem of determining properties that imply metrizability is called the *metrization problem*. In this section we shall prove the famous Urysohn Metrization Theorem and state without proof a more recently discovered collection of necessary and sufficient conditions for metrizability.

Definition: *A topological space $(X, \mathcal{T})$ is **metrizable** provided that there is a metric d for X for which the metric topology generated by d is identical with the original topology $\mathcal{T}$.*

Theorem 8.15:   *The product of a countable collection of metric spaces is metrizable.*

Proof: *This result has been previously assigned as an exercise for both the finite case (Theorem 7.6) and the infinite case (Problem 13, Exercise 7.2). The proof presented here is for the infinite case. Let $\{(X_n, d_n)\}_{n=1}^{\infty}$ be a countably infinite collection of metric spaces. It must be shown that the product topology for $X = \prod_{n=1}^{\infty} X_n$ is generated by a metric. The metric to be used is similar to the product metric of Section 3.6, with allowance made for the infinite number of coordinates involved.*

As a preliminary step, note first that every metric space (Y, ρ) has an equivalent metric ρ' in which Y is bounded: ρ' is defined by

$$\rho'(y_1, y_2) = \frac{\rho(y_1, y_2)}{1 + \rho(y_1, y_2)}, \quad y_1, y_2 \in Y.$$

It is left as an exercise to show that ρ' is a metric equivalent to ρ and that (Y, ρ') has diameter at most 1. (One could also use the metric ρ'' defined by

$$\rho''(y_1, y_2) = minimum \ \{\rho(y_1, y_2), 1\}$$

to accomplish the same purpose.) This justifies choosing the metric d_n for X_n, $n \geq 1$, in such a way that X_n has diameter at most 1.

For $x = (x_1, x_2, \ldots)$ and $y = (y_1, y_2, \ldots)$ in X, define

$$d(x, y) = \left(\sum_{n=1}^{\infty} \left(\frac{d_n(x_n, y_n)}{n} \right)^2 \right)^{1/2}$$

Since $d_n(x_n, y_n) \leq 1$, the series under the radical sign is dominated by the convergent series $\sum_{n=1}^{\infty} 1/n^2$ and is itself convergent. Thus $d(x, y)$ has been meaningfully defined. Since each d_n is a metric, it follows easily that d is a metric.

Let $\mathcal{T}$ denote the product topology for X and $\mathcal{T}'$ the metric topology generated by d. It will be shown that $\mathcal{T} = \mathcal{T}'$ by showing that $\mathcal{T} \subset \mathcal{T}'$ and $\mathcal{T}' \subset \mathcal{T}$. Let O be a member of $\mathcal{T}$ and $x = (x_1, x_2, \ldots)$ a member of O. By the definition of the product topology, there is a positive integer N and, for $1 \leq n \leq N$, an open ball $B(x_n, r_n)$ of positive radius r_n in X_n such that

$$x \in \bigcap_{n=1}^{N} p_n^{-1}(B(x_n, r_n)) \subset O.$$

Here $p_n\colon X \to X_n$ denotes the projection map on the nth coordinate space X_n. If r is the minimum of the numbers $r_1, r_2/2, \ldots r_N/N$, it is easy to see that

$$x \in B_d(x, r) \subset \bigcap_{n=1}^{N} p_n^{-1}(B(x_n, r_n)) \subset O.$$

Then O must be a union of open balls generated by d, so $O \in \mathcal{T}'$. Thus $\mathcal{T} \subset \mathcal{T}'$.

To see that $\mathcal{T}' \subset \mathcal{T}$, let U' be a member of $\mathcal{T}'$ and $y = (y_1, y_2, \ldots)$ a member of U'. There is a positive number ϵ for which the open ball $B_d(y, \epsilon)$ is contained in U'. Let M be a positive integer for which

$$\sum_{n=M+1}^{\infty} 1/n^2 < \epsilon^2/2.$$

Then

$$V = \bigcap_{n=1}^{M} p_n^{-1}(B(y_n, \epsilon/\sqrt{2M}))$$

is an open set containing y in the product topology $\mathcal{T}$. The reason for having the radii of the balls in $X_1, \ldots, X_N$ be the rather bizarre number $\epsilon/\sqrt{2M}$ will become clear in a moment. For $z = (z_1, z_2, \ldots)$ in V,

$$
\begin{aligned}
d(y, z) &= \left(\sum_{n=1}^{\infty} \left(\frac{d_n(y_n, z_n)}{n} \right)^2 \right)^{1/2} \\
&= \left(\sum_{n=1}^{M} \left(\frac{d_n(x_n, y_n)}{n} \right)^2 + \sum_{n=M+1}^{\infty} \left(\frac{d_n(x_n, y_n)}{n} \right)^2 \right)^{1/2} \\
&< \left(\sum_{n=1}^{M} \left(\frac{\epsilon/\sqrt{2M}}{n} \right)^2 + \frac{\epsilon^2}{2} \right)^{1/2} < \left(\sum_{n=1}^{M} \left(\frac{\epsilon^2}{2M} \right) + \frac{\epsilon^2}{2} \right)^{1/2} \\
&= \left(\frac{\epsilon^2}{2} + \frac{\epsilon^2}{2} \right)^{1/2} = \epsilon.
\end{aligned}
$$

Thus $V \subset U'$, so U' is a union of members of $\mathcal{T}$ and is therefore a member of $\mathcal{T}$. Thus $\mathcal{T}' \subset \mathcal{T}$, so $\mathcal{T}' = \mathcal{T}$.   □

Infinite dimensional Euclidean space $\mathbb{R}^\infty$ and the Hilbert cube I^∞ were introduced in Example 7.2.1. It is an immediate consequence of Theorem 8.15 that both $\mathbb{R}^\infty$ and I^∞ are metrizable. This fact was previously shown in Example 7.2.1 by exhibiting embeddings of $\mathbb{R}^\infty$ and I^∞ in Hilbert space H. A similar embedding technique is the basis of the most famous partial solution of the metrization problem, the Urysohn Metrization Theorem, which is proved next. Urysohn's remarkable idea shows that certain spaces can be embedded in H by associating with each point of the domain space a sequence of real numbers determined by Urysohn functions.

Theorem 8.16:   The Urysohn Metrization Theorem   *Every second countable regular space is metrizable.*

Proof: *Let X be a second countable regular space with countable basis $\mathcal{B} = \{B_n\}_{n=1}^{\infty}$. By Theorem 8.8, X is normal. Consider the collection of all ordered pairs (i, j) of integers for which $\bar{B}_i \subset B_j$. By Urysohn's Lemma (Theorem 8.13), there is for each such pair (i, j) a Urysohn function $f: X \rightarrow [0, 1]$ such that*

$$f(\bar{B}_i) = 0, \quad f(X \backslash B_j) = 1.$$

Let $\mathcal{F}$ denote such a collection of Urysohn functions having one member for each ordered pair (i, j) for which $\bar{B}_i \subset B_j$. Since $\mathcal{F}$ is countable, then it can be indexed by the set of positive integers, $\mathcal{F} = \{f_n\}_{n=1}^{\infty}$.

Define a function $F: X \rightarrow H$ from X into Hilbert space H by

$$F(x) = \left(f_1(x), \frac{f_2(x)}{2}, \frac{f_3(x)}{3}, \ldots \right), \quad x \in X.$$

Thus the coordinates of $F(x)$ are determined by the values of the members of $\mathcal{F}$ at x; each value $f_n(x)$ is divided by n only to insure that $F(x)$ is a member of H:

$$\sum_{n=1}^{\infty} \left(\frac{f_n(x)}{n} \right)^2 \leq \sum_{n=1}^{\infty} 1/n^2,$$

so the sum of the squares of the coordinates of $F(x)$ is a convergent series of real numbers.

To show that F is an embedding, it is sufficient to show that F is a one-to-one, continuous and open function to the subspace $F(X)$ of H. Then the metrizability of X will follow from the fact that X is homeomorphic to a subspace of the metric space H.

Let x and y be distinct points of X. Since X is Hausdorff, there is a basic open set B_j in $\mathcal{B}$ for which

$$x \in B_j, \quad y \notin B_j.$$

By regularity, there is a basic open set B_i for which

$$x \in B_i, \quad \bar{B}_i \subset B_j.$$

Then the set $\mathcal{F}$ of Urysohn functions has a member f_n corresponding to (i, j) for which

$$f_n(\bar{B}_i) = 0, \quad f_n(X \backslash B_j) = 1.$$

Then

$$f_n(x) = 0, \quad f_n(y) = 1,$$

so $F(x)$ and $F(y)$ differ in their nth coordinates. Then $F(x) \neq F(y)$, so F is a one-to-one function.

To prove the continuity of F, let $x \in X$ and let $B(F(x), \epsilon)$ be an open ball in H of positive radius ϵ centered at x. It must be shown that there is an open set V in X containing x for which

$$F(V) \subset B(F(x), \epsilon).$$

Let N be a positive integer for which

$$\sum_{n=N+1}^{\infty} 1/n^2 < \epsilon^2/2.$$

Since each function f_n, $1 \leq n \leq N$, is continuous, there exist open sets V_n such that for y in V_n,

$$|f_n(x) - f_n(y)| < \epsilon/\sqrt{2N}, \quad 1 \leq n \leq N.$$

Then $V = \bigcap_{n=1}^{N} V_n$ is an open set containing x. A calculation similar to that used in the proof of continuity in the Tietze Extension Theorem (Theorem 8.14) shows that $F(V)$ is a subset of $B(F(x), \epsilon)$. This calculation is left to the reader.

It remains to be shown that F is an open mapping from X onto $F(X)$. Let W be open in X. It must be shown that there is an open set U in H for which

$$F(W) = U \cap F(X).$$

Consider a point x in W. There is a pair B_i, B_j of basic open sets for which

$$x \in B_i \subset \bar{B}_i \subset B_j \subset W.$$

Then there is a member f_n of $\mathcal{F}$ for which

$$f_n(x) = 0, \quad f_n(X \backslash W) = 1.$$

Then for any point F(y) in B(F(x), 1/n) ∩ F(X),

$$d(F(x), F(y)) < 1/n$$

so $f_n(y)$ cannot possibly be 1. Then y cannot belong to X\W, so y belongs to W. This means that

$$B(F(x), 1/n) \cap F(X) \subset F(W).$$

Since x was arbitrary, we conclude that F(W) is the union of such relatively open sets and is therefore an open set in the subspace topology of F(X). This completes the proof that F is an embedding, and we conclude that X is metrizable.   □

The next theorem is an important consequence of the Urysohn Metrization Theorem. It shows that the property of being a compact metric space is a continuous invariant provided that the range space is Hausdorff.

Theorem 8.17:   *Let X be a compact metric space, Y a Hausdorff space, and f: X → Y a continuous function from X onto Y. Then Y is metrizable.*

Proof: *Since the continuous image of a compact space is compact, then $Y = f(X)$ is compact. As a compact Hausdorff space, Y is normal by Theorem 8.8 and therefore regular. It remains to be proved that Y is second countable.*

As the reader proved in Chapter 6 or should prove now, the compact metric space X has a countable basis $\mathcal{B}$. Let $\mathcal{A}$ be the collection of all finite unions of members of $\mathcal{B}$. Then $\mathcal{A}$ is countable since it is the union of a countable collection of countable sets. For A in $\mathcal{A}$, let

$$A^* = Y \setminus f(X \setminus A).$$

Then $\mathcal{A}^* = \{A^*: A \in \mathcal{A}\}$ is a countable collection of subsets of Y. Showing that each member of $\mathcal{A}^*$ is an open set is left as an exercise.

To see that $\mathcal{A}^*$ is a basis for Y, let U be open in Y and y a member of U. Then $f^{-1}(y)$ is a compact subset of X, $f^{-1}(U)$ is an open subset of X, and

$$f^{-1}(y) \subset f^{-1}(U).$$

Since $f^{-1}(y)$ is compact, there is a finite number of members $B_1, \ldots, B_n$ of $\mathcal{B}$ such that

$$f^{-1}(y) \subset \bigcup_{i=1}^{n} B_i \subset f^{-1}(U).$$

Then $A = \bigcup_{i=1}^{n} B_i$ is a member of $\mathcal{A}$, and the corresponding set A^* is a member of $\mathcal{A}^*$. Since $f^{-1}(y)$ is a subset of A, then

$$y \in (Y \backslash f(X \backslash A)) = A^*.$$

For any point $f(x) \in A^*$, $f(x)$ is not in $f(X \backslash A)$, so x cannot be in $X \backslash A$. Thus $x \in A$. Since

$$A \subset f^{-1}(U),$$

then $f(x) \in U$. Thus, for each open set U in Y and $y \in U$, there is a member A^* of $\mathcal{A}^*$ such that

$$y \in A^* \subset U.$$

This shows that $\mathcal{A}^*$ is a countable basis for Y. The fact that Y is metrizable follows from the Urysohn Metrization Theorem (Theorem 8.16).   □

Theorem 8.18:   *The following conditions are equivalent for a topological space X:*

(a)   *X is regular and second countable.*
(b)   *X can be embedded in Hilbert space H.*
(c)   *X is metrizable and separable.*

Proof: *An argument that (a) implies (b) is given by the proof of the Urysohn Metrization Theorem (Theorem 8.16). For the proof that (b) implies (c), note that any space satisfying (b) is metrizable since it is homeomorphic to a subspace of the metric space H. Since it was shown in Chapter 4 that H is second countable, then each of its subspaces is second countable as well. Thus X is second countable, and therefore separable by Theorem 4.6. To show that (c) implies (a) and complete the circle of implications, recall that every metric space is normal (and therefore regular) and that every separable metric space is second countable (Theorem 4.8).*   □

Theorem 8.18 is a type of theorem to which topologists aspire. It completely characterizes the set of second countable regular spaces by showing that this collection is the same as the collection of (ostensibly nicer) separable metric spaces, which can all be considered subsets of Hilbert space. There are related criteria that are necessary and sufficient for a space X to be metrizable. Regularity of the space is one of the criteria, and the other is a condition involving open sets which is comparable to but considerably weaker than second countability. That condition and the complete solution to the metrization problem are stated next. The proof of the general metrization theorem is beyond the scope of this text.

Definition: *A family $\mathcal{A}$ of subsets of a space X is **locally finite** provided that for each x in X, there is an open set U_x containing x such that U_x intersects only finitely many members of $\mathcal{A}$. A family $\mathcal{B}$ of subsets of X is σ-**locally finite** provided that $\mathcal{B}$ is the union of a countable collection of locally finite families:*

$$\mathcal{B} = \bigcup_{n=1}^{\infty} \mathcal{B}_n$$

where each family $\mathcal{B}_n$ is locally finite.

Theorem 8.19:   The Nagata-Smirnov-Bing Metrization Theorem   *In order that a topological space X be metrizable, it is necessary and sufficient that X be regular and have a σ-locally finite basis.*

Suggestions for additional reading on the metrization problem and a proof of Theorem 8.19 are made at the end of the chapter.

EXERCISE 8.5

1. Show, as suggested in the proof of Theorem 8.15, that every metric space (Y, ρ) has an equivalent metric ρ' for which the diameter of Y does not exceed 1.

2. (a) Show that the function d in the proof of Theorem 8.15 is a metric.

    (b) Show that the metric d in the proof of Theorem 8.15 is equivalent to the metric d' defined for $x = (x_1, x_2, \ldots)$ in X by

    $$d'(x, y) = \sum_{n=1}^{\infty} \frac{d_n(x_n, y_n)}{n^2}.$$

3. Show by a direct proof that every second countable regular space can be embedded in $\mathbb{R}^{\infty}$.

4. In the proof of Theorem 8.17, show that each member of $\mathcal{A}^*$ is an open set.

5. Prove that a compact Hausdorff space is metrizable if and only if it is second countable.

6. Let (X, d) be a separable metric space with countable dense subset $A = \{a_n\}_{n=1}^{\infty}$. Then X has an equivalent metric under which X has a diameter less than or equal to 1, so we may assume that this property holds for the metric d. For x in X, let

    $$y_n = d(x, a_n)/n, \quad f(x) = (y_1, y_2, \ldots).$$

    Prove that f is an embedding of X in Hilbert space H.

7. Prove that every locally compact, second countable Hausdorff space is metrizable. Prove that every topological manifold is metrizable.

8.6 THE STONE-ČECH COMPACTIFICATION

It was shown in Section 8.5 that every second countable regular space is metrizable. The method of proof involved embedding in Hilbert space. This section exhibits a similar embedding for each completely regular space in a compact Hausdorff space.

It is left as an exercise to show that every subspace of a compact Hausdorff space is completely regular. The main result of this section shows more than the converse. Every completely regular space X can be embedded as a dense subspace of a compact Hausdorff space $\beta(X)$ having the remarkable property that every continuous, bounded, real-valued function on X has an extension to a continuous, bounded, real-valued function on $\beta(X)$. To see that this property is remarkable, consider the completely regular space $(0, 1]$ and the continuous, bounded, real-valued function $f(x) = \sin(1/x)$, whose values fluctuate ever more rapidly between -1 and $+1$ as x approaches 0. This function cannot be extended continuously to $[0, 1]$, yet $(0, 1]$ is a subspace of a compact Hausdorff space $\beta((0, 1])$ to which f can be extended continuously. As we shall see, this Stone-Čech compactification $\beta(X)$ is quite difficult to visualize, even for relatively simple cases. This discussion has shown, for example, that $\beta((0, 1])$ is definitely not $[0, 1]$.

Definition: *Let X be a Hausdorff space. A **compactification** of X is an ordered pair (Y, e) for which Y is a compact Hausdorff space and $e: X \to Y$ is an embedding whose image $e(X)$ is a dense subspace of Y.*

Note that the one-point compactification X_∞ of a non-compact, locally compact Hausdorff space X provides a compactification according to the preceding definition. The embedding $e: X \to X_\infty$ in this case is the inclusion map.

If (Y, e) is a compactification of X, it is common practice to identify X and $e(X)$ and to think of X as a subspace of Y. With this understanding, one often refers to Y as a compactification of X, suppressing the role of the embedding.

The next definition extends to general topological spaces the space $C(X, \mathbb{R})$ of bounded, continuous, real-valued functions defined for metric spaces in Chapter 3.

Definition: *For a given topological space X, the symbol $C(X, \mathbb{R})$ denotes the set of bounded, continuous, real-valued functions with domain X and topology determined by the **supremum metric** ρ defined by*

$$\rho(f, g) = lub \; \{|f(x) - g(x)| : x \in X\}, f, g \in C(X, \mathbb{R}).$$

The first connection between completely regular spaces and function spaces is revealed by the following theorem.

Theorem 8.20:   *If $(X, \mathcal{T})$ is a completely regular space, then the weak topology for X generated by $C(X, \mathbb{R})$ is the given topology $\mathcal{T}$.*

Proof: *Let $\mathcal{T}'$ denote the weak topology generated by $C(X, \mathbb{R})$. Since each member of $C(X, \mathbb{R})$ is continuous with respect to $\mathcal{T}$ and $\mathcal{T}'$ is the weakest topology having this property, then $\mathcal{T}' \subset \mathcal{T}$.*

To show the reverse inclusion, consider an open set U in X and let $C = X \backslash U$ denote its complement. For x in U, complete regularity guarantees the existence of a continuous function $f: X \to [0, 1]$ such that

$$f(x) = 0, \quad f(C) = 1.$$

Then x belongs to the $\mathcal{T}'$-open set $f^{-1}([0, 1/2))$ and this set is disjoint from C. Thus

$$x \in f^{-1}([0, 1/2)) \subset U,$$

and U must be a union of $\mathcal{T}'$-open sets. Thus U is open in $\mathcal{T}'$, so $\mathcal{T} \subset \mathcal{T}'$. Thus $\mathcal{T} = \mathcal{T}'$, and the proof is complete. $\square$

Theorem 8.21:   The Stone-Čech Theorem   *Let X be a completely regular space. Then there is a compact Hausdorff space $\beta(X)$ which contains X as a dense subspace and for which every member of $C(X, \mathbb{R})$ can be extended to a member of $C(\beta(X), \mathbb{R})$.*

Proof: *Before proving the theorem, the following interpretive remarks may be in order. To say that $\beta(X)$ contains X as a dense subspace means that there is an embedding $e: X \to \beta(X)$ for which $e(X)$ is dense in $\beta(X)$. To say that every member of $C(X, \mathbb{R})$ can be extended to a member of $C(\beta(X), \mathbb{R})$ means that for each member f of $C(X, \mathbb{R})$, there is a member F of $C(\beta(X), \mathbb{R})$ such that*

$$F(e(x)) = f(x), \quad x \in X.$$

The long-awaited compactification $\beta(X)$ is defined as follows: For f in $C(X, \mathbb{R})$, let I_f denote the smallest closed interval containing $f(X)$. Let

$$Y = \prod_{f \in C(X, \mathbb{R})} I_f$$

be the product of all the intervals I_f, $f \in C(X, \mathbb{R})$. Then Y is compact and Hausdorff since each interval I_f is compact and Hausdorff and any product of compact Hausdorff spaces is a compact Hausdorff space. Define $e: X \to Y$ by

$$e(x)(f) = f(x), \quad x \in X, f \in C(X, \mathbb{R}).$$

This seemingly abstruse notation is explained as follows: For x in X, $e(x)$ is to be a member of the product space Y, whose members are functions from the index set

C(X, ℝ) into the union of all the sets I_f such that the image of f is a member of I_f. Thus we define

$$e(x): C(X, ℝ) \to \cup I_f$$

to be the function whose value at f is the point $e(x)(f) = f(x)$ of I_f. To prove continuity for e, consider the composition $p_f e: X \to I_f$ with an arbitrary projection map p_f. For x in X,

$$p_f e(x) = e(x)(f) = f(x),$$

so $p_f e = f$, and $p_f e$ is continuous since f is. Thus the composition $p_f e$ of e with each projection map is continuous, so e is continuous by Theorem 7.8.

To see that e is one-to-one, consider distinct points x_1 and x_2 in X. Complete regularity insures the existence of a continuous function $f: X \to [0, 1]$ such that

$$f(x_1) = 0, \quad f(x_2) = 1.$$

Then $e(x_1)$ and $e(x_2)$ must be different since their values at f, namely $e(x_1)(f) = f(x_1)$ and $e(x_2)(f) = f(x_2)$, are different. Thus we conclude that e is one-to-one.

Establishing $e: X \to Y$ as an embedding now hinges on proving that e is an open mapping from X onto e(X). Since e is one-to-one, it is sufficient to show that there is a subbasis $\mathcal{S}$ for the topology of X for which e maps each member of $\mathcal{S}$ to an open set in e(X). By Theorem 8.20, the weak topology for X generated by C(X, ℝ) equals the given topology on X. This means that the family of open sets $f^{-1}(U), f \in C(X, ℝ)$ and U open in ℝ, is a subbasis $\mathcal{S}$ for the given topology of X. For such a subbasic open set $f^{-1}(U)$, note that

$$e(f^{-1}(U)) = p_f^{-1}(U) \cap e(X),$$

an open set in the subspace topology for e(X). Thus e is an open mapping from X onto e(X) and embeds X as a subspace of Y.

Let β(X) denote the closure of e(X) in Y,

$$β(X) = \overline{e(X)}.$$

Since closed subspaces of compact spaces are compact, then β(X) is compact. Since the Hausdorff property is hereditary, β(X) is a compact Hausdorff space containing X, which we identify with e(X), as a dense subspace.

It remains to be proved that each member of C(X, ℝ) can be extended to a member of C(β(X), ℝ). This is accomplished as follows: For f in C(X, ℝ) and x in X

$$f(x) = p_f e(x).$$

Thus the projection map p_f provides the desired extension. □

For a completely regular space X, the compact Hausdorff space $\beta(X)$ and embedding $e: X \to \beta(X)$ described in the Stone-Čech Theorem and its proof form a compactification $(\beta(X), e)$ of X. This compactification is called the *Stone-Čech compactification* of X. It has the remarkable property that every member of $C(X, \mathbb{R})$ can be extended to a member of $C(\beta(X), \mathbb{R})$.

The Stone-Čech compactification $(\beta(X), e)$ is unique in the following sense: If (Z, i) is a compactification of X for which each member of $C(X, \mathbb{R})$ can be extended to a member of $C(Z, \mathbb{R})$, then Z is homeomorphic to $\beta(X)$ under a homeomorphism $h: \beta(X) \to Z$ which is the identity map on X. In other words, the homeomorphism h satisfies the property $he = i$ in the diagram below.

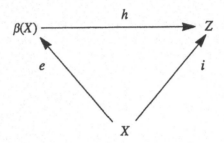

A proof of the uniqueness of the Stone-Čech compactification is outlined in the exercise for this section.

EXERCISE 8.6

1. Prove that every subspace of a compact Hausdorff space is completely regular.

2. Prove that every compact Hausdorff space can be embedded as a closed subspace of a product of intervals.

3. For a compact Hausdorff space X, prove that X and $\beta(X)$ are homeomorphic.

4. For a topological space X, show that $C(X, \mathbb{R})$ is a complete metric space.

5. (a) Prove that every product of intervals is completely regular.

   (b) Prove that every completely regular space can be embedded in a product of intervals.

6. Give an example of a completely regular space X whose Stone-Čech compactification is not homeomorphic to its one-point compactification.

7. For any completely regular space X and f in $C(X, \mathbb{R})$, prove that the extension of f to a member of $C(\beta(X), \mathbb{R})$ is unique.

8. Prove that in any product space, the subspace topology for a subspace equals the weak topology for the subspace generated by the restrictions of the projection maps to the subspace.

9. Let X be a normal space, A a closed subspace of X, Y a completely regular space, and $f: A \to Y$ a continuous function. Show that f has a continuous extension $F: X \to Z$,

where Z is a compact Hausdorff space which contains Y as a subspace. (*Hint:* Use the Tietze Extension Theorem (Theorem 8.14).)

10. The definition below extends to general topological spaces the concept of category defined for metric spaces in Section 3.7.

    Definition: *A topological space X that is the union of a countable family of nowhere dense sets is said to be of the **first category**. A space that is not of the first category is of the **second category**.*

    Prove that every compact Hausdorff space is of the second category.

11. This exercise is intended to show the uniqueness of the Stone-Čech compactification. The following terminology will simplify the discussion.

    Definition: *A space X is **C*-embedded** in a space Z provided that there is an embedding $i: X \to Z$ for which every member of $C(i(X), \mathbb{R})$ can be extended to a member of $C(Z, \mathbb{R})$.*

    The Stone-Čech Theorem asserts that for every completely regular space X, there is a compactification $(\beta(X), e)$ in which X is C^*-embedded by e. Prove the following to show that $\beta(X)$ is unique up to homeomorphism.

    (a) Let X and Y be completely regular spaces and $f: X \to Y$ a continuous function. Show that there is a unique continuous map $\hat{f}: \beta(X) \to \beta(Y)$ for which

    $$\hat{f}e_X = e_Y f.$$

    (b) Let Y be a compact Hausdorff space. Show that each continuous map $f: X \to Y$ has a unique continuous extension $\hat{f}: \beta(X) \to Y$.

    (c) Let (Z, i) be compactification of X where $i: X \to Z$ is an embedding such that every continuous map $f: X \to Y$ from X to a compact Hausdorff space Y has an extension to a continuous map $F: Z \to Y$. Then there is a homeomorphism $h: \beta(X) \to Z$ for which $he = i$. (Here $e: X \to \beta(X)$ is the usual embedding.)

    (d) Let (Z, i) be a compactification of X in which X is C^*-embedded by the embedding $i: X \to Z$. Then there is a homeomorphism $h: \beta(X) \to Z$ for which $he = i$. (*Hint:* Let $f: X \to Y$ be a continuous map from X into a compact Hausdorff space Y. Embed Y in a product T of closed intervals and show that $f: X \to Y$ can be continuously extended to $F: Z \to T$. Show that $F(Z)$ is actually a subset of Y and apply (c).)

SUGGESTIONS FOR FURTHER READING

For a more detailed treatment of separation and metrization, see Willard's *General Topology* or Munkres' *Topology: A First Course*. For an introduction to dimension theory, the classic treatise *Dimension Theory* by Hurewicz and Wallman is recommended.

HISTORICAL NOTES FOR CHAPTER 8

The numbering scheme for separation axioms or "Trennungsaxiomen" was introduced by Heinrich Tietze in 1923 to bring order to the various separation properties that had been proposed in the preceding two decades. The T_0 property is due to A. N. Kolmogorov; T_1-spaces were defined by Fréchet in 1907 under the name *accessible spaces;* and the T_2 property was used by Hausdorff in 1914 as one of his defining axioms for topological spaces. Regular spaces were first considered by Vietoris in 1921 and Tietze in 1923; completely regular spaces by Urysohn in 1924 and Tychonoff in 1930; normal spaces by Vietoris in 1921, Tietze in 1923, and Alexandroff and Urysohn in 1924; completely normal spaces by Tietze in 1923; and perfectly normal spaces by Urysohn in 1924 and Čech in 1932.

The normality of regular Lindelöf spaces (Theorem 8.9) was proved by Tychonoff in 1925. Theorem 8.10, which is widely used in examples, is due to F. B. Jones. The Sorgenfrey plane (Example 8.3.1) was defined by R. H. Sorgenfrey in 1947; the Moore plane or Niemytzki plane (Problem 6, Exercise 8.3) was considered independently by R. L. Moore, V. Niemytzki, and D. van Dantzig. The space of Example 8.2.1 is attributed to R. L. Moore and is sometimes called the Moore plane.

The important results of Sections 8.4 and 8.5 are due largely to the remarkable Russian mathematician Paul Urysohn (1898–1924). Urysohn's Lemma and the Urysohn Metrization Theorem date from 1924. Urysohn's preliminary ideas on embeddings in Hilbert space are illustrated by Problem 6 of Exercise 8.5, which he proved in 1923. The conclusion of the Tietze Extension Theorem (Theorem 8.14) was proved for closed subsets of $\mathbb{R}^2$ by Lebesgue in 1907, extended from the plane to metric spaces by Tietze in 1915, and proved for general normal spaces by Urysohn in 1924. In addition to the contributions already mentioned, Urysohn was one of the founders of the branch of topology known as *dimension theory,* which is not considered in this text. Urysohn's remarkable career was cut short by accidental drowning at the age of 26.

The general metrization theorem (Theorem 8.19) was proved independently by J. Nagata, Y. Smirnov, and R. H. Bing in 1950.

As was mentioned in the Historical Notes for Chapter 6, the Stone-Čech compactification was developed independently by M. H. Stone and E. Čech.

9 The Fundamental Group

9.1 THE NATURE OF ALGEBRAIC TOPOLOGY

In the first eight chapters we have dealt almost exclusively with point-set topology. This chapter introduces the fundamental group which, as the term "group" suggests, is an algebraic concept. Those who are not familiar with the basic properties of groups and homomorphisms should consult the Appendix or one of the standard textbooks on the subject before proceeding. Several excellent algebra texts are included in the supplementary reading list for this chapter.

The purpose of algebraic topology is to describe the structure of topological spaces by algebraic means, usually groups or rings. The algebraic structures involved are topological invariants in the sense that homeomorphic spaces are associated with isomorphic algebraic structures. Although algebraic topology and point-set topology share the common goal of classifying spaces by topological properties, the subjects are quite distinct in their historical development, emphasis, and methods. The development of point-set topology has been summarized in Chapter 1 and in the historical notes to the succeeding chapters. As we have seen, Cantor, Fréchet, and Hausdorff deserve the major credit for bringing together an amorphous and disparate collection of ideas about sets and continuous functions to form the subject of point-set topology. Algebraic topology, on the other hand, was introduced in the years 1895–1901 in remarkably modern form by the great French mathematician Henri Poincaré. Additional information about Poincaré and preliminary developments in algebraic topology are given in the historical notes at the end of the chapter.

Algebraic topology developed in response to specific geometric problems in Euclidean spaces. The purpose of the theory, roughly speaking, is to describe the connectivity or the "holes in the space" by algebraic methods. Connectivity properties are reflected in the algebraic properties of the associated groups. This chapter is restricted to the fundamental group, the first algebraic structure associated by Poincaré with topological spaces, and to its applications. The reader should be aware that algebraic topology is a very broad subject and that one chapter can give only a brief introduction to a few of its major aspects. Those interested in a more complete treatment should consult the supplementary reading list at the end of the chapter.

9.2 THE FUNDAMENTAL GROUP

The following examples, which deal with integration on multiply connected domains and with the classification of surfaces, are intended to illustrate the kind of analysis that led to the development of the fundamental group.

Example 9.2.1

(a)   Consider an annulus in the plane with closed paths A, B, C as shown in Figure 9.1.

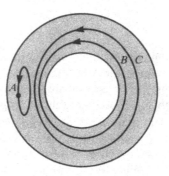

FIGURE 9.1

Let $F(x, y) = (p(x, y), q(x, y))$ be a continuous vector field defined on an open set containing the annulus and satisfying the exactness condition

$$\frac{\partial p}{\partial y} = \frac{\partial q}{\partial x}$$

According to Green's Theorem, $\int_A p\,dx + q\,dy = 0$ since the region interior to curve A is contained within the annulus. Since the region bounded by curves B and C is completely within the annulus, it follows also from Green's theorem that the curve integrals of the vector field over B and C are equal:

$$\int_B p\,dx + q\,dy = \int_C p\,dx + q\,dy.$$

Thus from the point of view of integrating exact vector fields, path A is trivial in the sense that integrals over it equal 0, and paths B and C are equivalent.

The geometric property producing these phenomena can be described intuitively as follows: Since the region inside curve A is completely contained within the annulus, then A can be "shrunk to a point" in the annulus. It is said that A is *homotopic to a constant path*. Analogously, paths B and C are *homotopic paths* since each can be continuously deformed into the other over a series of paths staying within the annulus. Note, however, that paths B and C are not homotopic to a constant path since they cannot be pulled across the "hole" that they enclose. The idea of homotopic paths will be defined rigorously in this section.

(b) Consider the problem of explaining the difference between a two-dimensional sphere S^2 and a torus. The difference, of course, is rather

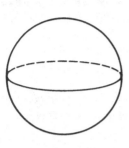

2-Sphere, S^2

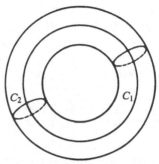

Torus, T

FIGURE 9.2

obvious. The torus encloses an inner region and has a "doughnut hole" while the sphere only encloses an inner region. In addition, the inner region enclosed by the sphere is different from the one enclosed by the torus, although this difference is difficult to describe rigorously. As we shall see later in the chapter, the idea of homotopic paths explains the difference clearly and rigorously. We shall see that every closed path in S^2 is homotopic to a constant path while the torus T has two basic types of paths, meridian circle C_1 and longitudinal circle C_2, which are not homotopic to constant paths. These facts, which seem intuitively plausible, are difficult to prove rigorously without a considerable amount of preliminary work.

Paths and path connected spaces were introduced in Section 5.5. This section extends these ideas to describe the concepts of simple and multiple connectedness for general topological spaces. Throughout this chapter, the closed unit interval [0, 1] is denoted by I.

Definition: Let α, $\beta: I \to X$ be paths with common initial point $\alpha(0) = \beta(0)$ and common terminal point $\alpha(1) = \beta(1)$. Then α and β are **equivalent** or **homotopic modulo endpoints** provided that there is a continuous function $F: I \times I \to X$ such that

$$F(t, 0) = \alpha(t), \quad F(t, 1) = \beta(t), \quad t \in I,$$
$$F(0, s) = \alpha(0) = \beta(0), \quad F(1, s) = \alpha(1) = \beta(1), \quad s \in I.$$

*The function F is called a **homotopy** between the paths α and β. For s in I, the restriction of F to I × {s}, denoted F(•, s), is called the **s-level** of the homotopy.*

Definition: *A **loop** in a topological space X is a path in X with common initial and terminal point. The common value of the initial point and terminal point is the **base point** of the loop. Two loops α and β with common base point x_0 are **equivalent** or **homotopic modulo** x_0, denoted α $\sim_{x_0}$ β, provided that they are equivalent as paths. In other words, α $\sim_{x_0}$ β if there is a homotopy F: I × I → X for which*

$$F(•, 0) = α, \quad F(•, 1) = β, \quad F(0, s) = F(1, s) = x_0, \quad s \in I.$$

Since F(0, s) = F(1, s) = x_0 for all s in I, it is said that F is a "base point preserving homotopy" or that the base point "stays fixed throughout the homotopy."

The usual practice in studying the loops in a space X is to specify a point x_0 in X to serve as the base point for the loops under consideration. This point x_0 is called the *base point* of X.

Definition: *The loop c: I → X whose only value is the base point x_0 of X is called the **constant loop** at x_0. A loop that is equivalent to a constant loop is said to be **null-homotopic**.*

Example 9.2.2

Consider the annulus X shown in Figure 9.3 with base point x_0 and closed curves A, B, C which are the images of paths α, β, and γ, respectively. Here, α, β, and γ are vector-valued functions, and we assume that the parametrizations for β and γ are chosen in such a way that the line segment from β(t) to γ(t), t ∈ I, lies in the annulus.

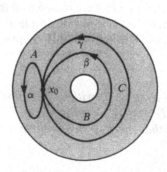

FIGURE 9.3

Then α is null-homotopic by the homotopy $F: I \times I \rightarrow X$ defined by

$$F(t, s) = (1 - s)\alpha(t) + sx_0, \quad (t, s) \in I \times I.$$

To see this, note that for fixed t, $F(t, s)$ defines the line segment joining x_0 and $\alpha(t)$. Thus for fixed s, the s-level $F(\bullet, s)$ is a path intermediate between $F(\bullet, 0) = \alpha$ and $F(\bullet, 1) = c$, the constant loop at x_0.

Note also that β and γ are equivalent by the homotopy $G: I \times I \rightarrow X$ defined by

$$G(t, s) = (1 - s)\beta(t) + s\gamma(t), \quad (s, t) \in I \times I.$$

Theorem 9.1:

(a) The relation of equivalence for paths is an equivalence relation.
(b) The relation of equivalence for loops is an equivalence relation.

Proof: *The following proof is for part (a); the obvious modifications needed to prove (b) are left as an exercise.*
If α is a path in a space X, the homotopy $F: I \times I \rightarrow X$ defined by

$$F(t, s) = \alpha(t), \quad (t, s) \in I \times I,$$

shows that α is equivalent to itself. Thus the relation is reflexive.
If α is equivalent to β by homotopy G with $G(\bullet, 0) = \alpha$ and $G(\bullet, 1) = \beta$, then $H: I \times I \rightarrow X$ defined by

$$H(t, s) = G(t, 1 - s), \quad (t, s) \in I \times I,$$

is a homotopy with 0-level β and 1-level α, which shows that β is equivalent to α. Thus equivalence of paths is a symmetric relation.
To prove the transitive property, suppose that α is equivalent to β by homotopy F and that β is equivalent to γ by homotopy G. Then the homotopy $K: I \times I \rightarrow X$ defined by

$$K(t, s) = \begin{cases} F(t, 2s) & 0 \le s \le 1/2, \quad t \in I \\ G(t, 2s - 1) & 1/2 \le s \le 1, \quad t \in I \end{cases}$$

has 0-level

$$K(\bullet, 0) = F(\bullet, 0) = \alpha$$

and 1-level

$$K(\bullet, 1) = G(\bullet, 1) = \gamma.$$

The continuity of K when s = 1/2 is shown by the Gluing Lemma since F(•, 1) = G(•, 0) = β. Thus equivalence of paths is a transitive relation and satisfies all the requirements of an equivalence relation.   □

We shall be concerned with the product of loops, as defined more generally for paths in Chapter 5. In order to facilitate the statements of several definitions and related properties, we assume for the remainder of this section that X represents a space with base point x_0 and that all loops mentioned are loops in X with x_0 as base point.

Definition: *The **product** of loops* α, β *in X with base point* x_0 *is the loop* $\alpha * \beta$ *defined by*

$$\alpha * \beta(t) = \begin{cases} \alpha(2t) & 0 \le t \le 1/2 \\ \beta(2t - 1) & 1/2 \le t \le 1. \end{cases}$$

Note that the product of loops α and β is simply their path product, which was defined in Chapter 5.

The following lemma asserts that equivalence of loops is preserved by the product operation. Its proof is left as an exercise.

Lemma: *If* $\alpha \sim_{x_0} \alpha'$ *and* $\beta \sim_{x_0} \beta'$, *then* $\alpha * \beta \sim_{x_0} \alpha' * \beta'$.

Definition: *For a loop* α *in X based at* x_0, *the **equivalence class** or **homotopy class** of* α, *denoted* $[\alpha]$, *is the set of all loops in X based at* x_0 *which are equivalent to* α. *The set of such equivalence classes is denoted by* $\prod_1 (X, x_0)$ *and is called the **fundamental group**, the **Poincaré group**, or the **first homotopy group** of X at* x_0. *For* $[\alpha]$, $[\beta]$ *in* $\prod_1 (X, x_0)$, *the group operation* ∘, *called the **product**, is defined by*

$$[\alpha] \cdot [\beta] = [\alpha * \beta].$$

It is, of course, necessary to prove that the set $\prod_1 (X, x_0)$ is actually a group. This is accomplished in several steps by the proof of the next theorem. Note that the lemma preceding the definition shows that the product of homotopy classes is a well-defined operation.

Theorem 9.2:   *If X is a space and* x_0 *a point of X, then* $\prod_1 (X, x_0)$ *is a group under the* ∘ *operation.*

Proof: *In order to conclude that* $\prod_1 (X, x_0)$ *is a group, it must be shown that (A) the operation* • *is associative, (B) there is a loop c for which* $[c]$ *is an identity element,* *and (C) each homotopy class [α] has an inverse [ᾱ] = [α]⁻¹ under the ∘ operation. Each of these is proved as a separate lemma.*

Lemma A: *The operation ∘ is associative.*

Proof: *It must be demonstrated that for all [α], [β], [γ] in $\Pi_1 (X, x_0)$,*

$$([α] \circ [β]) \circ [γ] = [α] \circ ([β] \circ [γ])$$

or, equivalently,

$$[(α * β) * γ] = [α * (β * γ)].$$

*The definition of the * operation shows that*

$$((α * β) * γ)(t) = \begin{cases} α(4t) & 0 \le t \le 1/4 \\ β(4t - 1) & 1/4 \le t \le 1/2 \\ γ(2t - 1) & 1/2 \le t \le 1 \end{cases}$$

and

$$(α * (β * γ))(t) = \begin{cases} α(2t) & 0 \le t \le 1/2 \\ β(4t - 2) & 1/2 \le t \le 3/4 \\ γ(4t - 3) & 3/4 \le t \le 1. \end{cases}$$

*The desired homotopy is obtained from Figure 9.4, by the following considerations. We desire a homotopy H: I × I → X which agrees with (α * β) * γ on the bottom and with α * (β * γ) on the top of the square. This can be accomplished by*

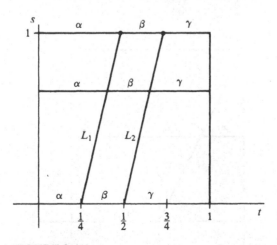

FIGURE 9.4

defining the s-level H(•, s) to follow the route of α from t = 0 horizontally across to the line L_1, the path of β from L_1 horizontally across to L_2, and the path of γ from L_2 horizontally across to t = 1.

Note that L_1 has equation s = 4t − 1 and that L_2 has equation s = 4t − 2. The reader should supply the elementary geometry necessary to derive the homotopy

$$H(t, s) = \begin{cases} \alpha(4t/(s + 1)) & 0 \le t \le (s + 1)/4 \\ \beta(4t - 1 - s) & (s + 1)/4 \le t \le (s + 2)/4 \\ \gamma((4t - 2 - s)/(2 - s)) & (s + 2)/4 \le t \le 1 \end{cases}$$

and verify that H is a homotopy modulo x_0 between (α ∗ β) ∗ γ and α ∗ (β ∗ γ). Note that the continuity of H is shown by applying the Gluing Lemma twice, to the union of three closed sets. This completes the proof of Lemma A.   □

Lemma B: *The homotopy class [c], where c is the constant loop whose only value is x_0, is the identity element for $\prod_1 (X, x_0)$.*

Proof: *Let us show first that*

$$[c] \cdot [\alpha] = [\alpha]$$

or, equivalently,

$$[c * \alpha] = [\alpha], \quad \alpha \in \prod_1 (X, x_0).$$

The reader should apply the method of Lemma A to Figure 9.5 to obtain the homotopy

$$K(t, s) = \begin{cases} x_0 & 0 \le t \le (1 - s)/2 \\ \alpha((2t + s - 1)/(s + 1)) & (1 - s)/2 \le t \le 1 \end{cases}$$

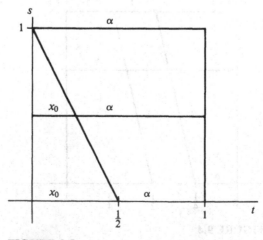

FIGURE 9.5

whose 0-level is c∗α and whose 1-level is α, thus demonstrating that [c] is an identity element for the ∘ operation when multiplied on the left. The proof that [c] also acts as an identity element when multiplied on the right is left as an exercise. Since the identity element in any group is unique, then [c] is the identity for $\prod_1 (X, x_0)$.   □

Lemma C: *For [α] in $\prod_1 (X, x_0)$, the inverse $[\alpha]^{-1}$ is the class of the reverse path $\bar{\alpha}(t) = \alpha(1 - t)$, $t \in I$.*

Proof: *Note that*

$$[\alpha] \cdot [\bar{\alpha}] = [\alpha * \bar{\alpha}]$$

and

$$\alpha * \bar{\alpha}(t) = \begin{cases} \alpha(2t) & 0 \le t \le 1/2 \\ \alpha(2 - 2t) & 1/2 \le t \le 1. \end{cases}$$

Thus α∗ᾱ follows the route of α and then reverses to follow the route of ᾱ back to the starting point x_0. We visualize a homotopy L for which the s-level L(∙, s) follows the route of α out to α(s) and then reverses its route back to x_0. This homotopy is defined by

$$L(t, s) = \begin{cases} \alpha(2ts) & 0 \le t \le 1/2 \\ \alpha(2s - 2ts) & 1/2 \le t \le 1. \end{cases}$$

It is easily observed that L is a homotopy between α∗ᾱ and c, showing that [ᾱ] is a right inverse of [α]. Since $\bar{\bar{\alpha}} = \alpha$, the above argument also shows that

$$[\bar{\alpha}] \cdot [\alpha] = [\bar{\alpha}] \cdot [\bar{\bar{\alpha}}] = [c],$$

so that $[\bar{\alpha}] = [\alpha]^{-1}$ is a two-sided inverse for [α]. This completes the proof of Lemma C and the proof that $\prod_1 (X, x_0)$ is a group.   □

Let X and Y be spaces with respective base points x_0 and y_0 and $f: X \to Y$ a continuous function for which $f(x_0) = y_0$. Then any loop α in X based at x_0 corresponds to the loop $f\alpha$ in Y based at y_0. It is left as an easy exercise for the reader to show that the function

$$f_*: \prod_1 (X, x_0) \to \prod_1 (Y, y_0)$$

defined by

$$f_*([\alpha]) = [f\alpha], \quad [\alpha] \in \prod_1 (X, x_0),$$

is a homomorphism. This function is called the *induced homomorphism* of the function *f*. Note also that if *f* is a homeomorphism, then the induced homomorphism *f*∗ is an isomorphism. Thus, the fundamental group is a topological invariant in the sense that homeomorphic spaces have isomorphic fundamental groups at base points which correspond under the homeomorphism.

In the next section we shall compute the fundamental group for several interesting spaces. First, however, it is necessary to establish some preliminary theorems.

Theorem 9.3:   *Let X be a path connected space and x_0, x_1 points of X. Then the fundamental groups $\prod_1 (X, x_0)$ and $\prod_1 (X, x_1)$ are isomorphic.*

Proof: *Let $\gamma: I \to X$ be a path from $\gamma(0) = x_0$ to $\gamma(1) = x_1$. Then for $[\alpha]$ in $\prod_1 (X, x_0)$, $(\bar{\gamma} * \alpha) * \gamma$ is a loop based at x_1. Thus we define a function f: $\prod_1 (X, x_0) \to \prod_1 (X, X_1)$ by*

$$f([\alpha]) = [(\bar{\gamma} * \alpha) * \gamma], \quad [\alpha] \in \prod_1 (X, x_0).$$

It is an easy exercise to show that the image of $[\alpha]$ under f is independent of the representative α in $[\alpha]$ so that f is well defined.

Analogously, define $g: \prod_1 (X, X_1) \to \prod_1 (X, x_0)$ by

$$g([\beta]) = [(\gamma * \beta) * \bar{\gamma}], \quad [\beta] \in \prod_1 (X, x_1).$$

Minor revisions in the proof of Lemma A show that $a * (b * c)$ and $(a * b) * c$ are equivalent paths for any paths a, b, c for which the indicated products are defined. Thus, in $[(\bar{\gamma} * \alpha) * \gamma]$, we may ignore the parentheses and write $[\bar{\gamma} * \alpha * \gamma]$ since the equivalence class is unaffected by the way in which the terms are associated. Similarly, Lemma C essentially shows that $[\gamma * \bar{\gamma}]$ and $[\bar{\gamma} * \gamma]$ are the identity elements of $\prod_1 (X, x_0)$ and $\prod_1 (X, x_1)$, respectively. With these observations, we are prepared to show that f is an isomorphism whose inverse is g.

To show that f is a homomorphism, note that for $[\alpha_1]$, $[\alpha_2]$ in $\prod_1 (X, x_0)$,

$$f([\alpha_1] \circ [\alpha_2]) = f([\alpha_1 * \alpha_2]) = [\bar{\gamma} * \alpha_1 * \alpha_2 * \gamma]$$
$$= [\bar{\gamma} * \alpha_1 * \gamma * \bar{\gamma} * \alpha_2 * \gamma] = [\bar{\gamma} * \alpha_1 * \gamma] \circ [\bar{\gamma} * \alpha_2 * \gamma]$$
$$= f([\alpha_1]) \circ f([\alpha_2]).$$

Thus f is a homomorphism.

For $[\alpha] \in \prod_1 (X, x_0)$,

$$gf([\alpha]) = g([\bar{\gamma} * \alpha * \gamma]) = [\gamma * \bar{\gamma} * \alpha * \gamma * \bar{\gamma}] = [\alpha].$$

Thus gf is the identity map on $\prod_1 (X, x_0)$, and a completely symmetric argument shows that fg is the identity map on $\prod_1 (X, x_1)$. Thus f is an isomorphism between $\prod_1 (X, x_0)$ and $\prod_1 (X, x_1)$. □

Because of the preceding theorem, it is common practice to omit mention of the base point for the fundamental group of a path connected space. Following this practice, we shall sometimes refer to $\prod_1 (X)$, the *fundamental group of X*, since the same abstract group results for each choice of the base point when X is path connected. Note, however, that Theorem 9.3 does not guarantee that the isomorphism between $\prod_1 (X, x_0)$ and $\prod_1 (X, x_1)$ is unique; different paths can produce different isomorphisms. In some applications, specification of the base point is important. For example, when comparing fundamental groups of spaces X and Y on the basis of a continuous map $f: X \rightarrow Y$, it is usually necessary to specify base points x_0 in X and y_0 in Y and to assume that $f(x_0) = y_0$.

Definition: *A **simply connected** space is a path connected space X whose fundamental group $\prod_1 (X)$ is the trivial group consisting only of an identity element.*

Definition: *A space X is **contractible to a point** x_0 in X provided that there is a continuous function F: $X \times I \rightarrow X$ such that*

$$F(x, 0) = x, \quad F(x, 1) = x_0, \quad F(x_0, s) = x_0, \quad x \in X, \quad s \in I.$$

*The function F is called a **contraction** of X to the point x_0. A space X is **contractible** if there is a point x_0 in X for which X is contractible to x_0.*

Intuitively speaking, a contractible space is one that can be "shrunk to a point" through a continuous deformation proceeding in stages from $t = 0$ to $t = 1$. Thinking of t as time, the 0-level $F(\bullet, 0)$ of the contraction is the identity map on X; successive levels $F(\bullet, t)$ shrink the space to the final stage $F(\bullet, 1)$ which maps all of X to the one point x_0. Our definition requires that $F(x_0, s) = x_0$ at each stage, so that the point x_0 stays fixed throughout the contraction process. The exercises for this section show how to relax this requirement.

Example 9.2.3

An interval $[a, b]$ on the real line is contractible to a. To see this, define F: $[a, b] \times I \rightarrow [a, b]$ by

$$F(x, t) = ta + (1 - t)x, \quad (x, t) \in [a, b] \times I.$$

Note that

$$F(x, 0) = x, \quad F(x, 1) = a, \quad x \in X,$$

so that over the "time interval" $0 \leq t \leq 1$ $F(x, t)$ moves from x at time 0 to a at time 1 along the line segment from x to a. Thus the contraction may be thought of as "squeezing" the interval to the point a. An analogous argument shows that $[a, b]$ is contractible to each of its points.

It is left as an exercise for the reader to modify the ideas of Example 9.2.3 and produce a proof of the following theorem:

Theorem 9.4:   *A convex set A in $\mathbb{R}^n$ is contractible to each point x_0 in A.*

Theorem 9.5:   *Every contractible space is simply connected.*

Proof: *Let X be a space contractible to a point x_0 in X by contraction F: $X \times I \to X$ satisfying*

$$F(x, 0) = x, \quad F(x, 1) = x_0, \quad F(x_0, t) = x_0, \quad x \in X, \quad t \in I.$$

To see first that X is path connected note that for a given point x, the function $f_x: I \to X$ defined by

$$f_x(t) = F(x, t), \quad t \in I,$$

is a path in X from x to x_0. Thus for x, y in X, the product $f_x * \bar{f}_y$ of the path f_x and the reverse of f_y is a path from x to y.
For $[\alpha] \in \prod_1 (X, x_0)$, define a homotopy $H: X \times I \to X$ by

$$H(t, s) = F(\alpha(t), s), \quad (t, s) \in I \times I.$$

The properties of the contraction F insure that H is a homotopy between $H(\cdot, 0) = \alpha$ and $H(\cdot, 1) = c$, the constant loop at x_0, demonstrating that $[\alpha] = [c]$. Thus $\prod_1 (X, x_0)$ consists only of an identity element, so X is simply connected.   □

From Theorems 9.4 and 9.5 we conclude that a single point, an interval, the real line, a disk, a rectangle, Euclidean n-space, and all other convex subspaces of Euclidean n-space have trivial fundamental group. Our first non-trivial example of a fundamental group occurs for the unit circle S^1 in the next section.

EXERCISE 9.2

1. Prove that equivalence of loops is an equivalence relation.

2. Prove Lemma B: The identity element for $\prod_1 (X, x_0)$ is the homotopy class $[c]$ determined by the constant loop c.

3. Prove that equivalence of loops is preserved by the $*$ product: If $\alpha \sim_{x_0} \alpha'$ and $\beta \sim_{x_0} \beta'$, then $\alpha * \beta \sim_{x_0} \alpha' * \beta'$.

4. Let γ be a loop in X with base point x_0 so that γ defines an isomorphism f: $\prod_1 (X, x_0) \to \prod_1 (X, x_0)$ as in the proof of Theorem 9.3. Show that f is the identity isomorphism if and only if $[\gamma]$ belongs to the center of $\prod_1 (X, x_0)$.

5. Let γ_1 and γ_2 be equivalent paths in X from x_0 to x_1. Show that the isomorphisms f_1 and f_2 determined by γ_1 and γ_2 in the proof of Theorem 9.3 are identical.

6. Give an example of a simply connected space that is not contractible.

7. (a) Let α and β be paths in a space X having common initial point x_0 and common terminal point x_1. Prove that α and β are equivalent if and only if the product $\alpha * \bar{\beta}$ is equivalent to the constant loop c whose only value is x_0.

   (b) Let X be a path connected space. Prove that X is simply connected if and only if each pair of paths in X having the same initial point and same terminal point are equivalent.

8. Prove Theorem 9.4.

9. Let $f: X \to Y$ be a continuous function on the spaces indicated and let x_0 be a point of X. Show that the function $f_*: \prod_1 (X, x_0) \to \prod_1 (Y, f(x_0))$ defined by

$$f_*([\alpha]) = [f\alpha], \quad [\alpha] \in \prod_1 (X, x_0),$$

is a homomorphism. Show in particular that f_* is a well-defined function.

10. Let $f: X \to Y$ be a homeomorphism from X onto Y and x_0 a point of X. Prove that $\prod_1 (X, x_0)$ and $\prod_1 (Y, f(x_0))$ are isomorphic.

11. Prove that the following spaces are contractible.

   (a) The upper hemisphere U of S^n: $U = \{x = (x_1 \ldots x_{n+1}) \in S^n: x_{n+1} \geq 0\}$.

   (b) The "punctured n-sphere" $S^n \backslash \{p\}$, where p is any particular point in S^n.

   (c) The topologist's comb (Example 5.6.1). (Find all base points to which the comb is contractible.)

12. **Definition:** *A space X is **weakly contractible** provided that there is a point x_0 in X and a continuous function $G: X \times I \to X$ such that*

$$G(x, 0) = x, \quad G(x, 1) = x_0, \quad x \in X.$$

*The function G is called a **weak contraction**.*

Thus the difference between a contraction on X and a weak contraction on X is in the fact that a weak contraction is not required to leave the base point x_0 fixed.

(a)   Give an example of a weakly contractible space that is not contractible.

(b)   Prove that each weakly contractible space is simply connected.

9.3 THE FUNDAMENTAL GROUP OF S^1

This section shows that the fundamental group of a circle is isomorphic to the additive group $\mathbb{Z}$ of integers. The function $p: \mathbb{R} \to S^1$ defined by

$$p(t) = (\cos 2\pi t, \sin 2\pi t), \quad t \in \mathbb{R},$$

to which we shall refer as the *covering projection* of $\mathbb{R}$ over S^1, will be instrumental in our computation of $\prod_1 (S^1)$. Note that p maps each integer k in $\mathbb{R}$ to the point $(1, 0)$ in S^1 and wraps each interval $[k, k + 1]$ around S^1 exactly once in the counterclockwise direction.

We shall use the point $(1, 0)$ as the base point of S^1 for the remainder of this chapter. For brevity, this base point will be denoted 1.

Definition: *Let X be a space and $f: X \to S^1$ a continuous function. A function $\tilde{f}$: $X \to \mathbb{R}$ for which $p\tilde{f} = f$ is called a **covering function** of f or a **lifting** of f to the real line $\mathbb{R}$.*

We shall be particularly interested in lifting a path $\alpha: I \to S^1$ to a covering path $\tilde{\alpha}: I \to \mathbb{R}$ and a homotopy $F: I \times I \to S^1$ to a covering homotopy $\tilde{F}$: $I \times I \to \mathbb{R}$.

The primary properties of the covering projection $p: \mathbb{R} \to S^1$ needed to produce liftings are developed in the following lemma.

Lemma: *There is a pair U_1, U_2 of path connected, open subsets of S^1 whose union is S^1 and for which $p: \mathbb{R} \to S^1$ maps each path component of $p^{-1}(U_i)$ homeomorphically onto U_i, $i = 1, 2$.*

Proof: *There are many possible ways to choose U_1 and U_2, one of which is the following. Let U_1 be the open arc on S^1 beginning at $(-1, 0)$ and extending counterclockwise to $(0, 1)$, and let U_2 be the corresponding open arc beginning at $(1, 0)$ and extending counterclockwise to $(0, -1)$.*

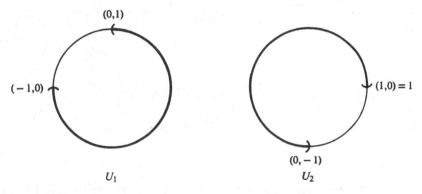

FIGURE 9.6

Then U_1 and U_2 are clearly path connected open subsets of S^1 whose union equals S^1. From the definition of the covering projection p, it follows easily that

$$p^{-1}(U_1) = \bigcup_{k=-\infty}^{\infty} (k - 1/2, k + 1/4),$$

$$p^{-1}(U_2) = \bigcup_{k=-\infty}^{\infty} (k, k + 3/4).$$

Note that the path components of $p^{-1}(U_1)$ are the intervals $(k - 1/2, k + 1/4)$, k an integer, each of which is mapped by a p homeomorphically onto U_1. Similarly, p maps the path components $(k, k + 3/4)$ of $p^{-1}(U_2)$ homeomorphically onto U_2. This completes the proof of the lemma. ☐

Theorem 9.6: The Covering Path Property *If $\alpha: I \to S^1$ is a path with initial point 1, then there is a lifting of α to a unique covering path $\tilde{\alpha}: I \to \mathbb{R}$ with initial point 0.*

Proof: *The proof rests on the following intuitive idea. Subdivide the range of the path α into connected sections so that each section is contained either in U_1 or U_2, the sets prescribed in the proof of the lemma. If a certain section is contained in U_1, we choose one of the intervals $A = (k - 1/2, k + 1/4)$ and consider the restriction $p|_A$ of p to this interval. Since this restriction is a homeomorphism, we can compose the inverse of $p|_A$ with the given section of α to "lift" this section to a section of a path in $\mathbb{R}$. Sections lying in U_2 are lifted similarly. Being careful to have the terminal point of one lifted section agree with the initial point of the next lifted section will insure a continuous lifting of the entire path.*

Proceeding to the proof, let ϵ be a Lebesgue number for the open cover $\{\alpha^{-1}(U_1),$ $\alpha^{-1}(U_2)\}$ of I. Choose a finite sequence

$$0 = t_0 < t_1 < \cdots < t_n = 1$$

of numbers in I with successive terms differing by less than ϵ. Then each set $\alpha([t_i, t_{i+1}])$, $0 \le i \le n - 1$, must be contained either in U_1 or in U_2.

Consider the first section $\alpha([t_0, t_1])$. This section must be contained in U_1 since

$$\alpha(t_0) = \alpha(0) = 1$$

is not in U_2. We wish the lifted section to have initial point 0, so we take $A_1 = (-1/2, 1/4)$ (i.e., $k = 0$) and define $\tilde{\alpha}$ on $[t_0, t_1]$ by

$$\tilde{\alpha}(t) = (p\,|_{A_1})^{-1}\alpha(t), \quad t \in [t_0, t_1].$$

This defines the lifted path on the first subinterval $[t_0, t_1]$.

Proceeding by induction, suppose that $\tilde{\alpha}$ has been defined on $[t_0, t_i]$. Then $\alpha([t_i, t_{i+1}])$ is contained in U, where U is either U_1 or U_2. Let A_{i+1} be the path component of $p^{-1}(U)$ to which $\tilde{\alpha}(t_i)$ belongs. Since A_{i+1} is one of the intervals $(k - 1/2, k + 1/4)$ or $(k, k + 3/4)$, then p maps A_{i+1} homeomorphically onto U. The desired extension of $\tilde{\alpha}$ to $[t_i, t_{i+1}]$ is defined by

$$\tilde{\alpha}(t) = (p\,|_{A_{i+1}})^{-1}\alpha(t), \quad t \in [t_i, t_{i+1}].$$

Continuity is assured by the fact that lifted sections agree at the endpoints. This inductive argument defines the lifting $\tilde{\alpha}$ on the entire interval $[t_0, t_n] = I$.

The fact that $\tilde{\alpha}$ is the only covering path of α with initial point 0 can be gleaned from what has already been done. Since $p\,|_{A_1}$ is a homeomorphism, the lifting of the first section of α to a section of a path beginning at 0 is unique. Since the second lifted section must begin where the first ends and $p\,|_{A_2}$ is a homeomorphism, then the definition of $\tilde{\alpha}$ on the second section is also unique. This argument extends by an obvious inductive step to show the uniqueness of the covering path beginning at 0. ☐

Theorem 9.7:   The Covering Homotopy Property   *If $H: I \times I \to S^1$ is a homotopy such that $H(0, 0) = 1$, then there is a lifting of H to a unique covering homotopy $\tilde{H}: I \times I \to \mathbb{R}$ for which $\tilde{H}(0, 0) = 0$.*

Proof: *Since the proof is similar to that of the Covering Path Property, some of the details are left to the reader. Using the same open sets U_1, U_2 defined for the Covering Path Property and a Lebesgue number argument, subdivide $I \times I$ into rectangles*

$$[t_k, t_{k+1}] \times [s_i, s_{i+1}], \quad 0 \le k \le n - 1, 0 \le i \le m - 1,$$

where

$$0 = t_0 < t_1 < \cdots < t_n = 1, \quad 0 = s_0 < s_1 < \cdots < s_m = 1$$

so that H *maps any of the prescribed rectangles into either* U_1 *or* U_2. *Since* $H(0, 0) = 1$ *is not in* U_2, *then* H *must map the first rectangle* $[t_0, t_1] \times [s_0, s_1]$ *into* U_1. *Letting* $A_1 = (-1/2, 1/4)$ *as before, define* $\check{H}$ *on* $[t_0, t_1] \times [s_0, s_1]$ *by*

$$\check{H}(t, s) = (p|_{A_1})^{-1} H(t, s).$$

The definition of $\check{H}$ *is extended over the rectangles* $[t_k, t_{k+1}] \times [s_0, s_1]$ *as in the proof of the Covering Path Property, being sure that the definitions agree on the edges between rectangles. This defines* $\check{H}$ *on the strip* $[0, 1] \times [s_0, s_1]$. *Next,* H *is defined in an analogous way on the strip* $[0, 1] \times [s_1, s_2]$, *with the definitions agreeing on the edges between rectangles. This argument extends inductively in a straightforward way to complete the proof.*   □

Definition: *For a loop* α *in* S^1 *with base point 1, the Covering Path Property specifies a unique covering path* $\tilde{\alpha}$ *of* α *with* $\tilde{\alpha}(0)$ *equal to 0. Since*

$$(\cos 2\pi\tilde{\alpha}(1), \sin 2\pi\tilde{\alpha}(1)) = p\tilde{\alpha}(1) = \alpha(1) = 1,$$

it follows that $\tilde{\alpha}(1)$ *must be an integer. This integer is called the* **degree** *of the loop* α *and is denoted* $\deg(\alpha)$.

    Intuitively, one thinks of the degree of a loop α as the net number of times that α "wraps" the interval $[0, 1]$ around S^1. Counterclockwise wrappings are counted as positive and clockwise ones as negative. The next theorem shows that the degree of a loop completely determines its equivalence class in $\prod_1(S^1, 1)$.

Theorem 9.8:   *For loops* α, β *in* S^1 *with base point 1,* $[\alpha] = [\beta]$ *if and only if* $\deg(\alpha) = \deg(\beta)$.

Proof: *Suppose first that* $[\alpha] = [\beta]$ *so that* α *and* β *are equivalent loops in* S^1. *Let* $F: I \times I \to S^1$ *be a homotopy demonstrating the equivalence of* α *and* β:

$$F(\bullet, 0) = \alpha, \quad F(\bullet, 1) = \beta,$$

$$F(0, s) = F(1, s) = 1, \quad s \in I.$$

The Covering Homotopy Property insures the existence of a unique covering homotopy $\tilde{F}$ *of* F *such that* $\tilde{F}(0, 0) = 0$. *For* s *in* I,

$$p\tilde{F}(0, s) = F(0, s) = 1,$$

so $\tilde{F}(0, s)$ must be an integer. Since I is connected, the same integral value must occur for each value of s. Since $\tilde{F}(0, 0) = 0$, then $\tilde{F}(0, s) = 0$ for all s in I. The same argument shows that $\tilde{F}(1, s)$ must have the constant integral value $\tilde{F}(1, 0)$ for all s in I. The uniqueness of covering paths insures that $\tilde{\alpha} = \tilde{F}(\cdot, 0)$ and $\tilde{\beta} = \tilde{F}(\cdot, 1)$ are the unique covering paths of α and β which begin at 0. Thus

$$deg(\alpha) = \tilde{\alpha}(1) \doteq \tilde{F}(1, 0) = \tilde{F}(1, 1) = \tilde{\beta}(1) = deg(\beta).$$

Thus equivalent loops must have the same degree.

The other part of the proof is easier. Suppose that α and β are loops in S^1 with base point 1 having the same degree. This means that the covering paths $\tilde{\alpha}$ and $\tilde{\beta}$ beginning at common initial point 0 also have common terminal point $\tilde{\alpha}(1) = \tilde{\beta}(1)$. Then the homotopy $G: I \times I \rightarrow \mathbb{R}$ defined by

$$G(t, s) = (1 - s)\tilde{\alpha}(t) + s\tilde{\beta}(t), \quad (t, s) \in I \times I,$$

demonstrates the equivalence of $\tilde{\alpha}$ and $\tilde{\beta}$. It follows easily that loops α and β are equivalent by the homotopy pG.   □

Theorem 9.8 shows how to associate each homotopy class of loops in $\prod_1 (S^1, 1)$ with an integer. The next theorem demonstrates that this correspondence is an isomorphism.

Theorem 9.9:   *The fundamental group $\prod_1 (S^1)$ is isomorphic to the additive group $\mathbb{Z}$ of integers.*

Proof: *Consider the degree function $d: \prod_1 (S^1, 1) \rightarrow \mathbb{Z}$ which assigns the integer degree of α to each equivalence class $[\alpha]$:*

$$d([\alpha]) = deg(\alpha), \quad [\alpha] \in \prod_1 (S^1, 1).$$

Theorem 9.8 shows that d is well-defined and one-to-one. To see that d is surjective, let k be an integer. The loop μ_k defined by

$$\mu_k(t) = p(kt), \quad t \in I,$$

has as covering path the function

$$\tilde{\mu}_k(t) = kt, \quad t \in I,$$

and thus has degree $\tilde{\mu}_k(1) = k$. Thus $d([\mu_k]) = k$.

It remains to be proved that d is a homomorphism. For [σ], [τ] in $\prod_1 (S^1, 1)$, let $\tilde{\sigma}$ and $\tilde{\tau}$ denote the unique covering paths of σ and τ beginning at 0. Then the path $g: I \to \mathbb{R}$ defined by

$$g(t) = \begin{cases} \tilde{\sigma}(2t) & 0 \le t \le 1/2 \\ \tilde{\sigma}(1) + \tilde{\tau}(2t - 1) & 1/2 \le t \le 1 \end{cases}$$

is the covering path of σ∗τ with initial point 0. Thus

$$deg(\sigma * \tau) = g(1) = \tilde{\sigma}(1) + \tilde{\tau}(1) = deg(\sigma) + deg(\tau).$$

Hence

$$d([\sigma] \circ [\tau]) = d([\sigma * \tau]) = deg(\sigma * \tau)$$
$$= deg(\sigma) + deg(\tau) = d([\sigma]) + d([\tau]).$$

Thus d is an isomorphism from $\prod_1(S^1, 1)$ onto $\mathbb{Z}$. □

The covering projection $p: \mathbb{R} \to S^1$ has been instrumental in our computation of $\prod_1(S^1)$. The relevant properties of this map have been generalized to define an important class of such functions $p: E \to B$ from a *covering space E* to a *base space B* for which analogues of the Covering Path Property and Covering Homotopy Property can be established. The fundamental group is used to determine which spaces are covering spaces for a given space B. More complete information about covering spaces can be found in the Suggestions for Further Reading at the end of the chapter.

EXERCISE 9.3

1. Explain in detail why the loop $\mu_k: I \to S^1$ defined by

$$\mu_k(t) = p(kt), \quad t \in I,$$

has degree k, for each integer k.

2. Complete the inductive definition of the covering homotopy in the proof of the Covering Homotopy Property (Theorem 9.7).

3. Consider S^1 as the set $z = x + iy$ of complex numbers having modulus 1. Then the covering projection $p: \mathbb{R} \to S^1$ is, by definition of the exponential function for complex variables,

$$p(t) = \cos 2\pi t + i \sin 2\pi t = e^{2\pi it}, t \in \mathbb{R}.$$

Use this representation to prove the uniqueness assertion of the Covering Path Property (Theorem 9.6) by showing the following:

(a) For a given loop α in S^1 with base point 1, let $\tilde{\alpha}_1$ and $\tilde{\alpha}_2$ be covering paths of α with initial point 0. Show that the composition of p with $\tilde{\alpha}_1 - \tilde{\alpha}_2$ is a constant path.

(b) Conclude from part (a) that $\tilde{\alpha}_1 - \tilde{\alpha}_2$ has only integral values. Use the connectedness of $\mathbb{R}$ to show that $\tilde{\alpha}_1 - \tilde{\alpha}_2$ has only the value 0.

9.4 ADDITIONAL EXAMPLES OF FUNDAMENTAL GROUPS

Our work in this chapter has revealed that the fundamental group of S^1 is the additive group of integers and that the fundamental group of a contractible space is trivial. It should be clear by now that the fundamental group is difficult to determine rigorously. This section presents several theorems that are useful in determining fundamental groups and some additional examples.

Definition: *Let X be a space with subspace D. Then D is a **deformation retract** of X provided that there is a continuous function $H: X \times I \rightarrow X$ for which*

$$H(x, 0) = x, \quad H(x, 1) \in D, \quad x \in X$$
$$H(y, t) = y, \quad y \in D, \quad t \in I.$$

*The homotopy H is called a **deformation retraction**. Since for each y in D, $H(y, t)$ has the constant value y as t varies from 0 to 1, it is sometimes said that the points of D **stay fixed** throughout the deformation retraction.*

Example 9.4.1

Consider the annulus

$$A = \{(x_1, x_2) \in \mathbb{R}^2 : 1 \leq x_1^2 + x_2^2 \leq 4\}$$

shown in Figure 9.7(a). The function $H: A \times I \rightarrow S^1$ defined by

$$H(x, t) = (1 - t)x + t\frac{x}{\|x\|}, \quad (x, t) \in A \times I,$$

is a deformation retraction from A onto its inner circle S^1. For Figure 9.7(b), a similar deformation retraction can be defined on any annulus X to show that a circle like D is also a deformation retract.

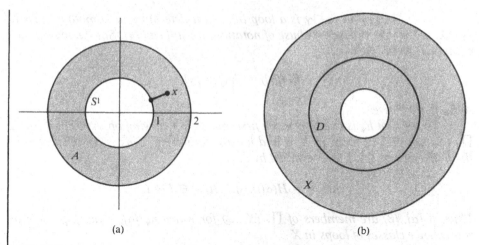

(a)   (b)

FIGURE 9.7

Example 9.4.2

A comparison of definitions will reveal that a space X is contractible to a point x_0 in X if and only if the single-point subspace $\{x_0\}$ is a deformation retract of X. Thus any point of $\mathbb{R}^n$ is a deformation retract of $\mathbb{R}^n$, any point of the n-ball B^n is a deformation retract of B^n, and any point of the n-cube I^n is a deformation retract of I^n.

Theorem 9.10:   *If D is a deformation retract of a space X and x_0 is a point of D, then $\prod_1 (X, x_0)$ and $\prod_1 (D, x_0)$ are isomorphic.*

Proof: *Let $H: X \times I \to X$ be a deformation retraction, as specified in the definition. For brevity of notation, let $h: X \to D$ denote the 1-level of H:*

$$h(x) = H(x, 1), \quad x \in X.$$

Let $h_*: \prod_1(X, x_0) \to \prod_1(D, x_0)$ be the homomorphism induced by h:

$$h_*([\alpha]) = [h\alpha], \quad [\alpha] \in \prod_1 (X, x_0).$$

The fact that h is a continuous map insures that h_* is well-defined, and the definition of the operation $*$ reveals it to be a homomorphism. For $[\alpha], [\beta]$ in $\prod_1 (X, x_0)$,

$$h_*([\alpha] \cdot [\beta]) = h_*([\alpha * \beta]) = [h(\alpha * \beta)] = [h\alpha * h\beta]$$
$$= h_*([\alpha]) \cdot h_*([\beta]).$$

For $[\gamma]$ in $\prod_1 (D, x_0)$, γ is a loop in X so it determines a homotopy class in $\prod_1 (X, x_0)$ which, with some abuse of notation, we still call $[\gamma]$. Since h leaves each point of D fixed, then

$$h_*([\gamma]) = [h\gamma] = [\gamma],$$

so h_* is surjective.

To see that h_* is one-to-one, we first make the following observation. If $[\alpha] \in \prod_1 (X, x_0)$, then as loops in X, α and $h\alpha$ are equivalent. This is demonstrated by the homotopy $K: I \times I \to X$ defined by

$$K(t, s) = H(\alpha(t), s), \quad (t, s) \in I \times I.$$

Thus, if $[\alpha]$, $[\beta]$ are members of $\prod_1 (X, x_0)$ for which $h_*([\alpha]) = h_*([\beta])$, then as equivalence classes of loops in X

$$[\alpha] = [h\alpha] = h_*([\alpha]) = h_*([\beta]) = [h\beta] = [\beta],$$

so $[\alpha] = [\beta]$ and h_* must be one-to-one. This completes the proof that h_* is an isomorphism. □

Example 9.4.3

For an annulus A, both its inner and outer circles are deformation retracts. From Theorem 9.10 we conclude that $\prod_1 (A)$ is isomorphic to the group $\mathbb{Z}$ of integers.

Example 9.4.4

The *punctured plane* $\mathbb{R}^2\backslash\{p\}$ consists of all points of $\mathbb{R}^2$ except a particular point p. It is left as an easy exercise for the reader to show that a circle containing p in its inner region is a deformation retract of $\mathbb{R}^2\backslash\{p\}$ and hence that the fundamental group of $\mathbb{R}^2\backslash\{p\}$ is the group of integers.

Theorem 9.11: *Let X_1 and X_2 be spaces with base points x_1 and x_2, respectively. Then*

$$\prod_1 (X_1 \times X_2, (x_1, x_2)) \cong \prod_1 (X_1, x_1) \otimes \prod_1 (X_2, x_2).$$

Proof: *Before beginning the proof, we note that the symbol "$\cong$" denotes isomorphism of groups and that $\prod_1 (X_1, x_1) \otimes \prod_1 (X_2, x_2)$ denotes the direct product of the indicated groups. The direct product consists of all ordered pairs $([\alpha_1], [\alpha_2])$ where $[\alpha_1] \in \prod_1 (X_1, x_1)$ and $[\alpha_2] \in \prod_1 (X_2, x_2)$ with group operation $\otimes$ defined by*

$$([\alpha_1], [\alpha_2]), \otimes ([\beta_1], [\beta_2]) = ([\alpha_1 * \beta_1], [\alpha_2 * \beta_2]).$$

Those not familiar with direct products should consult the Appendix on groups before proceeding.

Any loop α in $X_1 \times X_2$ based at (x_1, x_2) defines loops

$$\alpha_1 = p_1\alpha, \quad \alpha_2 = p_2\alpha$$

in X_1 and X_2 based at x_1 and x_2, respectively, where $p_1: X_1 \times X_2 \to X_1$ and $p_2: X_1 \times X_2 \to X_2$ are the projection maps from the product space onto the coordinate spaces. Also, any pair of loops α_1, α_2 in X_1 and X_2 based at x_1 and x_2, respectively, determine a loop $\alpha = (\alpha_1, \alpha_2)$ in $X_1 \times X_2$ with base point (x_1, x_2). It is a straightforward exercise to show that the function

$$f: \prod{}_{\!1} (X_1 \times X_2, (x_1, x_2)) \to \prod{}_{\!1} (X_1, x_1) \otimes \prod{}_{\!1} (X_2, x_2)$$

defined by

$$f([\alpha]) = ([\alpha_1], [\alpha_2]), \quad [\alpha] \in \prod{}_{\!1} (X_1 \times X_2, (x_1, x_2))$$

is an isomorphism. □

Example 9.4.5

(a) Since the torus T is homeomorphic to the product space $S^1 \times S^1$, then

$$\prod{}_{\!1} (T) \cong \prod{}_{\!1} (S^1) \otimes \prod{}_{\!1} (S^1) \cong \mathbb{Z} \otimes \mathbb{Z}.$$

(b) An *n-dimensional torus* T^n is the product of n factors of S^1. Thus $\prod_1 (T^n)$ is isomorphic to the direct product of n copies of the group $\mathbb{Z}$ of integers.

(c) Since a closed cylinder C is the product of S^1 with an interval $[a, b]$,

$$\prod{}_{\!1} (C) \cong \prod{}_{\!1} (S^1) \otimes \prod{}_{\!1} ([a, b]) \cong \mathbb{Z} \otimes \{0\} \cong \mathbb{Z}.$$

Theorem 9.12: *For $n \geq 2$, the n-sphere S^n is simply connected.*

Proof: *Let*

$$V_1 = \{x = (x_1, \ldots, x_{n+1}) \in S^n : x_{n+1} < 1/2\}$$

$$V_2 = \{x = (x_1, \ldots, x_{n+1}) \in S^n : x_{n+1} > -1/2\}.$$

Note that V_1, V_2 are simply connected open subsets of S^n whose union is S^n and whose intersection is non-empty and path connected. The proof will show that any space having such an open covering must be simply connected.

Let $1 = (1, 0, \ldots, 0) \in S^n$ be the base point and consider a homotopy class $[\alpha]$ in $\prod_1 (S^n, 1)$. Then $\alpha: I \to S^n$ is a continuous map and $\{\alpha^{-1}(V_1), \alpha^{-1}(V_2)\}$ is an open cover of I. Since I is compact, this open cover has a Lebesgue number ϵ. Choosing successive terms of a sequence

$$0 = s_0 < s_1 < \cdots < s_m = 1$$

to differ by less than ϵ insures that for $0 \le i \le m - 1$ the image $\alpha([s_i, s_{i+1}])$ will be a subset either of V_1 or of V_2. Defining

$$\alpha_i(t) = \alpha((1 - t)s_i + ts_{i+1}), \quad t \in I,$$

produces a sequence $\{\alpha_i\}_{i=0}^{m-1}$ of paths in S^n for which $\alpha_i(I)$ is a subset of a simply connected set U_i, which equals V_1 or V_2, and

$$[\alpha] = [\alpha_0 * \alpha_1 * \cdots * \alpha_{m-1}].$$

Since $V_1 \cap V_2$ is path connected, there is for $1 \le i \le m - 1$ a path γ_i from 1 to $\alpha(s_i)$ lying entirely within $U_{i-1} \cap U_i$. Letting $\bar{\gamma}_i$ denote the reverse of γ_i, it follows that $\bar{\gamma}_i * \gamma_i$ is equivalent to a constant loop, $1 \le i \le m - 1$, so that

$$[\alpha] = [\alpha_0 * \bar{\gamma}_1 * \gamma_1 * \alpha_1 * \bar{\gamma}_2 * \gamma_2 * \alpha_2 * \cdots * \bar{\gamma}_{m-1} * \gamma_{m-1} * \alpha_{m-1}]$$
$$= [\alpha_0 * \bar{\gamma}_1] \circ [\gamma_1 * \alpha_1 * \bar{\gamma}_2] \circ \cdots \circ [\bar{\gamma}_{m-1} * \alpha_{m-1}].$$

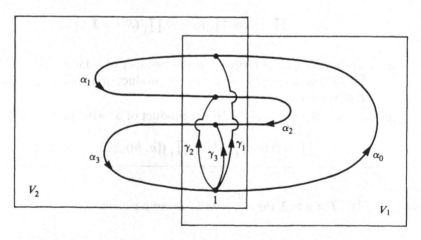

FIGURE 9.8   Schematic diagram for Theorem 9.12 with $m = 4$.

Each equivalence class in this product is determined by a loop based at 1 in either V_1 or V_2, a simply connected set in either case. Thus each term in the product is *the identity class, so [α] must also be the identity class. Hence* Π_1 *(S^n, 1) = {0}* *and S^n is simply connected for n ≥ 2.*   □

Question: *Where in the preceding proof was the assumption n ≥ 2 used?*

The examples of fundamental groups given in this chapter are all abelian. There are relatively simple topological spaces which have non-abelian fundamental groups; the doubly punctured plane (plane with two points removed) and the sub-space of the plane consisting of two tangent circles (a figure eight) are two examples. Showing that the fundamental groups of these spaces are non-abelian would require a considerable departure from the mainstream of this chapter, so these demonstrations are omitted. Additional information on this subject can be found in the Suggestions for Further Reading at the end of the chapter.

EXERCISE 9.4

1. Show that S^1 is a deformation retract of the cylinder $S^1 \times I$. Use this to prove that the fundamental group of a cylinder is isomorphic to Z.

2. Explain in detail where the assumption $n \geq 2$ was used in the proof of Theorem 9.12.

3. Generalize the proof of Theorem 9.12 to prove the following:

    Theorem: *Suppose X is a space with an open cover {V$_i$} such that*

    (a) $\cap V_i \neq \emptyset$,

    (b)   *each V$_i$ is simply connected,*

    (c)   *for i ≠ j, V$_i \cap$ V$_j$ is path connected.*

    Then X is simply connected.

4. Determine the fundamental group of the Möbius strip.

5. (a)   Prove that S^{n-1} is a deformation retract of $\mathbb{R}^n \backslash \{\theta\}$.

    (b)   Use part (a) to prove that punctured n-space $\mathbb{R}^n \backslash \{p\}$ is simply connected for $n \geq 3$.

6. Let X be a space consisting of two spheres S^m and S^n, where $m, n \geq 2$, joined at a point. Prove that X is simply connected.

9.5 THE BROUWER FIXED POINT THEOREM AND RELATED RESULTS

Recall from Chapter 5 that a *fixed point* of a given function $f: X \rightarrow X$ from a space X into itself is a point x in X for which $f(x) = x$. A space X has the *fixed-*

point property provided that every continuous function $f: X \to X$ has at least one fixed point. The main result of this section shows that the closed unit ball

$$B^2 = \{(x_1, x_2) \in \mathbb{R}^2: x_1^2 + x_2^2 \leq 1\}$$

has the fixed-point property. The proof is based on an algebraic comparison of $\Pi_1(B^2)$ and $\Pi_1(S^1)$ and illustrates the power of the fundamental group in geometry. Actually, the closed unit n-ball

$$B^n = \{(x_1, x_2, \ldots, x_n) \in \mathbb{R}^n: x_1^2 + x_2^2 + \cdots + x_n^2 \leq 1\}$$

has the fixed-point property for every positive integer n; this celebrated result is the Brouwer Fixed Point Theorem. The case $n = 1$ was proved in Theorem 5.9, and the case $n = 2$ is our object in this section. Those interested in the general result should consult the supplementary reading list.

Two intuitively plausible results, Theorems 9.13 and 9.14, will be needed to prove the $n = 2$ case of Brouwer's theorem.

Theorem 9.13:   *The unit circle S^1 is not contractible.*

Proof: *Any contractible space has trivial fundamental group. Since $\Pi_1(S^1) \cong \mathbb{Z}$, then S^1 is not contractible.*   □

Definition: *A subspace A of a topological space X is a **retract** of X provided that there is a continuous function $f: X \to A$ for which $f(a) = a$ for each a in A. The function f is called a **retraction** of X onto A.*

Example 9.5.1

   (a)   In an annulus, both the inner and outer circles are retracts.

   (b)   A closed subinterval $[c, d]$ of a given interval $[a, b]$ is a retract of $[a, b]$.

   (c)   The set of endpoints $A = \{a, b\}$ is not a retract of a closed interval $[a, b]$, where $a < b$, for the following reason: Since $[a, b]$ is connected and A is not, there cannot be a continuous function from $[a, b]$ onto A.

It should be clear that every deformation retract is a retract but that not every retract is a deformation retract. This relation is pursued in the exercises for this section.

Theorem 9.14:   The Brouwer No Retraction Theorem   *The unit circle S^1 is not a retract of the closed unit ball B^2.*

Proof: *The proof is by contradiction. Suppose, contrary to the theorem, that S^1 is a retract of the unit ball B^2 and let $f: B^2 \to S^1$ be a retraction. Define a homotopy $h: S^1 \times I \to S^1$ by*

$$h(x, t) = f(t + (1 - t)x_1, (1 - t)x_2), \quad x = (x_1, x_2) \in S^1, \quad t \in I.$$

It is not difficult to see that h is a contraction of S^1 to the point $1 = (1, 0)$, contradicting the fact that S^1 is not contractible.   □

Theorem 9.15:   The Brouwer Fixed Point Theorem   *Every continuous function $f: B^2 \to B^2$ from the two-dimensional ball to itself has at least one fixed point.*

Proof: *Suppose to the contrary that $f: B^2 \to B^2$ is a continuous map having no fixed point. Then for all x in B^2, $f(x)$ and x are distinct points. For x in B^2, consider the half-line from $f(x)$ through x, and let $h(x)$ denote the intersection of this half-line with the bounding circle S^1, as shown in Figure 9.9.*

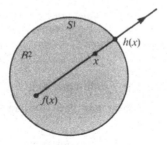

FIGURE 9.9

Then $h: B^2 \to S^1$ is easily seen to be a retraction from B^2 onto S^1, contradicting Theorem 9.14. Thus we conclude that the assumption that a continuous map $f: B^2 \to B^2$ can have no fixed point must be false   □

The sequence of results just used to establish the Brouwer Fixed Point Theorem in dimension two generalizes to higher dimensions in a very natural way. It is intuitively plausible that the n-sphere S^n is not contractible for any positive integer n. Once this fact is established, the Brouwer No Retraction Theorem and the Brouwer Fixed Point Theorem can be proved with only minor modifications to the arguments given above. Thus, the main result needed for the Brouwer Fixed Point

Theorem is the non-contractibility of S^n. Since this is established for S^1 by the fundamental group, it seems natural to think that some similar algebraic structure might do the same for S^n, $n \geq 2$. This structure should describe the "hole" in S^n in the way that the fundamental group describes the equivalence classes of loops in S^1. This is, in fact, the case. There is a sequence $\prod_n (X, x_0)$, $n = 1, 2, 3, \ldots$, of groups defined for general topological spaces which generalize the fundamental group to higher dimensions. These groups establish the non-contractibility of S^n and allow a proof of the Brouwer Fixed Point Theorem, as well as many other beautiful geometric theorems. The interested reader will find relatively easy access to these higher homotopy groups in the suggested reading list for this chapter.

EXERCISE 9.5

1. Let X be a space. Prove that every deformation retract of X is also a retract of X.

2. (a) Prove that a single point in S^1 is a retract but not a deformation retract of S^1.

    (b) Prove that a meridian or longitudinal circle on the torus T is a retract but not a deformation retract of T.

3. Fill in the details in the proof of Theorem 9.15 by proving that the function h is continuous and leaves each point of S^1 fixed.

4. **Definition:** *Let $f: X \rightarrow Y$ be a continuous function on the indicated spaces. Then f is* **null-homotopic** *provided that there is a continuous function $H: X \times I \rightarrow Y$ such that $H(\cdot, 0) = f$ and $H(\cdot, 1)$ is a constant map.*

    Prove the following:

    (a) If Y is contractible, then every continuous map $f: X \rightarrow Y$ is null-homotopic.

    (b) If $g: X \rightarrow S^n$, $n \geq 1$, is a continuous map whose image $g(X)$ is a proper subset of S^n, then g is null-homotopic.

    (c) If X is contractible, then every continuous map $h: X \rightarrow Y$ is null-homotopic.

5. Assuming that the n-sphere S^n is not contractible, prove the Brouwer No Retraction Theorem (Theorem 9.14) and the Brouwer Fixed Point Theorem (Theorem 9.15) for $n > 2$.

6. (a) Show that for the Cantor set C, every closed subset is a retract.

    (b) Which subsets of C are deformation retracts?

7. Let X be a space, A a retract of X, x_0 a point of A, and $f: X \rightarrow A$ a retraction. Show that the induced homomorphism

$$f_*: \prod_1 (X, x_0) \rightarrow \prod_1 (A, x_0)$$

    is a surjection.

8. Show that S^1 is not weakly contractible. (Weak contractibility is defined in Problem 12 of Exercise 9.2.)

9.6 CATEGORIES AND FUNCTORS

The theme of the present chapter has been the use of algebraic structures, in the form of fundamental groups, to answer topological and geometric questions. We think of translating a topological problem to an algebraic one and solving the latter to reveal an insight into the former. This method of attack is formalized in mathematical language by the theory of categories, the subject of this section. Roughly speaking, a *category* is a class of sets, called *objects*, which share some particular structure and a collection of functions or *morphisms* from one object to another. For example, it would be said that a problem in the category of topological spaces and continuous maps can be translated to the category of groups and homomorphisms. The mechanism of translation between categories is called a *functor*. Precise definitions and examples of these concepts are the object of this section.

Definition: *A **category** $\mathcal{C}$ is a class whose members are the **objects** of the category, together with sets of **maps** or **morphisms** from each object to every other object. The set of all morphisms from object A to object B is denoted Hom(A, B). If f is a member of Hom(A, B), then we write f: $A \to B$ or $A \xrightarrow{f} B$ and call A and B the **domain** and **range** of f, respectively. The morphisms are required to satisfy the following properties:*

(a) *For each triple A, B, C of objects and morphisms f: $A \to B$ and g: $B \to C$, there is a **composite morphism***

$$gf = g \circ f : A \to C.$$

(b) *Composition of morphisms is associative, in the following sense. If f: $A \to B$, g: $B \to C$ and h: $C \to D$ are morphisms, then*

$$h \circ (g \circ f) = (h \circ g) \circ f : A \to D.$$

(c) *For each object A there is an **identity morphism** $1_A: A \to A$ such that for each object X*

$$1_A f = f, \quad f \in Hom(X, A),$$

$$g 1_A = g, \quad g \in Hom(A, X).$$

Example 9.6.1

The system whose objects are topological spaces X and for which $Hom(X, Y)$ is the family of continuous functions from X to Y is a category. The composite operation on morphisms is simply composition of functions, and the identity morphism on an object X is the identity function. Following the usual practice, we shall refer to categories by specifying the class of objects first and the allowable morphisms second, without further designation. Thus, the category of this example is called the *category of topological spaces and continuous functions*.

Example 9.6.2

As additional examples of categories, note the following:

    (a)   The category of groups and homomorphisms.

    (b)   The category of metric spaces and continuous functions.

    (c)   The category of metric spaces and isometries.

    (d)   The category of sets and functions.

    (e)   The category of sets and injective functions.

    (f)   The category whose objects are ordered pairs (X, x_0), where X is a space with base point x_0, and whose morphisms are continuous maps $f: (X, x_0) \to (Y, y_0)$ which satisfy $f(x_0) = y_0$. This category is the category of *pointed topological spaces and base point preserving continuous maps*.

Definition: *Let $\mathcal{C}$ and $\mathcal{C}'$ be categories. A* **covariant functor** *from $\mathcal{C}$ to $\mathcal{C}'$ consists of two functions, an* **object function** *which maps each object A of $\mathcal{C}$ to an object $T(A)$ of $\mathcal{C}'$ and a* **morphism function** *which maps each morphism f of $\mathcal{C}$ to a morphism $T(f)$ of $\mathcal{C}'$ subject to:*

    (a)   *If f is a morphism of $\mathcal{C}$ from A to B, then $T(f)$ is a morphism from $T(A)$ to $T(B)$.*

    (b)   *If $f: A \to B$ and $g: B \to C$ are morphisms of $\mathcal{C}$, then*

$$T(g \cdot f) = T(g) \cdot T(f).$$

    (c)   *For each object A of $\mathcal{C}$,*

$$T(1_A) = 1_{T(A)}.$$

    The definition of the term **contravariant functor** *differs from the preceding definition as follows. A contravariant functor satisfies (c), but instead of (a) and (b) it satisfies*

    (a')   *If f is a morphism of $\mathcal{C}$ from A to B, then $T(f)$ is a morphism of $\mathcal{C}'$ from $T(B)$ to $T(A)$.*

    (b')   *If $f: A \to B$ and $g: B \to C$ are morphisms of $\mathcal{C}$, then*

$$T(g \cdot f) = T(f) \cdot T(g).$$

Example 9.6.3

The purpose of this example is to describe the *fundamental group functor*. Consider the category $\mathcal{P}$ of pointed topological spaces and continuous base point preserving maps and the category $\mathcal{G}$ of groups and homomorphisms. As we know, the fundamental group assigns to each object (X, x_0) of $\mathcal{P}$ an object $\prod_1 (X, x_0)$ of $\mathcal{G}$. For a morphism $f: (X, x_0) \to (Y, y_0)$, the induced homomorphism f_*, which is denoted $\prod_1 (f)$ in functorial notation, is defined from $\prod_1 (X, x_0)$ to $\prod_1 (Y, y_0)$ by composition of functions:

$$f_*([\alpha]) = [f\alpha], \quad [\alpha] \in \prod_1 (X, x_0).$$

This functor is covariant since $f_*: \prod_1 (X, x_0) \to \prod_1 (Y, y_0)$ whenever $f: (X, x_0) \to (Y, y_0)$.

Example 9.6.4

As an easier example of a covariant functor, consider the category $\mathcal{T}$ of topological spaces and continuous functions and the category $\mathcal{S}$ of sets and functions. The "forgetful functor" from $\mathcal{T}$ to $\mathcal{S}$ assigns to each topological space X the underlying set X and to each continuous function $f: X \to Y$ on topological spaces the same function f from set X to set Y. The name of this functor is derived from its property of "forgetting" the structure of the category $\mathcal{T}$.

Example 9.6.5

For an example of a contravariant functor, consider the category $\mathcal{T}$ of topological spaces and continuous functions and the category $\mathcal{M}$ of metric spaces and continuous functions. For each object X in $\mathcal{T}$, let $C(X, \mathbb{R})$ denote the space of bounded, continuous real-valued functions $f: X \to \mathbb{R}$. With the supremum metric, $C(X, \mathbb{R})$ is a metric space. We define a functor from $\mathcal{T}$ to $\mathcal{M}$ whose object function assigns to each member X of $\mathcal{T}$ the corresponding metric space $T(X) = C(X, \mathbb{R})$ and assigns to a morphism $f: X \to Y$ the morphism $T(f): C(Y, \mathbb{R}) \to C(X, \mathbb{R})$ defined as follows:

$$T(f)(h) = hf, \quad h \in C(Y, \mathbb{R}).$$

Note that if $h: Y \to \mathbb{R}$, then $hf: X \to \mathbb{R}$, so the functor T is contravariant.

EXERCISE 9.6

1. Show in detail that Example 9.6.3 defines a covariant functor.

2. Show in detail that Example 9.6.5 defines a contravariant functor.

3. Let $\mathcal{T}$ be the category of topological spaces and continuous maps and $\mathcal{S}$ the category of sets and functions. Define a functor from $\mathcal{T}$ to $\mathcal{S}$ which assigns to each object X in $\mathcal{T}$ its set of components. Describe the action of this functor on morphisms. Is this functor covariant or contravariant?

4. Let $\mathcal{C}$ be the category of completely regular spaces and continuous functions and $\mathcal{D}$ the category of compact Hausdorff spaces. The *Stone-Čech functor* assigns to each object X of $\mathcal{C}$ its Stone-Čech compactification $\beta(x)$ in $\mathcal{D}$. Use Problem 11 of Exercise 8.6 to show that this assignment does define a functor and describe its action on morphisms.

5. Let $\mathcal{C}$ be a category and A an object of $\mathcal{C}$. Show that the identity morphism $1_A: A \to A$ is unique.

6. **Definition:** *Let $\mathcal{C}$ be a category. If $f: X \to Y$ and $g: Y \to X$ are morphisms for which $gf = 1_X$, then f is called a **right inverse** of g and g is called **left inverse** of f. A **two-sided inverse** or **inverse** g of f is a morphism which is both a left and right inverse for f. If f has an inverse, then f is called an **equivalence,** and the objects X and Y are said to be **equivalent.***

    Prove:

    (a) Let $f: X \to Y$ be a morphism with left inverse $g: Y \to X$ and right inverse $h: Y \to X$. Then $g = h$ and f is an equivalence.

    (b) Equivalence of objects in a category is an equivalence relation.

SUGGESTIONS FOR FURTHER READING

For an introduction to various aspects of algebraic topology at a level accessible to undergraduates, see *Basic Concepts of Algebraic Topology* by Croom or *Algebraic Topology* by Massey. Several other introductory textbooks on the subject are included in the Bibliography.

Introduction to Knot Theory by Crowell and Fox has an excellent exposition of the fundamental group. *Algebraic Topology* by Spanier, at a higher level, gives an excellent overview of algebraic topology; it is particularly recommended for categories and functors.

For the background in group theory required for Chapter 9, Herstein's *Topics in Algebra*, Jacobson's *Basic Algebra I*, and Shapiro's *Introduction to Abstract Algebra* are recommended.

HISTORICAL NOTES FOR CHAPTER 9

As was mentioned in the introduction to this chapter, algebraic topology was introduced in a series of papers during the years 1895–1901 by the French mathematician Henri Poincaré (1854–1912). Algebraic topology did not develop as an outgrowth of point-set topology. Poincaré's first paper preceded Fréchet's work on general metric spaces by 11 years and Hausdorff's definition of general topological spaces by 19 years. Furthermore, Poincaré's work on the fundamental group and other aspects of algebraic topology was not influenced by Cantor's theory of sets. Indeed, Poincaré was unimpressed by the theory of sets and referred to Cantor's work as a "disease" from which mathematics would eventually recover.

Henri Poincaré made significant mathematical contributions in the fields of algebraic topology, algebra, non-Euclidean geometry, differential equations, complex variables, algebraic geometry, astronomy, mathematical physics, and celestial mechanics. As the author of 30 books and over 500 papers on new mathematics, Poincaré was the second most prolific writer of mathematics, being surpassed in volume only by Leonard Euler. He is often considered the last universal mathematician, in the sense of having a real grasp of most of the significant mathematics of his time.

The Brouwer Fixed Point Theorem is due to the Dutch mathematician L. E. J. Brouwer (1881–1966). The proof given in the text, using the fundamental group, is not the one originally given by Brouwer. The higher homotopy groups, which were mentioned in the discussion of the Brouwer Fixed Point Theorem in higher dimensions, were defined during the years 1932–1935 by Eduard Čech (1893–1960) and Witold Hurewicz (1904–1956).

Appendix: Introduction to Groups

This brief appendix is intended to refresh the reader's memory with the basic facts about groups essential for Chapter 9 of the text. Proofs of propositions are not included, and the presentation here is not to be considered an adequate introduction to the theory of groups in any sense. Detailed exposition can be found in many standard textbooks for algebra. The books *Basic Algebra I* by Jacobson and *Introduction to Abstract Algebra* by Shapiro are recommended.

Definition: *A **binary operation** on a set X is a function $f: X \times X \rightarrow X$. For x, y, in X, it is customary to replace $f(x, y)$ by xy or $x \cdot y$ (multiplicative notation) or $x + y$ (additive notation).*

Definition: *A **group** is an ordered pair $(G, \cdot)$ consisting of a set G and a binary operation on G satisfying the following properties:*

- *(a) (The Associative Property) $(ab)c = a(bc)$ for all a, b, c in G.*
- *(b) There is an element e of G, called the **identity element,** satisfying $ae = ea - a$ for all a in G.*
- *(c) For each element a in G, there is an element a^{-1}, called the **inverse** of a, for which $aa^{-1} = a^{-1}a = e$.*

In additive notation, the identity element is usually denoted by 0 and the inverse of a by $-a$. A *trivial group* is a group $\{0\}$ consisting only of an identity element 0. It is a simple consequence of the definition that the identity element of a group is unique and that each element has a unique inverse.

Definition: *A group G satisfying $ab = ba$ for all a, b in G is said to be **commutative** or **abelian.***

Definition: *A subset H of a group G is a **subgroup** of G if H is a group under the operation of G.*

The group $\mathbb{Z}$ of integers under addition is the group most frequently encountered in the text. The subgroups of $\mathbb{Z}$ are the groups H_n composed of all integral multiples of a given integer n.

Definition: *A function f: G → H from a group G to a group H is a* **homomorphism** *if for all a, b in G*

$$f(ab) = f(a)f(b).$$

The **kernel** *of f is the set of all members of G mapped by f to the identity of H. A homomorphism which is also a one-to-one function from G onto H is called an* **isomorphism;** *in this case the groups G and H are called* **isomorphic.**

The relation of isomorphism for groups is an equivalence relation.

Definition: *For groups G and H, the* **direct product** *G ⊗ H is the set G × H with group operation ⊗ defined by*

$$(a_1, b_1) \otimes (a_2, b_2) = (a_1 \cdot a_2, b_1 \cdot b_2)$$

for all (a₁, b₁), (a₂, b₂) in G × H. (In order not to complicate the notation, the symbol · is used here for the operations in both G and H.)

The concept of direct product extends in the usual way to any number of factor groups.

BIBLIOGRAPHY

Apostol, T.M., *Calculus,* 2nd ed., Blaisdell, Waltham, MA, 1967.

Armstrong, M.A., *Basic Topology,* Springer-Verlag, New York, 1983.

Arnold, B.H., *Intuitive Concepts in Elementary Topology,* Prentice-Hall, Englewood Cliffs, NJ, 1962.

Bing, R. H., "Elementary Point-Set Topology," Slaught Memorial Papers, Number 8, *American Mathematical Monthly,* Vol. 67, No. 7, 1960.

Bourbaki, N., *General Topology,* Addison-Wesley, Reading, MA, 1966.

Brand, L., *Differential and Difference Equations,* Wiley, New York, 1966.

Courant, R., and Robbins, H., *What is Mathematics?* Oxford University Press, London, 1941.

Croom, F.H., *Basic Concepts of Algebraic Topology,* Springer-Verlag, New York, 1978.

Crowell, R.H., and Fox, R.H., *Introduction to Knot Theory,* Springer-Verlag, New York, 1977.

Dugundji, J., *Topology,* Allyn and Bacon, Boston, 1965.

Eisenberg, M., *Topology,* Holt, Rinehart and Winston, New York, 1974.

Gamelin, T.W., and Greene, R.E., *Introduction to Topology,* CBS College Publishing, New York, 1983.

Guillemin, V., and Pollack, A., *Differential Topology,* Prentice-Hall, Englewood Cliffs, NJ, 1974.

Hamilton, A.G., *Numbers, Sets and Axioms,* Cambridge University Press, Cambridge, 1982.

Hausdorff, F., *Grundzüge der Mengenlehre,* 2nd ed., Walter de Gruyter, Leipzig, 1914.

Herstein, I.N., *Topics in Algebra,* Xerox, Lexington, MA, 1964.

Hocking, J.G., and Young, G.S., *Topology,* Addison-Wesley, Reading, MA, 1961.

Hurewicz, W., and Wallman, H., *Dimension Theory,* Princeton University Press, Princeton, NJ, 1969.

Jacobson, N., *Basic Algebra I,* W.H. Freeman, San Francisco, 1974.

Jänich, K., *Topology,* Springer-Verlag, New York, 1984.

Kaplansky, I., *Set Theory and Metric Spaces,* Allyn and Bacon, Boston, 1972.

Kelley, J.L., *General Topology,* Springer-Verlag, New York, 1975.

Kosniowski, C., *A First Course in Algebraic Topology,* Cambridge University Press, Cambridge, 1980.

Lefschetz, S., *Introduction to Topology,* Princeton University Press, Princeton, NJ, 1949.

Manheim, J., *The Genesis of Point-Set Topology,* Pergamon, London, 1964.

Massey, W.S., *Algebraic Topology: An Introduction,* Harcourt, Brace and World, New York, 1967: Springer-Verlag, 1977.

Milnor, J., *Topology from a Differential Viewpoint,* University of Virginia Press, Charlottesville, 1965.

Moise, E.E., *Geometric Topology in Dimensions Two and Three,* Springer-Verlag, New York, 1977.

Munkres, J.R., *Topology: A First Course,* Prentice-Hall, Englewood Cliffs, NJ, 1975.

Newman, M.H.A., *Elements of the Topology of Plane Sets of Points,* 2nd ed., Cambridge University Press, London, 1964.

Rucker, R., *Infinity and the Mind,* Birkhäuser, Boston, 1982.

Rudin, W., *Principles of Mathematical Analysis,* McGraw Hill, New York, 1964.

Schurle, A.W., *Topics in Topology,* Elsevier/North Holland, New York, 1979.

Shapiro, L., *Introduction to Abstract Algebra,* McGraw-Hill, New York, 1975.

Simmons, G.F., *Introduction to Topology and Modern Analysis,* McGraw-Hill, New York, 1963.

Singer, I.M., and Thorpe, J.A., *Lecture Notes on Elementary Topology and Geometry,* Scott, Foresman, Greenview, IL, 1967.

Spanier, E.H., *Algebraic Topology,* McGraw Hill, New York, 1966.

Steen, L.A., and Seebach, J.A., *Counterexamples in Topology,* Holt, Rinehart and Winston, New York, 1970.

Steenrod, N., and Chinn, W.G., *First Concepts of Topology,* The New Mathematical Library, Number 18, The Mathematical Association of America, Washington, DC, 1966.

Tucker, A.W., and Bailey, H.S., "Topology," Scientific American, CLXXXII: 18–24, 1950.

Wall, C.T.C., *A Geometric Introduction to Topology,* Addison-Wesley, Reading, MA, 1972.

Willard, S., *General Topology,* Addison-Wesley, Reading, MA, 1970.

INDEX

abelian (or commutative) group, 301
accumulation point
  in a metric space, 66
  in a topological space, 101
  of a subset of the real line, 44
Alexander Subbasis Theorem, 203, 207, 229
Alexander, J. W., 229
Alexandroff, P. S., 193, 194, 229, 266
algebraic number, 53
algebraic topology, 267
analysis situs, 9
Anderson, R. D., 229
annulus, 223–224
antipodal point, 28
Arzela, C., 97
Ascoli, G., 97

Baer, R. W., 229
Baire Category Theorem (Theorem 3.17), 89
Baire, René, 97
ball,
  closed, 62
  open, 50, 62
  radius of, 62
Banach, Stephan, 97
base or basis for a topology, 109
  equivalent, 113
  local, 110
base point,
  of a loop, 270
  of a space, 270
basic open set, 109
Betti, Enrico, 229
bijection, 22
binary operation, 301
Bing, R. H., 266
Bolyai, Janos, 10
Bolzano, Bernard, 53
Bolzano-Weierstrass property, 175
Bolzano-Weierstrass Theorem (Theorem 2.13), 48
Borel, Emile, 53, 97, 193
boundary,
  of a manifold, 223, 227

boundary (*Continued*)
  of a set in a metric space, 73
  of a set in a topological space, 103
boundary point,
  in a metric space, 73
  in a topological space, 103
  of a manifold, 223, 227
bounded function, 93
bounded interval, 29
bounded metric space, 60
bounded set in a metric space, 60
bounded set on the real line, 40
box product space, 210
box topology, 210
broom space, 157–158
Brouwer Fixed Point Theorem (Theorem 9.15), 293
Brouwer No Retraction Theorem (Theorem 9.14), 293
Brouwer, L. E. J., 299

C*-embedding, 265
Cantor function, 192
Cantor set (Section 6.5), 187, 209
Cantor's Intersection Theorem, 95
Cantor's Nested Intervals Theorem (Theorem 2.11), 46, 164
Cantor's Theorem of Deduction (Theorem 6.2), 164
Cantor, Georg, 10, 39, 53, 97, 160, 194, 267
cardinally equivalent sets, 34
Cartesian product,
  finite case, 19
  general case, 205
category, 295
category,
  first, 89, 265
  second, 89, 265
Cauchy sequence, 87
Cauchy, Augustin Louis, 10, 97
Cauchy-Schwarz Inequality, 56
Čech, Eduard, 194, 266, 299
center of a ball, 62
characteristic function, 121

305

A CATALOG OF SELECTED
DOVER BOOKS
IN SCIENCE AND MATHEMATICS

Astronomy

CHARIOTS FOR APOLLO: The NASA History of Manned Lunar Spacecraft to 1969, Courtney G. Brooks, James M. Grimwood, and Loyd S. Swenson, Jr. This illustrated history by a trio of experts is the definitive reference on the Apollo spacecraft and lunar modules. It traces the vehicles' design, development, and operation in space. More than 100 photographs and illustrations. 576pp. 6 3/4 x 9 1/4. 0-486-46756-2

EXPLORING THE MOON THROUGH BINOCULARS AND SMALL TELESCOPES, Ernest H. Cherrington, Jr. Informative, profusely illustrated guide to locating and identifying craters, rills, seas, mountains, other lunar features. Newly revised and updated with special section of new photos. Over 100 photos and diagrams. 240pp. 8 1/4 x 11. 0-486-24491-1

WHERE NO MAN HAS GONE BEFORE: A History of NASA's Apollo Lunar Expeditions, William David Compton. Introduction by Paul Dickson. This official NASA history traces behind-the-scenes conflicts and cooperation between scientists and engineers. The first half concerns preparations for the Moon landings, and the second half documents the flights that followed Apollo 11. 1989 edition. 432pp. 7 x 10. 0-486-47888-2

APOLLO EXPEDITIONS TO THE MOON: The NASA History, Edited by Edgar M. Cortright. Official NASA publication marks the 40th anniversary of the first lunar landing and features essays by project participants recalling engineering and administrative challenges. Accessible, jargon-free accounts, highlighted by numerous illustrations. 336pp. 8 3/8 x 10 7/8. 0-486-47175-6

ON MARS: Exploration of the Red Planet, 1958-1978–The NASA History, Edward Clinton Ezell and Linda Neuman Ezell. NASA's official history chronicles the start of our explorations of our planetary neighbor. It recounts cooperation among government, industry, and academia, and it features dozens of photos from Viking cameras. 560pp. 6 3/4 x 9 1/4. 0-486-46757-0

ARISTARCHUS OF SAMOS: The Ancient Copernicus, Sir Thomas Heath. Heath's history of astronomy ranges from Homer and Hesiod to Aristarchus and includes quotes from numerous thinkers, compilers, and scholasticists from Thales and Anaximander through Pythagoras, Plato, Aristotle, and Heraclides. 34 figures. 448pp. 5 3/8 x 8 1/2. 0-486-43886-4

AN INTRODUCTION TO CELESTIAL MECHANICS, Forest Ray Moulton. Classic text still unsurpassed in presentation of fundamental principles. Covers rectilinear motion, central forces, problems of two and three bodies, much more. Includes over 200 problems, some with answers. 437pp. 5 3/8 x 8 1/2. 0-486-64687-4

BEYOND THE ATMOSPHERE: Early Years of Space Science, Homer E. Newell. This exciting survey is the work of a top NASA administrator who chronicles technological advances, the relationship of space science to general science, and the space program's social, political, and economic contexts. 528pp. 6 3/4 x 9 1/4. 0-486-47464-X

STAR LORE: Myths, Legends, and Facts, William Tyler Olcott. Captivating retellings of the origins and histories of ancient star groups include Pegasus, Ursa Major, Pleiades, signs of the zodiac, and other constellations. "Classic." – Sky & Telescope. 58 illustrations. 544pp. 5 3/8 x 8 1/2. 0-486-43581-4

A COMPLETE MANUAL OF AMATEUR ASTRONOMY: Tools and Techniques for Astronomical Observations, P. Clay Sherrod with Thomas L. Koed. Concise, highly readable book discusses the selection, set-up, and maintenance of a telescope; amateur studies of the sun; lunar topography and occultations; and more. 124 figures. 26 halftones. 37 tables. 335pp. 6 1/2 x 9 1/4. 0-486-42820-6

Browse over 9,000 books at www.doverpublications.com

Chemistry

MOLECULAR COLLISION THEORY, M. S. Child. This high-level monograph offers an analytical treatment of classical scattering by a central force, quantum scattering by a central force, elastic scattering phase shifts, and semi-classical elastic scattering. 1974 edition. 310pp. 5 3/8 x 8 1/2. 0-486-69437-2

HANDBOOK OF COMPUTATIONAL QUANTUM CHEMISTRY, David B. Cook. This comprehensive text provides upper-level undergraduates and graduate students with an accessible introduction to the implementation of quantum ideas in molecular modeling, exploring practical applications alongside theoretical explanations. 1998 edition. 832pp. 5 3/8 x 8 1/2. 0-486-44307-8

RADIOACTIVE SUBSTANCES, Marie Curie. The celebrated scientist's thesis, which directly preceded her 1903 Nobel Prize, discusses establishing atomic character of radioactivity; extraction from pitchblende of polonium and radium; isolation of pure radium chloride; more. 96pp. 5 3/8 x 8 1/2. 0-486-42550-9

CHEMICAL MAGIC, Leonard A. Ford. Classic guide provides intriguing entertainment while elucidating sound scientific principles, with more than 100 unusual stunts: cold fire, dust explosions, a nylon rope trick, a disappearing beaker, much more. 128pp. 5 3/8 x 8 1/2. 0-486-67628-5

ALCHEMY, E. J. Holmyard. Classic study by noted authority covers 2,000 years of alchemical history: religious, mystical overtones; apparatus; signs, symbols, and secret terms; advent of scientific method, much more. Illustrated. 320pp. 5 3/8 x 8 1/2. 0-486-26298-7

CHEMICAL KINETICS AND REACTION DYNAMICS, Paul L. Houston. This text teaches the principles underlying modern chemical kinetics in a clear, direct fashion, using several examples to enhance basic understanding. Solutions to selected problems. 2001 edition. 352pp. 8 3/8 x 11. 0-486-45334-0

PROBLEMS AND SOLUTIONS IN QUANTUM CHEMISTRY AND PHYSICS, Charles S. Johnson and Lee G. Pedersen. Unusually varied problems, with detailed solutions, cover of quantum mechanics, wave mechanics, angular momentum, molecular spectroscopy, scattering theory, more. 280 problems, plus 139 supplementary exercises. 430pp. 6 1/2 x 9 1/4. 0-486-65236-X

ELEMENTS OF CHEMISTRY, Antoine Lavoisier. Monumental classic by the founder of modern chemistry features first explicit statement of law of conservation of matter in chemical change, and more. Facsimile reprint of original (1790) Kerr translation. 539pp. 5 3/8 x 8 1/2. 0-486-64624-6

MAGNETISM AND TRANSITION METAL COMPLEXES, F. E. Mabbs and D. J. Machin. A detailed view of the calculation methods involved in the magnetic properties of transition metal complexes, this volume offers sufficient background for original work in the field. 1973 edition. 240pp. 5 3/8 x 8 1/2. 0-486-46284-6

GENERAL CHEMISTRY, Linus Pauling. Revised third edition of classic first-year text by Nobel laureate. Atomic and molecular structure, quantum mechanics, statistical mechanics, thermodynamics correlated with descriptive chemistry. Problems. 992pp. 5 3/8 x 8 1/2. 0-486-65622-5

ELECTROLYTE SOLUTIONS: Second Revised Edition, R. A. Robinson and R. H. Stokes. Classic text deals primarily with measurement, interpretation of conductance, chemical potential, and diffusion in electrolyte solutions. Detailed theoretical interpretations, plus extensive tables of thermodynamic and transport properties. 1970 edition. 590pp. 5 3/8 x 8 1/2. 0-486-42225-9

Engineering

FUNDAMENTALS OF ASTRODYNAMICS, Roger R. Bate, Donald D. Mueller, and Jerry E. White. Teaching text developed by U.S. Air Force Academy develops the basic two-body and n-body equations of motion; orbit determination; classical orbital elements, coordinate transformations; differential correction; more. 1971 edition. 455pp. 5 3/8 x 8 1/2. 0-486-60061-0

INTRODUCTION TO CONTINUUM MECHANICS FOR ENGINEERS: Revised Edition, Ray M. Bowen. This self-contained text introduces classical continuum models within a modern framework. Its numerous exercises illustrate the governing principles, linearizations, and other approximations that constitute classical continuum models. 2007 edition. 320pp. 6 1/8 x 9 1/4. 0-486-47460-7

ENGINEERING MECHANICS FOR STRUCTURES, Louis L. Bucciarelli. This text explores the mechanics of solids and statics as well as the strength of materials and elasticity theory. Its many design exercises encourage creative initiative and systems thinking. 2009 edition. 320pp. 6 1/8 x 9 1/4. 0-486-46855-0

FEEDBACK CONTROL THEORY, John C. Doyle, Bruce A. Francis and Allen R. Tannenbaum. This excellent introduction to feedback control system design offers a theoretical approach that captures the essential issues and can be applied to a wide range of practical problems. 1992 edition. 224pp. 6 1/2 x 9 1/4. 0-486-46933-6

THE FORCES OF MATTER, Michael Faraday. These lectures by a famous inventor offer an easy-to-understand introduction to the interactions of the universe's physical forces. Six essays explore gravitation, cohesion, chemical affinity, heat, magnetism, and electricity. 1993 edition. 96pp. 5 3/8 x 8 1/2. 0-486-47482-8

DYNAMICS, Lawrence E. Goodman and William H. Warner. Beginning engineering text introduces calculus of vectors, particle motion, dynamics of particle systems and plane rigid bodies, technical applications in plane motions, and more. Exercises and answers in every chapter. 619pp. 5 3/8 x 8 1/2. 0-486-42006-X

ADAPTIVE FILTERING PREDICTION AND CONTROL, Graham C. Goodwin and Kwai Sang Sin. This unified survey focuses on linear discrete-time systems and explores natural extensions to nonlinear systems. It emphasizes discrete-time systems, summarizing theoretical and practical aspects of a large class of adaptive algorithms. 1984 edition. 560pp. 6 1/2 x 9 1/4. 0-486-46932-8

INDUCTANCE CALCULATIONS, Frederick W. Grover. This authoritative reference enables the design of virtually every type of inductor. It features a single simple formula for each type of inductor, together with tables containing essential numerical factors. 1946 edition. 304pp. 5 3/8 x 8 1/2. 0-486-47440-2

THERMODYNAMICS: Foundations and Applications, Elias P. Gyftopoulos and Gian Paolo Beretta. Designed by two MIT professors, this authoritative text discusses basic concepts and applications in detail, emphasizing generality, definitions, and logical consistency. More than 300 solved problems cover realistic energy systems and processes. 800pp. 6 1/8 x 9 1/4. 0-486-43932-1

THE FINITE ELEMENT METHOD: Linear Static and Dynamic Finite Element Analysis, Thomas J. R. Hughes. Text for students without in-depth mathematical training, this text includes a comprehensive presentation and analysis of algorithms of time-dependent phenomena plus beam, plate, and shell theories. Solution guide available upon request. 672pp. 6 1/2 x 9 1/4. 0-486-41181-8

HELICOPTER THEORY, Wayne Johnson. Monumental engineering text covers vertical flight, forward flight, performance, mathematics of rotating systems, rotary wing dynamics and aerodynamics, aeroelasticity, stability and control, stall, noise, and more. 189 illustrations. 1980 edition. 1089pp. 5 5/8 x 8 1/4.   0-486-68230-7

MATHEMATICAL HANDBOOK FOR SCIENTISTS AND ENGINEERS: Definitions, Theorems, and Formulas for Reference and Review, Granino A. Korn and Theresa M. Korn. Convenient access to information from every area of mathematics: Fourier transforms, Z transforms, linear and nonlinear programming, calculus of variations, random-process theory, special functions, combinatorial analysis, game theory, much more. 1152pp. 5 3/8 x 8 1/2.   0-486-41147-8

A HEAT TRANSFER TEXTBOOK: Fourth Edition, John H. Lienhard V and John H. Lienhard IV. This introduction to heat and mass transfer for engineering students features worked examples and end-of-chapter exercises. Worked examples and end-of-chapter exercises appear throughout the book, along with well-drawn, illuminating figures. 768pp. 7 x 9 1/4.   0-486-47931-5

BASIC ELECTRICITY, U.S. Bureau of Naval Personnel. Originally a training course; best nontechnical coverage. Topics include batteries, circuits, conductors, AC and DC, inductance and capacitance, generators, motors, transformers, amplifiers, etc. Many questions with answers. 349 illustrations. 1969 edition. 448pp. 6 1/2 x 9 1/4.
0-486-20973-3

BASIC ELECTRONICS, U.S. Bureau of Naval Personnel. Clear, well-illustrated introduction to electronic equipment covers numerous essential topics: electron tubes, semiconductors, electronic power supplies, tuned circuits, amplifiers, receivers, ranging and navigation systems, computers, antennas, more. 560 illustrations. 567pp. 6 1/2 x 9 1/4.   0-486-21076-6

BASIC WING AND AIRFOIL THEORY, Alan Pope. This self-contained treatment by a pioneer in the study of wind effects covers flow functions, airfoil construction and pressure distribution, finite and monoplane wings, and many other subjects. 1951 edition. 320pp. 5 3/8 x 8 1/2.   0-486-47188-8

SYNTHETIC FUELS, Ronald F. Probstein and R. Edwin Hicks. This unified presentation examines the methods and processes for converting coal, oil, shale, tar sands, and various forms of biomass into liquid, gaseous, and clean solid fuels. 1982 edition. 512pp. 6 1/8 x 9 1/4.   0-486-44977-7

THEORY OF ELASTIC STABILITY, Stephen P. Timoshenko and James M. Gere. Written by world-renowned authorities on mechanics, this classic ranges from theoretical explanations of 2- and 3-D stress and strain to practical applications such as torsion, bending, and thermal stress. 1961 edition. 560pp. 5 3/8 x 8 1/2.   0-486-47207-8

PRINCIPLES OF DIGITAL COMMUNICATION AND CODING, Andrew J Viterbi and Jim K. Omura. This classic by two digital communications experts is geared toward students of communications theory and to designers of channels, links, terminals, modems, or networks used to transmit and receive digital messages. 1979 edition. 576pp. 6 1/8 x 9 1/4.   0-486-46901-8

LINEAR SYSTEM THEORY: The State Space Approach, Lotfi A. Zadeh and Charles A. Desoer. Written by two pioneers in the field, this exploration of the state space approach focuses on problems of stability and control, plus connections between this approach and classical techniques. 1963 edition. 656pp. 6 1/8 x 9 1/4.
0-486-46663-9

Mathematics–Bestsellers

HANDBOOK OF MATHEMATICAL FUNCTIONS: with Formulas, Graphs, and Mathematical Tables, Edited by Milton Abramowitz and Irene A. Stegun. A classic resource for working with special functions, standard trig, and exponential logarithmic definitions and extensions, it features 29 sets of tables, some to as high as 20 places. 1046pp. 8 x 10 1/2.   0-486-61272-4

ABSTRACT AND CONCRETE CATEGORIES: The Joy of Cats, Jiri Adamek, Horst Herrlich, and George E. Strecker. This up-to-date introductory treatment employs category theory to explore the theory of structures. Its unique approach stresses concrete categories and presents a systematic view of factorization structures. Numerous examples. 1990 edition, updated 2004. 528pp. 6 1/8 x 9 1/4.   0-486-46934-4

MATHEMATICS: Its Content, Methods and Meaning, A. D. Aleksandrov, A. N. Kolmogorov, and M. A. Lavrent'ev. Major survey offers comprehensive, coherent discussions of analytic geometry, algebra, differential equations, calculus of variations, functions of a complex variable, prime numbers, linear and non-Euclidean geometry, topology, functional analysis, more. 1963 edition. 1120pp. 5 3/8 x 8 1/2.   0-486-40916-3

INTRODUCTION TO VECTORS AND TENSORS: Second Edition--Two Volumes Bound as One, Ray M. Bowen and C.-C. Wang. Convenient single-volume compilation of two texts offers both introduction and in-depth survey. Geared toward engineering and science students rather than mathematicians, it focuses on physics and engineering applications. 1976 edition. 560pp. 6 1/2 x 9 1/4.   0-486-46914-X

AN INTRODUCTION TO ORTHOGONAL POLYNOMIALS, Theodore S. Chihara. Concise introduction covers general elementary theory, including the representation theorem and distribution functions, continued fractions and chain sequences, the recurrence formula, special functions, and some specific systems. 1978 edition. 272pp. 5 3/8 x 8 1/2.
0-486-47929-3

ADVANCED MATHEMATICS FOR ENGINEERS AND SCIENTISTS, Paul DuChateau. This primary text and supplemental reference focuses on linear algebra, calculus, and ordinary differential equations. Additional topics include partial differential equations and approximation methods. Includes solved problems. 1992 edition. 400pp. 7 1/2 x 9 1/4.   0-486-47930-7

PARTIAL DIFFERENTIAL EQUATIONS FOR SCIENTISTS AND ENGINEERS, Stanley J. Farlow. Practical text shows how to formulate and solve partial differential equations. Coverage of diffusion-type problems, hyperbolic-type problems, elliptic-type problems, numerical and approximate methods. Solution guide available upon request. 1982 edition. 414pp. 6 1/8 x 9 1/4.   0-486-67620-X

VARIATIONAL PRINCIPLES AND FREE-BOUNDARY PROBLEMS, Avner Friedman. Advanced graduate-level text examines variational methods in partial differential equations and illustrates their applications to free-boundary problems. Features detailed statements of standard theory of elliptic and parabolic operators. 1982 edition. 720pp. 6 1/8 x 9 1/4.   0-486-47853-X

LINEAR ANALYSIS AND REPRESENTATION THEORY, Steven A. Gaal. Unified treatment covers topics from the theory of operators and operator algebras on Hilbert spaces; integration and representation theory for topological groups; and the theory of Lie algebras, Lie groups, and transform groups. 1973 edition. 704pp. 6 1/8 x 9 1/4.
0-486-47851-3

A SURVEY OF INDUSTRIAL MATHEMATICS, Charles R. MacCluer. Students learn how to solve problems they'll encounter in their professional lives with this concise single-volume treatment. It employs MATLAB and other strategies to explore typical industrial problems. 2000 edition. 384pp. 5 3/8 x 8 1/2. 0-486-47702-9

NUMBER SYSTEMS AND THE FOUNDATIONS OF ANALYSIS, Elliott Mendelson. Geared toward undergraduate and beginning graduate students, this study explores natural numbers, integers, rational numbers, real numbers, and complex numbers. Numerous exercises and appendixes supplement the text. 1973 edition. 368pp. 5 3/8 x 8 1/2. 0-486-45792-3

A FIRST LOOK AT NUMERICAL FUNCTIONAL ANALYSIS, W. W. Sawyer. Text by renowned educator shows how problems in numerical analysis lead to concepts of functional analysis. Topics include Banach and Hilbert spaces, contraction mappings, convergence, differentiation and integration, and Euclidean space. 1978 edition. 208pp. 5 3/8 x 8 1/2. 0-486-47882-3

FRACTALS, CHAOS, POWER LAWS: Minutes from an Infinite Paradise, Manfred Schroeder. A fascinating exploration of the connections between chaos theory, physics, biology, and mathematics, this book abounds in award-winning computer graphics, optical illusions, and games that clarify memorable insights into self-similarity. 1992 edition. 448pp. 6 1/8 x 9 1/4. 0-486-47204-3

SET THEORY AND THE CONTINUUM PROBLEM, Raymond M. Smullyan and Melvin Fitting. A lucid, elegant, and complete survey of set theory, this three-part treatment explores axiomatic set theory, the consistency of the continuum hypothesis, and forcing and independence results. 1996 edition. 336pp. 6 x 9. 0-486-47484-4

DYNAMICAL SYSTEMS, Shlomo Sternberg. A pioneer in the field of dynamical systems discusses one-dimensional dynamics, differential equations, random walks, iterated function systems, symbolic dynamics, and Markov chains. Supplementary materials include PowerPoint slides and MATLAB exercises. 2010 edition. 272pp. 6 1/8 x 9 1/4. 0-486-47705-3

ORDINARY DIFFERENTIAL EQUATIONS, Morris Tenenbaum and Harry Pollard. Skillfully organized introductory text examines origin of differential equations, then defines basic terms and outlines general solution of a differential equation. Explores integrating factors; dilution and accretion problems; Laplace Transforms; Newton's Interpolation Formulas, more. 818pp. 5 3/8 x 8 1/2. 0-486-64940-7

MATROID THEORY, D. J. A. Welsh. Text by a noted expert describes standard examples and investigation results, using elementary proofs to develop basic matroid properties before advancing to a more sophisticated treatment. Includes numerous exercises. 1976 edition. 448pp. 5 3/8 x 8 1/2. 0-486-47439-9

THE CONCEPT OF A RIEMANN SURFACE, Hermann Weyl. This classic on the general history of functions combines function theory and geometry, forming the basis of the modern approach to analysis, geometry, and topology. 1955 edition. 208pp. 5 3/8 x 8 1/2. 0-486-47004-0

THE LAPLACE TRANSFORM, David Vernon Widder. This volume focuses on the Laplace and Stieltjes transforms, offering a highly theoretical treatment. Topics include fundamental formulas, the moment problem, monotonic functions, and Tauberian theorems. 1941 edition. 416pp. 5 3/8 x 8 1/2. 0-486-47755-X

Mathematics-Logic and Problem Solving

PERPLEXING PUZZLES AND TANTALIZING TEASERS, Martin Gardner. Ninety-three riddles, mazes, illusions, tricky questions, word and picture puzzles, and other challenges offer hours of entertainment for youngsters. Filled with rib-tickling drawings. Solutions. 224pp. 5 3/8 x 8 1/2.						0-486-25637-5

MY BEST MATHEMATICAL AND LOGIC PUZZLES, Martin Gardner. The noted expert selects 70 of his favorite "short" puzzles. Includes The Returning Explorer, The Mutilated Chessboard, Scrambled Box Tops, and dozens more. Complete solutions included. 96pp. 5 3/8 x 8 1/2.						0-486-28152-3

THE LADY OR THE TIGER?: and Other Logic Puzzles, Raymond M. Smullyan. Created by a renowned puzzle master, these whimsically themed challenges involve paradoxes about probability, time, and change; metapuzzles; and self-referentiality. Nineteen chapters advance in difficulty from relatively simple to highly complex. 1982 edition. 240pp. 5 3/8 x 8 1/2.						0-486-47027-X

SATAN, CANTOR AND INFINITY: Mind-Boggling Puzzles, Raymond M. Smullyan. A renowned mathematician tells stories of knights and knaves in an entertaining look at the logical precepts behind infinity, probability, time, and change. Requires a strong background in mathematics. Complete solutions. 288pp. 5 3/8 x 8 1/2.
						0-486-47036-9

THE RED BOOK OF MATHEMATICAL PROBLEMS, Kenneth S. Williams and Kenneth Hardy. Handy compilation of 100 practice problems, hints and solutions indispensable for students preparing for the William Lowell Putnam and other mathematical competitions. Preface to the First Edition. Sources. 1988 edition. 192pp. 5 3/8 x 8 1/2.						0-486-69415-1

KING ARTHUR IN SEARCH OF HIS DOG AND OTHER CURIOUS PUZZLES, Raymond M. Smullyan. This fanciful, original collection for readers of all ages features arithmetic puzzles, logic problems related to crime detection, and logic and arithmetic puzzles involving King Arthur and his Dogs of the Round Table. 160pp. 5 3/8 x 8 1/2.
						0-486-47435-6

UNDECIDABLE THEORIES: Studies in Logic and the Foundation of Mathematics, Alfred Tarski in collaboration with Andrzej Mostowski and Raphael M. Robinson. This well-known book by the famed logician consists of three treatises: "A General Method in Proofs of Undecidability," "Undecidability and Essential Undecidability in Mathematics," and "Undecidability of the Elementary Theory of Groups." 1953 edition. 112pp. 5 3/8 x 8 1/2.						0-486-47703-7

LOGIC FOR MATHEMATICIANS, J. Barkley Rosser. Examination of essential topics and theorems assumes no background in logic. "Undoubtedly a major addition to the literature of mathematical logic." – *Bulletin of the American Mathematical Society.* 1978 edition. 592pp. 6 1/8 x 9 1/4.						0-486-46898-4

INTRODUCTION TO PROOF IN ABSTRACT MATHEMATICS, Andrew Wohlgemuth. This undergraduate text teaches students what constitutes an acceptable proof, and it develops their ability to do proofs of routine problems as well as those requiring creative insights. 1990 edition. 384pp. 6 1/2 x 9 1/4.	0-486-47854-8

FIRST COURSE IN MATHEMATICAL LOGIC, Patrick Suppes and Shirley Hill. Rigorous introduction is simple enough in presentation and context for wide range of students. Symbolizing sentences; logical inference; truth and validity; truth tables; terms, predicates, universal quantifiers; universal specification and laws of identity; more. 288pp. 5 3/8 x 8 1/2.						0-486-42259-3

Mathematics–Algebra and Calculus

VECTOR CALCULUS, Peter Baxandall and Hans Liebeck. This introductory text offers a rigorous, comprehensive treatment. Classical theorems of vector calculus are amply illustrated with figures, worked examples, physical applications, and exercises with hints and answers. 1986 edition. 560pp. 5 3/8 x 8 1/2. 0-486-46620-5

ADVANCED CALCULUS: An Introduction to Classical Analysis, Louis Brand. A course in analysis that focuses on the functions of a real variable, this text introduces the basic concepts in their simplest setting and illustrates its teachings with numerous examples, theorems, and proofs. 1955 edition. 592pp. 5 3/8 x 8 1/2. 0-486-44548-8

ADVANCED CALCULUS, Avner Friedman. Intended for students who have already completed a one-year course in elementary calculus, this two-part treatment advances from functions of one variable to those of several variables. Solutions. 1971 edition. 432pp. 5 3/8 x 8 1/2. 0-486-45795-8

METHODS OF MATHEMATICS APPLIED TO CALCULUS, PROBABILITY, AND STATISTICS, Richard W. Hamming. This 4-part treatment begins with algebra and analytic geometry and proceeds to an exploration of the calculus of algebraic functions and transcendental functions and applications. 1985 edition. Includes 310 figures and 18 tables. 880pp. 6 1/2 x 9 1/4. 0-486-43945-3

BASIC ALGEBRA I: Second Edition, Nathan Jacobson. A classic text and standard reference for a generation, this volume covers all undergraduate algebra topics, including groups, rings, modules, Galois theory, polynomials, linear algebra, and associative algebra. 1985 edition. 528pp. 6 1/8 x 9 1/4. 0-486-47189-6

BASIC ALGEBRA II: Second Edition, Nathan Jacobson. This classic text and standard reference comprises all subjects of a first-year graduate-level course, including in-depth coverage of groups and polynomials and extensive use of categories and functors. 1989 edition. 704pp. 6 1/8 x 9 1/4. 0-486-47187-X

CALCULUS: An Intuitive and Physical Approach (Second Edition), Morris Kline. Application-oriented introduction relates the subject as closely as possible to science with explorations of the derivative; differentiation and integration of the powers of x; theorems on differentiation, antidifferentiation; the chain rule; trigonometric functions; more. Examples. 1967 edition. 960pp. 6 1/2 x 9 1/4. 0-486-40453-6

ABSTRACT ALGEBRA AND SOLUTION BY RADICALS, John E. Maxfield and Margaret W. Maxfield. Accessible advanced undergraduate-level text starts with groups, rings, fields, and polynomials and advances to Galois theory, radicals and roots of unity, and solution by radicals. Numerous examples, illustrations, exercises, appendixes. 1971 edition. 224pp. 6 1/8 x 9 1/4. 0-486-47723-1

AN INTRODUCTION TO THE THEORY OF LINEAR SPACES, Georgi E. Shilov. Translated by Richard A. Silverman. Introductory treatment offers a clear exposition of algebra, geometry, and analysis as parts of an integrated whole rather than separate subjects. Numerous examples illustrate many different fields, and problems include hints or answers. 1961 edition. 320pp. 5 3/8 x 8 1/2. 0-486-63070-6

LINEAR ALGEBRA, Georgi E. Shilov. Covers determinants, linear spaces, systems of linear equations, linear functions of a vector argument, coordinate transformations, the canonical form of the matrix of a linear operator, bilinear and quadratic forms, and more. 387pp. 5 3/8 x 8 1/2. 0-486-63518-X

CATALOG OF DOVER BOOKS

Mathematics–Probability and Statistics

BASIC PROBABILITY THEORY, Robert B. Ash. This text emphasizes the probabilistic way of thinking, rather than measure-theoretic concepts. Geared toward advanced undergraduates and graduate students, it features solutions to some of the problems. 1970 edition. 352pp. 5 3/8 x 8 1/2. 0-486-46628-0

PRINCIPLES OF STATISTICS, M. G. Bulmer. Concise description of classical statistics, from basic dice probabilities to modern regression analysis. Equal stress on theory and applications. Moderate difficulty; only basic calculus required. Includes problems with answers. 252pp. 5 5/8 x 8 1/4. 0-486-63760-3

OUTLINE OF BASIC STATISTICS: Dictionary and Formulas, John E. Freund and Frank J. Williams. Handy guide includes a 70-page outline of essential statistical formulas covering grouped and ungrouped data, finite populations, probability, and more, plus over 1,000 clear, concise definitions of statistical terms. 1966 edition. 208pp. 5 3/8 x 8 1/2. 0-486-47769-X

GOOD THINKING: The Foundations of Probability and Its Applications, Irving J. Good. This in-depth treatment of probability theory by a famous British statistician explores Keynesian principles and surveys such topics as Bayesian rationality, corroboration, hypothesis testing, and mathematical tools for induction and simplicity. 1983 edition. 352pp. 5 3/8 x 8 1/2. 0-486-47438-0

INTRODUCTION TO PROBABILITY THEORY WITH CONTEMPORARY APPLICATIONS, Lester L. Helms. Extensive discussions and clear examples, written in plain language, expose students to the rules and methods of probability. Exercises foster problem-solving skills, and all problems feature step-by-step solutions. 1997 edition. 368pp. 6 1/2 x 9 1/4. 0-486-47418-6

CHANCE, LUCK, AND STATISTICS, Horace C. Levinson. In simple, non-technical language, this volume explores the fundamentals governing chance and applies them to sports, government, and business. "Clear and lively ... remarkably accurate." – Scientific Monthly. 384pp. 5 3/8 x 8 1/2. 0-486-41997-5

FIFTY CHALLENGING PROBLEMS IN PROBABILITY WITH SOLUTIONS, Frederick Mosteller. Remarkable puzzlers, graded in difficulty, illustrate elementary and advanced aspects of probability. These problems were selected for originality, general interest, or because they demonstrate valuable techniques. Also includes detailed solutions. 88pp. 5 3/8 x 8 1/2. 0-486-65355-2

EXPERIMENTAL STATISTICS, Mary Gibbons Natrella. A handbook for those seeking engineering information and quantitative data for designing, developing, constructing, and testing equipment. Covers the planning of experiments, the analyzing of extreme-value data; and more. 1966 edition. Index. Includes 52 figures and 76 tables. 560pp. 8 3/8 x 11. 0-486-43937-2

STOCHASTIC MODELING: Analysis and Simulation, Barry L. Nelson. Coherent introduction to techniques also offers a guide to the mathematical, numerical, and simulation tools of systems analysis. Includes formulation of models, analysis, and interpretation of results. 1995 edition. 336pp. 6 1/8 x 9 1/4. 0-486-47770-3

INTRODUCTION TO BIOSTATISTICS: Second Edition, Robert R. Sokal and F. James Rohlf. Suitable for undergraduates with a minimal background in mathematics, this introduction ranges from descriptive statistics to fundamental distributions and the testing of hypotheses. Includes numerous worked-out problems and examples. 1987 edition. 384pp. 6 1/8 x 9 1/4. 0-486-46961-1

Browse over 9,000 books at www.doverpublications.com

Mathematics–Geometry and Topology

PROBLEMS AND SOLUTIONS IN EUCLIDEAN GEOMETRY, M. N. Aref and William Wernick. Based on classical principles, this book is intended for a second course in Euclidean geometry and can be used as a refresher. More than 200 problems include hints and solutions. 1968 edition. 272pp. 5 3/8 x 8 1/2. 0-486-47720-7

TOPOLOGY OF 3-MANIFOLDS AND RELATED TOPICS, Edited by M. K. Fort, Jr. With a New Introduction by Daniel Silver. Summaries and full reports from a 1961 conference discuss decompositions and subsets of 3-space; n-manifolds; knot theory; the Poincaré conjecture; and periodic maps and isotopies. Familiarity with algebraic topology required. 1962 edition. 272pp. 6 1/8 x 9 1/4. 0-486-47753-3

POINT SET TOPOLOGY, Steven A. Gaal. Suitable for a complete course in topology, this text also functions as a self-contained treatment for independent study. Additional enrichment materials make it equally valuable as a reference. 1964 edition. 336pp. 5 3/8 x 8 1/2. 0-486-47222-1

INVITATION TO GEOMETRY, Z. A. Melzak. Intended for students of many different backgrounds with only a modest knowledge of mathematics, this text features self-contained chapters that can be adapted to several types of geometry courses. 1983 edition. 240pp. 5 3/8 x 8 1/2. 0-486-46626-4

TOPOLOGY AND GEOMETRY FOR PHYSICISTS, Charles Nash and Siddhartha Sen. Written by physicists for physics students, this text assumes no detailed background in topology or geometry. Topics include differential forms, homotopy, homology, cohomology, fiber bundles, connection and covariant derivatives, and Morse theory. 1983 edition. 320pp. 5 3/8 x 8 1/2. 0-486-47852-1

BEYOND GEOMETRY: Classic Papers from Riemann to Einstein, Edited with an Introduction and Notes by Peter Pesic. This is the only English-language collection of these 8 accessible essays. They trace seminal ideas about the foundations of geometry that led to Einstein's general theory of relativity. 224pp. 6 1/8 x 9 1/4. 0-486-45350-2

GEOMETRY FROM EUCLID TO KNOTS, Saul Stahl. This text provides a historical perspective on plane geometry and covers non-neutral Euclidean geometry, circles and regular polygons, projective geometry, symmetries, inversions, informal topology, and more. Includes 1,000 practice problems. Solutions available. 2003 edition. 480pp. 6 1/8 x 9 1/4. 0-486-47459-3

TOPOLOGICAL VECTOR SPACES, DISTRIBUTIONS AND KERNELS, François Trèves. Extending beyond the boundaries of Hilbert and Banach space theory, this text focuses on key aspects of functional analysis, particularly in regard to solving partial differential equations. 1967 edition. 592pp. 5 3/8 x 8 1/2.
0-486-45352-9

INTRODUCTION TO PROJECTIVE GEOMETRY, C. R. Wylie, Jr. This introductory volume offers strong reinforcement for its teachings, with detailed examples and numerous theorems, proofs, and exercises, plus complete answers to all odd-numbered end-of-chapter problems. 1970 edition. 576pp. 6 1/8 x 9 1/4. 0-486-46895-X

FOUNDATIONS OF GEOMETRY, C. R. Wylie, Jr. Geared toward students preparing to teach high school mathematics, this text explores the principles of Euclidean and non-Euclidean geometry and covers both generalities and specifics of the axiomatic method. 1964 edition. 352pp. 6 x 9. 0-486-47214-0

Mathematics–History

THE WORKS OF ARCHIMEDES, Archimedes. Translated by Sir Thomas Heath. Complete works of ancient geometer feature such topics as the famous problems of the ratio of the areas of a cylinder and an inscribed sphere; the properties of conoids, spheroids, and spirals; more. 326pp. 5 3/8 x 8 1/2. 0-486-42084-1

THE HISTORICAL ROOTS OF ELEMENTARY MATHEMATICS, Lucas N. H. Bunt, Phillip S. Jones, and Jack D. Bedient. Exciting, hands-on approach to understanding fundamental underpinnings of modern arithmetic, algebra, geometry and number systems examines their origins in early Egyptian, Babylonian, and Greek sources. 336pp. 5 3/8 x 8 1/2. 0-486-25563-8

THE THIRTEEN BOOKS OF EUCLID'S ELEMENTS, Euclid. Contains complete English text of all 13 books of the Elements plus critical apparatus analyzing each definition, postulate, and proposition in great detail. Covers textual and linguistic matters; mathematical analyses of Euclid's ideas; classical, medieval, Renaissance and modern commentators; refutations, supports, extrapolations, reinterpretations and historical notes. 995 figures. Total of 1,425pp. All books 5 3/8 x 8 1/2.

Vol. I: 443pp. 0-486-60088-2
Vol. II: 464pp. 0-486-60089-0
Vol. III: 546pp. 0-486-60090-4

A HISTORY OF GREEK MATHEMATICS, Sir Thomas Heath. This authoritative two-volume set that covers the essentials of mathematics and features every landmark innovation and every important figure, including Euclid, Apollonius, and others. 5 3/8 x 8 1/2.
Vol. I: 461pp. 0-486-24073-8
Vol. II: 597pp. 0-486-24074-6

A MANUAL OF GREEK MATHEMATICS, Sir Thomas L. Heath. This concise but thorough history encompasses the enduring contributions of the ancient Greek mathematicians whose works form the basis of most modern mathematics. Discusses Pythagorean arithmetic, Plato, Euclid, more. 1931 edition. 576pp. 5 3/8 x 8 1/2.
0-486-43231-9

CHINESE MATHEMATICS IN THE THIRTEENTH CENTURY, Ulrich Libbrecht. An exploration of the 13th-century mathematician Ch'in, this fascinating book combines what is known of the mathematician's life with a history of his only extant work, the Shu-shu chiu-chang. 1973 edition. 592pp. 5 3/8 x 8 1/2.
0-486-44619-0

PHILOSOPHY OF MATHEMATICS AND DEDUCTIVE STRUCTURE IN EUCLID'S ELEMENTS, Ian Mueller. This text provides an understanding of the classical Greek conception of mathematics as expressed in Euclid's Elements. It focuses on philosophical, foundational, and logical questions and features helpful appendixes. 400pp. 6 1/2 x 9 1/4. 0-486-45300-6

BEYOND GEOMETRY: Classic Papers from Riemann to Einstein, Edited with an Introduction and Notes by Peter Pesic. This is the only English-language collection of these 8 accessible essays. They trace seminal ideas about the foundations of geometry that led to Einstein's general theory of relativity. 224pp. 6 1/8 x 9 1/4. 0-486-45350-2

HISTORY OF MATHEMATICS, David E. Smith. Two-volume history – from Egyptian papyri and medieval maps to modern graphs and diagrams. Non-technical chronological survey with thousands of biographical notes, critical evaluations, and contemporary opinions on over 1,100 mathematicians. 5 3/8 x 8 1/2.
Vol. I: 618pp. 0-486-20429-4
Vol. II: 736pp. 0-486-20430-8